T0297024

LONDON MATHEMATICAL SOCIETY LECTURE NOTE SERIES

Managing Editor: Professor M. Reid, Mathematics Institute,
University of Warwick, Coventry CV4 7AL, United Kingdom

The titles below are available from booksellers, or from Cambridge University Press at
www.cambridge.org/mathematics

London Mathematical Society Lecture Note Series: 397

Hyperbolic Geometry and Applications in Quantum Chaos and Cosmology

Edited by

JENS BOLTE
Royal Holloway, University of London

FRANK STEINER
Universität Ulm, Germany

CAMBRIDGE
UNIVERSITY PRESS

CAMBRIDGE
UNIVERSITY PRESS

University Printing House, Cambridge CB2 8BS, United Kingdom

One Liberty Plaza, 20th Floor, New York, NY 10006, USA

477 Williamstown Road, Port Melbourne, VIC 3207, Australia

314-321, 3rd Floor, Plot 3, Splendor Forum, Jasola District Centre, New Delhi - 110025, India

103 Penang Road, #05-06/07, Visioncrest Commercial, Singapore 238467

Cambridge University Press is part of the University of Cambridge.

It furthers the University's mission by disseminating knowledge in the pursuit of education, learning and research at the highest international levels of excellence.

www.cambridge.org
Information on this title: www.cambridge.org/9781107610491

© Cambridge University Press 2012

First published 2012

A catalogue record for this publication is available from the British Library

ISBN 978-1-107-61049-1 Paperback

Contents

List of Contributors

Aline Aigon-Dupuy
Ecole Polytechnique Fédérale de
Lausanne
Section de Mathématiques
SB-IGAT-Geom Station 8
CH-1015 Lausanne, Switzerland
aline.dupuy@bluewin.ch

Ralf Aurich
Institut für Theoretische Physik
Universität Ulm
Albert-Einstein-Allee 11
D-89081 Ulm, Germany
ralf.aurich@uni-ulm.de

Peter Buser
Ecole Polytechnique Fédérale de
Lausanne
Section de Mathématiques
SB-IGAT-Geom Station 8
CH-1015 Lausanne, Switzerland
peter.buser@epfl.ch

Dennis A. Hejhal
School of Mathematics
University of Minnesota
Minneapolis
MN 55455, USA
hejhal@math.umn.edu
and
Department of Mathematics

Uppsala University
Box 480
S-75106 Uppsala, Sweden
hejhal@math.uu.se

Jens Marklof
School of Mathematics
University of Bristol
Bristol BS8 1TW, UK
j.marklof@bristol.ac.uk

Dieter H. Mayer
Institut für Theoretische Physik
TU-Clausthal
D-38678 Clausthal-Zellerfeld,
Germany
dieter.mayer@tu-clausthal.de

Klaus-Dieter Semmler
Ecole Polytechnique Fédérale de
Lausanne
Section de Mathématiques
SB-IGAT-Geom Station 8
CH-1015 Lausanne, Switzerland
klaus-dieter.semmler@epfl.ch

Martin Sieber
School of Mathematics
University of Bristol
Bristol BS8 1TW, UK
m.sieber@bristol.ac.uk

Frank Steiner
Institut für Theoretische Physik
Universität Ulm
Albert-Einstein-Allee 11
D-89081 Ulm, Germany
frank.steiner@uni-ulm.de
and
Université Lyon 1
Centre de Recherche Astrophysique
de Lyon
CNRS UMR 5574
9 avenue Charles André
F-69230 Saint-Genis-Laval, France

Fredrik Strömberg
TU Darmstadt
Fachbereich Mathematik, AG AGF
Schlossgartenstraße 7
D-64289 Darmstadt, Germany
stroemberg@mathematik.tu-darmstadt.de

Holger Then
School of Mathematics
University of Bristol
Bristol BS8 1TW, UK
holger.then@bristol.ac.uk

Preface

Hyperbolic geometry is a classical subject of pure mathematics. It was invented as the first example of a non-euclidean geometry, and hence played a significant role in the development of modern geometry. Soon, however, its manifold connections with other branches of mathematics and science in general became apparent. Subsequently, it played a major role in such diverse fields as number theory, representation theory, complex functions, ergodic theory, dynamical systems, string theory, quantum chaos, and cosmology.

One of the earliest appearances of hyperbolic geometry in applications arose in connection with ergodic theory and dynamical systems, when in 1898 Hadamard realised that the negative curvature of hyperbolic surfaces produced an unstable and erratic behaviour of the geodesic motion. As a specific example, in 1924 Artin considered the geodesic billiard motion in an infinite triangle on the hyperbolic plane. He constructed a symbolic coding of the geodesics in terms of continued fractions and for the first time proved a quasi ergodic behaviour of the geodesics. Subsequently, many developments in ergodic theory and dynamical systems began with studies of geodesic flows on hyperbolic manifolds.

Beginning in the 1940s, a further far reaching development was initiated by the work of Maass, Selberg and others. Maass considered the spectral theory of the Laplace Beltrami operator on a hyperbolic surface, and Selberg discovered a close connection between the eigenvalues of the Laplacian and the closed geodesics on the surface, as expressed by the celebrated Selberg trace formula. This relation can be viewed as an analogue of the Riemann-von Mangoldt explicit formula that connects the non-trivial zeros of the Riemann zeta function with the prime numbers. The explicit formula plays an important role in studies of the distribution of prime numbers. On the other hand, it is also used to investigate properties of the Riemann zeros, most notably in connection with the famous Riemann hypothesis.

Roughly at the same time, but independently of these developments, in physics the statistical correlations of energy eigenvalues of complicated quantum systems, like heavy nuclei, were found to be the same as the eigenvalue correlations of certain random matrices. This observation was later extended to eigenvalue correlations of quantum Hamiltonians with only a few degrees of freedom, when their classical counterparts possessed a chaotic dynamical behaviour. In 1984 this connection was brought in a condensed form and since then is known as the conjecture of Bohigas, Giannoni and Schmit. Based on these developments the field of quantum chaos emerged which is, in general, concerned with the implications of classical chaos for the dynamical properties of quantum systems. In this context new light was also shed on the similarities between Laplace eigenvalues and Riemann zeros. Montgomery had proven that certain correlations of the Riemann zeros were the same as in random matrix theory. Subsequently, Odlydzko's extensive calculations of Riemann zeros revealed that these correlations follow the random matrix behaviour to an amazing extent.

In the context of quantum chaos, Gutzwiller soon realised that the force free motion of a particle on a hyperbolic surface provides an ideal model system. On the one hand, the quantum system is defined in terms of the Laplace Beltrami operator on the surface which, in suitable units, serves as the quantum Hamiltonian. On the other hand, the geodesic flow on the surface models the classical motion of a particle. Due to the negative curvature this is strongly chaotic, i.e., ergodic, mixing, and uniformly hyperbolic. In this context, the Selberg trace formula can be seen as a connection between the classical and the quantum motion, and hence it can serve as an ideal tool to relate both descriptions of the physical (model) system. Although in the form of the Gutzwiller trace formula a similar connection exists for practically every physical system, in general such a relation is only of an asymptotic (semiclassical) nature. It is an extremely remarkable feature of hyperbolic manifolds that the Selberg trace formula is an identity rather than an asymptotic relation. This property is related to a number of further peculiarities such as, e.g., the connection of the trace formula with Ruelle transfer operators that usually only appear in ergodic theory. These are some of the major reasons why the motion on hyperbolic manifolds emerged as a preferred model in the field of quantum chaos.

This volume now gathers lectures on various aspects of hyperbolic geometry and its applications, mainly to quantum chaos. The contributions are based on lectures held by experts in the field during the International School *Quantum Chaos on Hyperbolic Manifolds*. This school was organised in the framework of the European Research Training Network *Mathematical Aspects of Quantum Chaos* and took place on Schloß Reisensburg in Günzburg (Germany), 4–11 October 2003. The lectures are intended for graduate students and young researchers, as well as for those experienced scientists who want to learn about the subject in some detail. The first contribution is an extended introduction to the geometry of hyperbolic surfaces. This is followed by a detailed exposition of the Selberg trace formula for compact surfaces in the second contribution. An application of the trace formula to correlations of Laplace eigenvalues is the subject of the third contribution. An alternative approach to Laplace eigenvalues that is based on ergodic theory is explained in the following contribution. The subject of the fifth and the sixth contribution are computations of Laplace eigenfunctions on certain non-compact surfaces. The final contribution then deals with three-dimensional hyperbolic manifolds and their applications in cosmology.

We are indebted to the authors of the contributions gathered here, who spent so much effort in preparing their lectures with great care and enthusiasm. We would also like to thank them for their patience during the editorial process of this volume. In addition, we are grateful to Holger Then for his help in preparing this volume. Further thanks go to the European Commission for their financial support through the Research Training Network *Mathematical Aspects of Quantum Chaos*, no. HPRN-CT-2000-00103 of the IHP Programme, which initially brought us all together.

Egham and Lyon   *Jens Bolte*
                                                        Frank Steiner

I. Hyperbolic Geometry

Aline Aigon-Dupuy, Peter Buser, and Klaus-Dieter Semmler

Ecole Polytechnique Fédérale de Lausanne
Section de Mathématiques
SB-IGAT-Geom Station 8
CH-1015 Lausanne, Switzerland
aline.dupuy@bluewin.ch, peter.buser@epfl.ch,
klaus-dieter.semmler@epfl.ch

Summary. The aim of these lectures is to provide a self-contained introduction into the geometry of the hyperbolic plane and to develop computational tools for the construction and the study of discrete groups of non-Euclidean motions.

The first lecture provides a minimal background in the classical geometries: spherical, hyperbolic and Euclidean. In Lecture 2 we present the Poincaré model of the hyperbolic plane. This is the most widely used model in the literature, but possibly not always the most suitable one for effective computation, and so we present in Lectures 4 and 5 an independent approach based on what we call the *matrix model*. Between the two approaches, in Lecture 3, we introduce the concepts of a Fuchsian group, its fundamental domains and the hyperbolic surfaces. In the final lecture we bring everything together and study, as a special subject, the construction of hyperbolic surfaces of genus 2 based on geodesic octagons.

The course has been designed such that the reader may work himself linearly through it. The prerequisites in differential geometry are kept to a minimum and are largely covered, for example by the first chapters of the books by Do Carmo [8] or Lee [15]. The numerous exercises in the text are all of a computational rather than a problem-to-solve nature and are, hopefully, not too hard to do.

Lecture 1:

The Classical Geometries in Dimension 2

It was long believed that the homogeneity and isotropy of our physical space in which we live, are characteristic features of Euclidean geometry, and it was not until the works of Bolyai, Gauss and Lobatchewski in the early nineteenth century that it became clear that other equally homogeneous geometries exist. These, in the meanwhile classical, geometries will be introduced here in a unified form, where for simplicity we restrict ourselves to dimension two.

1.1 The ε-Hyperboloïd Model

The starting point is the following bilinear form in $\mathbb{R}^3$, where $\varepsilon \in \mathbb{R}$ is a constant, $\varepsilon \neq 0$,

$$h_\varepsilon(x, y) = x_1 y_1 + x_2 y_2 + \frac{1}{\varepsilon} x_3 y_3. \tag{1}$$

Here $x = (x_1, x_2, x_3)$ and $y = (y_1, y_2, y_3)$ may be interpreted either as points of $\mathbb{R}^3$ or as vectors at some point $p \in \mathbb{R}^3$. Along with this bilinear form comes the surface

$$\mathbb{S}_\varepsilon^2 = \{x \in \mathbb{R}^3 \mid \varepsilon(x_1^2 + x_2^2) + x_3^2 = 1\}. \tag{2}$$

When $\varepsilon > 0$, $\mathbb{S}_\varepsilon^2$ is an ellipsoid, when $\varepsilon < 0$, $\mathbb{S}_\varepsilon^2$ is a two-sheeted hyperboloïd with the two sheets

$$\mathbb{S}_\varepsilon^{2+} = \{(x_1, x_2, x_3) \in \mathbb{S}_\varepsilon^2 \mid x_3 > 0\}, \quad \mathbb{S}_\varepsilon^{2-} = \{(x_1, x_2, x_3) \in \mathbb{S}_\varepsilon^2 \mid x_3 < 0\}.$$

We will now verify that $\mathbb{S}_\varepsilon^2$ with h_ε is a Riemannian manifold, study its isometry group, and learn how the geodesics look like.

Isometries

An economic way to carry out this program is to begin with the transformations of $\mathbb{R}^3$ which preserve h_ε. We use the following notation. $M(3, \mathbb{R})$ is the set of all 3×3 matrices with real coefficients. Any $M \in M(3, \mathbb{R})$ has three columns m_1, m_2, m_3, and we write $M = (m_1, m_2, m_3) = (m_{ij})$. A special case is the identity matrix (e_1, e_2, e_3), where e_1, e_2, e_3 is the standard basis of $\mathbb{R}^3$. As the columns of M are members of $\mathbb{R}^3$ we may apply h_ε to them, and so the following definition makes sense,

$$O_\varepsilon(3) = \{(m_1, m_2, m_3) \in M(3, \mathbb{R}) \mid h_\varepsilon(m_i, m_j) = h_\varepsilon(e_i, e_j)\}. \tag{3}$$

Observing that $M e_i = m_i$, $i = 1, 2, 3$, and using that multiplication by a matrix is a linear operation we easily see that for $M \in M(3, \mathbb{R})$ we have

$$M \in O_\varepsilon(3) \iff h_\varepsilon(Mx, My) = h_\varepsilon(x, y), \ \forall x, y \in \mathbb{R}^3$$

(with x and y written as column vectors). From this we further see that if $L, M \in O_\varepsilon(3)$, then also $LM \in O_\varepsilon(3)$ and $M^{-1} \in O_\varepsilon(3)$. Hence, $O_\varepsilon(3)$ is a group. It is called the *orthogonal group of* h_ε.

Example 1.1. The following matrices belong to $O_\varepsilon(3)$,

$$S_1 = \begin{pmatrix} -1 & 0 & 0 \\ 0 & 1 & 0 \\ 0 & 0 & 1 \end{pmatrix}, \quad S_2 = \begin{pmatrix} 1 & 0 & 0 \\ 0 & -1 & 0 \\ 0 & 0 & 1 \end{pmatrix}, \quad S_3 = \begin{pmatrix} 1 & 0 & 0 \\ 0 & 1 & 0 \\ 0 & 0 & -1 \end{pmatrix},$$

$$R_\alpha = \begin{pmatrix} \cos\alpha & -\sin\alpha & 0 \\ \sin\alpha & \cos\alpha & 0 \\ 0 & 0 & 1 \end{pmatrix}, \quad N_t = \begin{pmatrix} \mathbf{C}_\varepsilon(t) & 0 & \mathbf{S}_\varepsilon(t) \\ 0 & 1 & 0 \\ -\varepsilon\mathbf{S}_\varepsilon(t) & 0 & \mathbf{C}_\varepsilon(t) \end{pmatrix}, \quad \alpha, t \in \mathbb{R},$$

where $\mathbf{C}_\varepsilon$ and $\mathbf{S}_\varepsilon$ are the generalized trigonometric functions

$$\mathbf{C}_\varepsilon(t) = \begin{cases} \cos\sqrt{\varepsilon}\,t, & \text{if } \varepsilon > 0; \\ \cosh\sqrt{|\varepsilon|}\,t, & \text{if } \varepsilon < 0; \end{cases} \qquad \mathbf{S}_\varepsilon(t) = \begin{cases} \dfrac{1}{\sqrt{\varepsilon}}\sin\sqrt{\varepsilon}\,t, & \text{if } \varepsilon > 0; \\ \dfrac{1}{\sqrt{|\varepsilon|}}\sinh\sqrt{|\varepsilon|}\,t, & \text{if } \varepsilon < 0. \end{cases}$$

The Taylor series expansion of these functions for $t \to 0$ is

$$\mathbf{C}_\varepsilon(t) = 1 - \frac{\varepsilon t^2}{2!} + \cdots, \qquad \mathbf{S}_\varepsilon(t) = t - \frac{\varepsilon t^3}{3!} + \cdots. \tag{4}$$

Exercise 1.2.
(a) Any $M \in O_\varepsilon(3)$ maps $\mathbb{S}_\varepsilon^2$ onto itself.
(b) For any $p, q \in \mathbb{S}_\varepsilon^2$ there exists $M \in O_\varepsilon(3)$ such that $Mp = q$.
(c) $R_\alpha R_\beta = R_{\alpha+\beta}$, for all $\alpha, \beta \in \mathbb{R}$.
(d) $N_s N_t = N_{s+t}$, for all $s, t \in \mathbb{R}$.
(e) S_1, S_2, S_3 and the R_α's and N_t's generate $O_\varepsilon(3)$.
(Hint: in (b) show first that there exists α and t such that $N_t R_\alpha p = \pm e_3$.)

We will now introduce a Riemannian metric on $\mathbb{S}_\varepsilon^2$ such that $O_\varepsilon(3)$ operates by isometries on it. (For the concept of a Riemannian metric and worked out examples, see e.g. the third chapter in Lee's book [15].)

For $p \in \mathbb{S}_\varepsilon^2$ we denote by $T_p\mathbb{S}_\varepsilon^2$ the space of all vectors at p which are tangent to $\mathbb{S}_\varepsilon^2$. In the special case $p = e_3$ this space is easy to describe:

$$v = (v_1, v_2, v_3) \in T_{e_3}\mathbb{S}_\varepsilon^2 \iff v_3 = 0.$$

For the other points in $\mathbb{S}_\varepsilon^2$ it will not be necessary to describe $T_p\mathbb{S}_\varepsilon^2$ explicitly; it suffices to know, by Exercise 1.2(b), that there exists $M \in O_\varepsilon(3)$ sending the tangent vectors at e_3 to the tangent vectors at p, and vice-versa.

Definition 1.3. For any given $p \in \mathbb{S}_\varepsilon^2$ and for all $v, w \in T_p\mathbb{S}_\varepsilon^2$ we define $g_\varepsilon(v, w) = h_\varepsilon(v, w)$.

Theorem 1.4. $\mathbb{S}_\varepsilon^2$ *endowed with the product* g_ε *is a Riemannian 2-manifold and* $O_\varepsilon(3)$ *acts by isometries. If* $v, w \in T_p\mathbb{S}_\varepsilon^2$ *and* $v', w' \in T_{p'}\mathbb{S}_\varepsilon^2$ *are orthonormal bases, then there exists a unique* $\phi \in O_\varepsilon(3)$ *satisfying* $\phi p = p'$, $\phi v = v'$, $\phi w = w'$.

Proof. To prove that g_ε is a Riemannian metric on $\mathbb{S}_\varepsilon^2$ it only remains to show that g_ε is positive definite on any $T_p\mathbb{S}_\varepsilon^2$. For $p = e_3$ this is clearly the case because for $v, w \in T_{e_3}\mathbb{S}_\varepsilon^2$ the product $g_\varepsilon(v, w)$ coincides with the ordinary scalar product. For any other $p \in \mathbb{S}_\varepsilon^2$ it now suffices to recall (Exercise 1.2(b)) that there exists $M \in O_\varepsilon(3)$ sending $T_p\mathbb{S}_\varepsilon^2$ to $T_{e_3}\mathbb{S}_\varepsilon^2$.

Since $O_\varepsilon(3)$ preserves h_ε it preserves g_ε, and this means that $O_\varepsilon(3)$ acts by isometries.

For the two orthonormal bases we take $M, M' \in O_\varepsilon(3)$ such that $Mp = e_3$ and $M'p' = e_3$. Now Mv, Mw and $M'v', M'w'$ are orthonormal bases of $T_{e_3}\mathbb{S}^2_\varepsilon$ with respect to g_ε and thus also with respect to the ordinary scalar product. We find therefore an isometry S either of the form $S = \mathbb{R}_\alpha$ or $S = \mathbb{R}_\alpha S_1$ such that $SMv = M'v'$ and $SMw = M'w'$. Setting $\phi = (M')^{-1}SM$ we get the desired isometry. The uniqueness is left as an exercise (use Theorem 1.5 below). $\square$

For any Riemannian manifold $\mathcal{M}$ we denote by $\mathrm{Isom}(\mathcal{M})$ its *isometry group* i.e. the group of all isometries $\phi : \mathcal{M} \to \mathcal{M}$.

We shall use the following standard fact from Riemannian geometry which we state here without giving a proof.

Theorem 1.5. *Let $\mathcal{M}$ be an n-dimensional connected Riemannian manifold, $p \in \mathcal{M}$ a point, and $v_1, \ldots, v_n$ a basis of $T_p\mathcal{M}$. If $\phi, \psi \in \mathrm{Isom}(\mathcal{M})$ send p to the same image point and $v_1, \ldots, v_n$ to the same image vectors, then $\phi = \psi$.*

(The theorem is actually seldom stated, but it is a simple consequence of the fact that the so-called exponential map is a local diffeomorphism (Proposition 2.9 in [8], or Lemma 5.10 in [15]).

The Hyperboloïd

If $\varepsilon < 0$, then $\mathbb{S}^2_\varepsilon$ has two isometric sheets $\mathbb{S}^{2+}_\varepsilon$ and $\mathbb{S}^{2-}_\varepsilon$. An isometry between the two is e.g. given by S_3 in Example 1.1. We shall therefore restrict ourselves from now on to one of the two sheets. To treat all cases in a uniform manner we use the following notation.

Definition 1.6.
$$H^2_\varepsilon = \begin{cases} \mathbb{S}^2_\varepsilon, & \text{if } \varepsilon > 0; \\ \mathbb{S}^{2+}_\varepsilon, & \text{if } \varepsilon < 0. \end{cases}$$

We call H^2_ε the *ε-hyperboloïd*. Together with g_ε it is a connected two-dimensional Riemannian manifold.

Geodesics

In this part we use some elements from Riemannian geometry such as the Levi-Civita connection to characterize the geodesics of H^2_ε. The reader who is not familiar with this may skip the proofs and use the conclusion of Theorem 1.8 as an ad hoc definition of a geodesic.

In what follows, all geodesics are understood to be parametrized by unit speed on a maximal possible interval that contains 0. Frequently such a geodesic will be identified with the set of its points.

Lemma 1.7. *The following curve is a geodesic of* H^2_ε,

$$\eta(t) = N_t \begin{bmatrix} 0 \\ 0 \\ 1 \end{bmatrix} = \begin{bmatrix} \mathbf{S}_\varepsilon(t) \\ 0 \\ \mathbf{C}_\varepsilon(t) \end{bmatrix}, \quad t \in \mathbb{R}.$$

Proof. We denote by $\dot{\eta}(t) = \frac{d}{dt}\eta(t)$ the tangent vectors of η. The isometry $S = S_2$ fixes η point-wise, and therefore $\nabla_{(S\circ\eta)'(t)}(S \circ \eta)'(t) = \nabla_{\dot{\eta}(t)}\dot{\eta}(t)$.

On the other hand, η satisfies $g_\varepsilon(\dot{\eta}(t), \dot{\eta}(t)) = \text{constant} = 1$, and the rule $\frac{d}{dt}g_\varepsilon(\dot{\eta}(t), \dot{\eta}(t)) = 2g_\varepsilon(\nabla_{\dot{\eta}(t)}\dot{\eta}(t), \dot{\eta}(t))$ implies $g_\varepsilon(\nabla_{\dot{\eta}(t)}\dot{\eta}(t), \dot{\eta}(t)) = 0$, i.e. $\nabla_{\dot{\eta}(t)}\dot{\eta}(t)$ is orthogonal to η and therefore $\nabla_{(S\circ\eta)'(t)}(S \circ \eta)'(t) = S(\nabla_{\dot{\eta}(t)}\dot{\eta}(t)) = -\nabla_{\dot{\eta}(t)}\dot{\eta}(t)$.

Altogether $\nabla_{\dot{\eta}(t)}\dot{\eta}(t) = 0$, i.e. η is a geodesic. □

Theorem 1.8. *The geodesics in* H^2_ε *are the curves* $\mathrm{H}^2_\varepsilon \cap G$*, where G is a plane through O that intersects* H^2_ε.

Proof. Let E be the plane through O, e_1, e_3. Then $\mathrm{H}^2_\varepsilon \cap E = \{\eta(t) \mid t \in \mathbb{R}\}$, where η is the geodesic from Lemma 1.7. If G is any plane intersecting H^2_ε, then by Theorem 1.4, there exists $\phi \in O_\varepsilon(3)$ such that $\phi(E) = G$. (Since ϕ is a linear map any plane through O is mapped to a plane through O !) It follows that $\mathrm{H}^2_\varepsilon \cap G = \phi(\mathrm{H}^2_\varepsilon \cap E) = \phi(\eta)$ is a geodesic.

Conversely, let γ be an arbitrary geodesic on H^2_ε. By Theorem 1.4, there exists $\psi \in O_\varepsilon(3)$ sending $\eta(0)$ to $\gamma(0)$ and $\dot{\eta}(0)$ to $\dot{\gamma}(0)$. The uniqueness theorem for geodesics then implies that $\gamma = \psi(\eta)$. It follows that $\gamma = \psi(\mathrm{H}^2_\varepsilon \cap E) = \mathrm{H}^2_\varepsilon \cap G$ where G is the plane $G = \psi(E)$. □

Corollary 1.9. *Any geodesic on* H^2_ε *is the image of η under some $\psi \in O_\varepsilon(3)$.*
□

Polar Coordinates

For the next consideration we set

$$d_\varepsilon = \begin{cases} \pi/\sqrt{\varepsilon}, & \text{if } \varepsilon > 0; \\ \infty, & \text{if } \varepsilon \leq 0. \end{cases} \tag{5}$$

For any given $\sigma \in [0, 2\pi]$, the curve

$$\rho \mapsto \gamma_\sigma(\rho) = R_\sigma\eta(\rho), \quad 0 \leq \rho < d_\varepsilon,$$

(where η is again as in Lemma 1.7) is a unit speed geodesic. For any given $\rho > 0$ the curve

$$\sigma \mapsto c_\rho(\sigma) = R_\sigma\eta(\rho) = \begin{bmatrix} \cos(\sigma)\mathbf{S}_\varepsilon(\rho) \\ \sin(\sigma)\mathbf{S}_\varepsilon(\rho) \\ \mathbf{C}_\varepsilon(\rho) \end{bmatrix}$$

has tangent vectors

$$\dot{c}_\rho(\sigma) = \begin{bmatrix} -\sin(\sigma)\mathbf{S}_\varepsilon(\rho) \\ \cos(\sigma)\mathbf{S}_\varepsilon(\rho) \\ 0 \end{bmatrix},$$

and $g_\varepsilon(\dot{c}_\rho(\sigma), \dot{c}_\rho(\sigma))^{1/2} = \mathbf{S}_\varepsilon(\rho)$. One also easily checks that $g_\varepsilon(\dot{c}_\rho(\sigma), \dot{\gamma}_\sigma(\rho)) = 0$. This shows that if we use ρ and σ as new coordinates for points in H_ε^2, then the expression for the Riemannian metric g_ε is as follows, where we use the differential notation ds_ε for the length of an infinitesimal segment with components $d\rho$ and $d\sigma$:

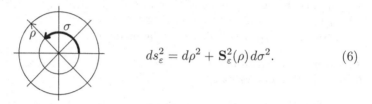

$$ds_\varepsilon^2 = d\rho^2 + \mathbf{S}_\varepsilon^2(\rho)\,d\sigma^2. \tag{6}$$

The coordinates ρ and σ are called *geodesic polar coordinates*. The figure illustrates them as ordinary polar coordinates in the Euclidean plane. The radial straight line of argument σ corresponds to γ_σ, the circle of radius ρ corresponds to c_ρ. The length of c_ρ is $2\pi\,\mathbf{S}_\varepsilon(\rho)$.

The polar coordinates here have been centered at point e_3; applying $\mathrm{O}_\varepsilon(3)$ we may center them at any other point of H_ε^2.

Exercise 1.10. Let $f(\rho, \sigma) = \gamma_\sigma(\rho)$.

(a) In the case $\varepsilon > 0$ the mapping $f : \,]0, d_\varepsilon[\times [0, 2\pi[\to \mathrm{H}_\varepsilon^2 \setminus \{\pm e_3\}$ is one-to-one and onto.

(b) In the case $\varepsilon < 0$ the mapping $f : \,]0, \infty[\times [0, 2\pi[\to \mathrm{H}_\varepsilon^2 \setminus \{e_3\}$ is one-to-one and onto.

Exercise 1.11. Use (6) to show that for any r in the interval $]0, d_\varepsilon[$ the restriction of the geodesic γ_σ to the interval $[0, r]$ is the shortest curve on $\mathbb{S}_\varepsilon^2$ from $\gamma_\sigma(0)$ to $\gamma_\sigma(r)$.

Use Corollary 1.9 to conclude that this is true for all geodesics in $\mathbb{S}_\varepsilon^2$.

Distances

Definition 1.12. For $p, q \in \mathrm{H}_\varepsilon^2$ the *distance* $d(p, q)$ is defined as the infimum over the lengths of all arcs from p to q on H_ε^2.

In the next theorem d_ε is again $\pi/\sqrt{\varepsilon}$ if $\varepsilon > 0$ and ∞ otherwise (see (5)).

Theorem 1.13. *For any $p, q \in \mathrm{H}_\varepsilon^2$, $p \neq q$, the following holds.*

(i) $d(p, q) \le d_\varepsilon$.

(ii) *There exists a geodesic through p, q, and on this geodesic an arc of length $d(p, q)$ from p to q.*

(iii) *If $d(p, q) < d_\varepsilon$, then this geodesic and the arc are unique.*

Proof. There exists $\phi \in O_\varepsilon(3)$ such that $\phi(p) = e_3$. By Exercise 1.10, $\phi(q) = \gamma_\sigma(r)$ for some $\sigma \in [0, 2\pi[$ and $r \leq d_\varepsilon$. By Exercise 1.11, $r = d(p, q)$. This yields (i) and (ii). If $r < d_\varepsilon$, then $\phi(q) \neq -e_3$ and so the plane through $O, e_3, \phi(q)$ is unique. By Theorem 1.8 this implies that the geodesic through e_3 and $\phi(q)$ is unique. If $\varepsilon < 0$, then γ_σ contains only one arc from $e_3 = \phi(p)$ to $\phi(q)$. If $\varepsilon > 0$, then there are two such arcs, one of length r, the other of length $2d_\varepsilon - r > r$. This yields (iii). $\square$

Displacement Lengths and Angles of Rotation

From the rule $N_s N_t = N_{s+t}$ (Exercise 1.2(d)) we see that N_s shifts any point $\eta(t) = N_t e_3$ to $N_{s+t} e_3 = \eta(t+s)$. This shows that η is an invariant geodesic of N_s, and for any point $\eta(t)$ on this geodesic the distance to its image is equal to $|s|$.

More generally, if $M = \phi N_s \phi^{-1}$, where $\phi \in O_\varepsilon(3)$ maps η to some geodesic γ_M of H_ε^2, then γ_M is invariant under M, and for any point $p \in \gamma$ the distance to its image has the same value $\ell_M := |s|$.

Definition 1.14. In the above situation we call γ_M the *axis* of M and ℓ_M its *displacement length*.

Exercise 1.15.
$$\text{trace}(M) = 1 + 2\mathbf{C}_\varepsilon(\ell_M).$$

In a similar way, if $p \in H_\varepsilon^2$, $\phi \in O_\varepsilon(3)$ such that $\phi(p) = e_3$, and $M = \phi^{-1} R_{\alpha_M} \phi$ for some $\alpha_M \in]-\pi, \pi]$, then we have

Exercise 1.16.
$$\text{trace}(M) = 1 + 2\cos(\alpha_M).$$

We call, in this case, M a *rotation* with center p and $|\alpha_M|$ is the (absolute) *angle of rotation* of M.

Exercise 1.17. If M is a rotation with angle α_M and center p, then for any unit vector $v \in T_p H_\varepsilon^2$,
$$g_\varepsilon(v, Mv) = \cos(\alpha_M).$$

The preceding exercise shows that our definition of an angle of rotation is compatible with the usual definition of the angle between vectors relatively to a given scalar product:

Definition 1.18. For any $p \in H_\varepsilon^2$ and any pair of unit vectors $v, w \in T_p H_\varepsilon^2$ the *angle* $\sphericalangle(v, w)$ is defined as the unique real number in the interval $[0, \pi]$ such that
$$\cos \sphericalangle(v, w) = g_\varepsilon(v, w).$$

If M is the rotation with center p and angle α_M sending v to w, then the comparison of this definition with Exercise 1.17 shows that indeed $|\alpha_M| = \sphericalangle(v, w)$.

1.2 The Case $\varepsilon = 0$

Interpreting the terms suitably, it is also possible to include the case $\varepsilon = 0$ into our considerations. The set

$$\mathbb{S}_0^2 = \{x \in \mathbb{R}^3 \mid x_3^2 = 1\}$$

is a pair of parallel planes, and H_0^2 becomes the plane $x_3 = 1$. For arbitrary vectors in $\mathbb{R}^3$ the definition of h_0 as given by (1) does not make sense, but for $p \in \mathbb{S}_0^2$ the tangent vectors have the third coordinate equal to zero, and we may well define

$$g_0(v, w) = v_1 w_1 + v_2 w_2, \quad v, w \in T_p \mathbb{S}_0^2.$$

Thus, H_0^2 together with g_0 becomes a model of the Euclidean plane.

The planes through O intersect H_0^2 in straight lines, and we get all straight lines like this. Hence, on H_0^2 the geodesics as described in Theorem 1.8 are precisely the straight lines of Euclidean geometry.

It is interesting to see how the group $O_0(3)$ comes out. Since $h_\varepsilon(e_3, e_3) \to \infty$ as $\varepsilon \to 0$, the definition of $O_\varepsilon(3)$ as given by (3) cannot be extended to $\varepsilon = 0$. But the matrices S_1, S_2, S_3, R_α in Example 1.1 are still defined, and the N_t too: inspecting the Taylor series expansion (4) we may extend the definitions of the functions $\mathbf{C}_\varepsilon$ and $\mathbf{S}_\varepsilon$ to $\varepsilon = 0$ as follows.

$$\mathbf{C}_0(t) = 1, \quad \mathbf{S}_0(t) = t, \quad t \in \mathbb{R}.$$

The matrices N_t then become

$$N_t = \begin{pmatrix} 1 & 0 & t \\ 0 & 1 & 0 \\ 0 & 0 & 1 \end{pmatrix},$$

and for $x \in \mathrm{H}_0^2$ we get

$$N_t \begin{bmatrix} x_1 \\ x_2 \\ 1 \end{bmatrix} = \begin{bmatrix} x_1 + t \\ x_2 \\ 1 \end{bmatrix}.$$

i.e. N_t operates as a translation. Guided by Exercise 1.2 we now define $O_0(3)$ as the group generated by S_1, S_2, S_3, the R_α's and the N_t's in the above form. The reader may then easily check the following, where $O(2)$ is the group of all orthogonal 2×2-matrices.

$$O_0(3) = \left\{ \begin{pmatrix} a_{11} & a_{12} & t_1 \\ a_{21} & a_{22} & t_2 \\ 0 & 0 & \pm 1 \end{pmatrix} \mid \begin{bmatrix} t_1 \\ t_2 \end{bmatrix} \in \mathbb{R}^2, \begin{bmatrix} a_{11} & a_{12} \\ a_{21} & a_{22} \end{bmatrix} \in O(2) \right\} \qquad (7)$$

Exercise 1.19. The subgroup $O_0^+(3)$ formed by all $M \in O_0(3)$ with last component equal to $+1$ preserves H_0^2 and coincides with the group of Euclidean motions in the plane.

1.3 Curvature

In Riemannian geometry of dimension 2 the *Gauss curvature* $K(p)$ at any point p measures the deviation from Euclidean geometry by comparing the lengths of small distance circles $c_\rho(p)$ of radius ρ and center p with the lengths of the corresponding circles in the Euclidean plane. The definition is as follows, where ℓ denotes the length,

$$K(p) \stackrel{\text{def}}{=} -\frac{3}{\pi} \lim_{\rho \to 0} \frac{1}{\rho^3} \left(\ell(c_\rho(p)) - 2\pi\rho \right). \tag{8}$$

For H_ε^2 the length of $c_\rho(p)$ may be computed in polar coordinates via (6) and is equal to $2\pi \mathbf{S}_\varepsilon(\rho)$. The Taylor series expansion (4) then yields

$$\mathbb{S}_\varepsilon^2 \text{ has constant curvature } \varepsilon. \tag{9}$$

Volume Growth

Expression (8) for the deviation from Euclidean geometry may give the impression that we have here merely a negligible third order effect. Yet, the contrary is true! Let us consider, e.g. the area of large discs. By (6), the area $d\omega_\varepsilon$ of an infinitesimal region whose description in polar coordinates (ρ, σ) is that of a rectangle of width $d\rho$ and height $d\sigma$, is given by

$$d\omega_\varepsilon = \mathbf{S}_\varepsilon(\rho) \, d\rho \, d\sigma. \tag{10}$$

For the circular disc $B_r(p) = \{x \in H_\varepsilon^2 \mid d(x, p) < r\}$ of radius $r < d_\varepsilon$ (see (5)), the area becomes

$$\text{area} B_r(p) = \int_0^{2\pi} \int_0^r \mathbf{S}_\varepsilon(\rho) \, d\rho \, d\sigma = \frac{-2\pi}{\varepsilon} (\mathbf{C}_\varepsilon(r) - 1). \tag{11}$$

We see here that in the case of constant negative curvature the area grows exponentially. This exponential growth, so radically different from the Euclidean case, is responsible for many unexpected spectral phenomena in hyperbolic geometry.

Trigonometry and the Gauss-Bonnet Formula for Triangles

In the following we consider geodesic triangles i.e. triangles formed by geodesic arcs of lengths $< d_\varepsilon$. By abuse of notation, sides and their lengths are named by the same symbol.

Theorem 1.20 (Right-angled triangles). *For any geodesic triangle in* H^2_ε *with sides* a, b, c, *opposite angles* α, β *and* a *orthogonal to* b *the following hold.*

(i) $\mathbf{C}_\varepsilon(c) = \mathbf{C}_\varepsilon(a)\,\mathbf{C}_\varepsilon(b)$,

(ii) $\cos(\beta) = \mathbf{C}_\varepsilon(b)\sin(\alpha)$.

Proof. We show (i) and leave (ii) as an exercise. The strategy is to move the triangles using different isometries and obtain the formulas by comparing the traces.

Consider the common axis η of the isometries N_t (Lemma 1.7), and define $\mu = R_{\pi/2}(\eta)$. Then μ is the common axis of all

$$M_t = R_{\pi/2} N_t R_{-\pi/2} = \begin{pmatrix} 1 & 0 & 0 \\ 0 & \mathbf{C}_\varepsilon(t) & \mathbf{S}_\varepsilon(t) \\ 0 & -\varepsilon \mathbf{S}_\varepsilon(t) & \mathbf{C}_\varepsilon(t) \end{pmatrix}. \tag{12}$$

Place the triangle such that a, b lie on μ, η as shown in Fig. 1 with vertices at $\mu(a)$ and $\eta(b)$, and consider a symmetric copy (shaded) across vertex $\mu(a)$.

Now shift the triangles along μ using M_{-2a} as shown in the figure so that in its new position the shaded one has side b on η, and then apply N_{2b}. Under this operation side c is shifted by $2c$ along the geodesic γ through $\mu(a), \eta(b)$, and so $N_{2b}M_{-2a}$ has axis γ and displacement length $2c$. Using Exercise 1.15 we compute

$$1 + 2\mathbf{C}_\varepsilon(2c) = \text{trace}(N_{2b}M_{-2a}) = (1 + \mathbf{C}_\varepsilon(2a))(1 + \mathbf{C}_\varepsilon(2b)) - 1,$$

and (i) follows from the rule $\mathbf{C}_\varepsilon(2t) = 2\mathbf{C}_\varepsilon(t)^2 - 1$.

For the proof of (ii) the exercise is to find a position of the triangle such that applying $R_{2\alpha-\pi}M_{-2b}$ to it is the same as rotating it by -2β around its vertex that has angle β. $\square$

Remark 1.21. It is interesting to observe what formula (i) yields in the limit $\varepsilon \to 0$. In paragraph 1.2 we defined $\mathbf{C}_0$ as the function which is constant equal

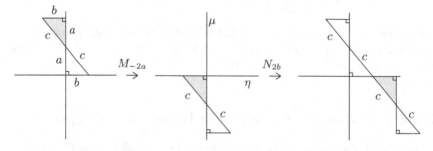

Fig. 1. The product $N_{2b}M_{-2a}$ of the vertical shift followed by the horizontal shift has displacement length $2c$.

to 1. With this definition the formula, as written, collapses to " $1 = 1$ ". If, however, we write out the formula using the Taylor series expansion of $\mathbf{C}_\varepsilon$ and subtract 1 on either side, then in the limit $\varepsilon \to 0$ we obtain

$$\text{`` } c^2 = a^2 + b^2 \text{ ''.}$$

Hence, formula (i) may be understood as a generalization to H_ε^2 of the classical Pythagorean theorem.

Another classical fact from Euclidean geometry, namely that the angle sum of a triangle equals π has the following generalization.

Theorem 1.22 (Gauss-Bonnet formula for triangles). *Let T be a geodesic triangle in* H_ε^2 *with interior angles* α, β, γ. *Then*

$$\varepsilon \operatorname{area}(T) = \alpha + \beta + \gamma - \pi.$$

Proof. Observe that any triangle can be either divided into two right-angled triangles or completed into a right-angled triangle. We may therefore restrict ourselves to right-angled triangles.

For $0 \le \sigma \le \alpha$, we consider the right-angled triangle with angle σ as in Fig. 2. According to Theorem 1.20, we have

$$\cos \beta(\sigma) = \mathbf{C}_\varepsilon(b) \sin \sigma$$
$$\cos \sigma = \mathbf{C}_\varepsilon(a(\sigma)) \sin \beta(\sigma).$$

Taking the derivative, $-\dot{\beta}(\sigma) \sin \beta(\sigma) = \mathbf{C}_\varepsilon(b) \cos \sigma = \mathbf{C}_\varepsilon(b)\mathbf{C}_\varepsilon(a(\sigma)) \sin \beta(\sigma)$ $= \mathbf{C}_\varepsilon(c(\sigma)) \sin \beta(\sigma)$, we get

$$-\dot{\beta}(\sigma) = \mathbf{C}_\varepsilon(c(\sigma)).$$

Using expression (10) for the area element we get

$$\operatorname{area}(T) = \int_0^\alpha \int_0^{c(\sigma)} \mathbf{S}_\varepsilon(\rho) d\rho d\sigma = \frac{-1}{\varepsilon} \int_0^\alpha (\mathbf{C}_\varepsilon(c(\sigma)) - 1)\, d\sigma$$
$$= \frac{1}{\varepsilon} \int_0^\alpha (1 + \dot{\beta}(\sigma))\, d\sigma.$$

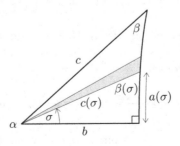

Fig. 2. Computing the area of a right-angled geodesic triangle.

Since $\beta(\alpha) = \beta$ and $\beta(0) = \pi/2$ this yields

$$\text{area}(T) = \frac{1}{\varepsilon}(\alpha + \beta - \pi/2).$$

$\square$

Remark 1.23 (The classical geometries). The spaces H_ε^2 are called the *classical geometries* (in dimension 2). One can show, e.g. using polar coordinates, that if ε and ε' have the same sign, then the Riemannian manifolds H_ε^2 and $\mathrm{H}_{\varepsilon'}^2$ differ only by a scaling factor. There are therefore essentially three classical geometries: the Euclidean plane H_0^2, the sphere H_1^2, and the space H_{-1}^2.

Definition 1.24. Any Riemannian manifold isometric to H_{-1}^2 is called (a model of) the *hyperbolic plane*.

Lecture 2:

The Poincaré Model of the Hyperbolic Plane

In this lecture we study the Poincaré model of the hyperbolic plane using Möbius transformations. In the first paragraph we obtain this model out of H_{-1}^2 using stereographic projection. After a brief account of the Möbius transformations in the second paragraph we will detail two versions of the Poincaré model, one in the unit disc and one in the upper half-plane.

2.1 Stereographic Projection

Let us again consider $\mathbb{S}_\varepsilon^2$ for arbitrary $\varepsilon \neq 0$ (c.f. (2)). The *stereographic projection* is defined as the central projection π_ε of $\mathbb{S}_\varepsilon^2$ onto the (u_1, u_2)-plane with center of projection $(0, 0, -1)$. It is given by the following formulas :

$$\pi_\varepsilon : \mathbb{S}_\varepsilon^2 \smallsetminus \{(0,0,-1)\} \longrightarrow \mathbb{R}^2$$

$$(x_1, x_2, x_3) \longmapsto (u_1, u_2) = \left(\frac{x_1}{1+x_3}, \frac{x_2}{1+x_3} \right). \tag{13}$$

In order to extend this mapping to all of $\mathbb{S}_\varepsilon^2$ we add a "point at infinity" ∞ to $\mathbb{R}^2$ setting $\widehat{\mathbb{R}}^2 = \mathbb{R}^2 \cup \{\infty\}$ and use the convention $\pi_\varepsilon(0,0,-1) = \infty$. The image $\mathbb{D}_\varepsilon = \pi_\varepsilon(\mathbb{S}_\varepsilon^2)$ becomes

$$\mathbb{D}_\varepsilon = \widehat{\mathbb{R}}^2, \text{ if } \varepsilon > 0, \qquad \mathbb{D}_\varepsilon = \widehat{\mathbb{R}}^2 \smallsetminus \left\{ u \in \mathbb{R}^2 \mid u_1^2 + u_2^2 = \frac{1}{|\varepsilon|} \right\}, \text{ if } \varepsilon < 0.$$

Observe that $\pi_\varepsilon : \mathbb{S}_\varepsilon^2 \longrightarrow \mathbb{D}_\varepsilon$ is one-to-one. For the inverse mapping $\pi_\varepsilon^{-1} : \mathbb{D}_\varepsilon \longrightarrow \mathbb{S}_\varepsilon^2$, we use that $\varepsilon(u_1^2 + u_2^2)(1 + x_3)^2 + x_3^2 = 1$ to get the following:

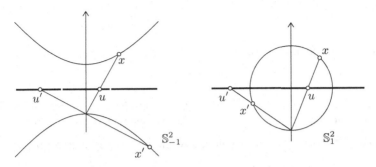

Fig. 3. Profile views of the stereographic projection for the cases $\varepsilon = -1$ and $\varepsilon = 1$. The second figure is drawn with a larger scale.

$$x_k = \frac{2u_k}{1+\rho}, \quad k = 1, 2, \quad x_3 = \frac{1-\rho}{1+\rho}, \quad \text{where } \rho = \varepsilon(u_1^2 + u_2^2). \tag{14}$$

Exercise 2.1. The induced metric on $\mathbb{D}_\varepsilon$ via π_ε^{-1} is

$$ds^2 = \frac{4(du_1^2 + du_2^2)}{(1+\rho)^2}.$$

Remark 2.2. The following are simple consequences of the above.

(i) $(\mathbb{S}_\varepsilon^2 \setminus \{(0,0,-1)\}, \pi_\varepsilon|_{\mathbb{S}_\varepsilon^2 \setminus \{(0,0,-1)\}})$ is a coordinate chart for $\mathbb{S}_\varepsilon^2$.
(ii) The angular measure for tangent vectors with respect to g_ε (Definition 1.3) is the same as in the coordinate chart.
(iii) If $\varepsilon = 1$, g_ε is the ordinary scalar product and π_ε the standard stereographic projection of the sphere $\mathbb{S}^2$. Hence, Exercise 2.1 proves that the stereographic projection preserves angles.

Definition 2.3. A *generalized circle of* $\widehat{\mathbb{R}}^2$ is either a circle in $\mathbb{R}^2$ or the union $L \cup \{\infty\}$, where L is a straight line in $\mathbb{R}^2$.

A *generalized circle of* $\mathbb{D}_\varepsilon$ is the intersection $\mathbb{D}_\varepsilon \cap C$, where C is a generalized circle of $\widehat{\mathbb{R}}^2$.

Theorem 2.4. *If a plane $E \subset \mathbb{R}^3$ intersects $\mathbb{S}_\varepsilon^2$ in more than one point, then $\pi_\varepsilon(E \cap \mathbb{S}_\varepsilon^2)$ is a generalized circle of $\mathbb{D}_\varepsilon$.*

Proof. Generalized circles are subsets of $\mathbb{D}_\varepsilon$ satisfying an equation of the form

$$m_1 u_1 + m_2 u_2 + m_3(u_1^2 + u_2^2) + m_4 = 0.$$

Using (14) and assuming that the image lies in a plane, i.e satisfies a linear equation

$$a_1 x_1 + a_2 x_2 + a_3 x_3 + a_4 = 0,$$

we obtain

$$(a_1u_1 + a_2u_2)(1 + x_3) + a_3\frac{1 - \rho}{1 + \rho} + a_4 = 0.$$

Since $(1 + \rho) \neq 0$, this is equivalent to

$$2\,(a_1u_1 + a_2u_2) + \varepsilon(u_1^2 + u_2^2)(a_4 - a_3) + (a_3 + a_4) = 0.$$

□

It is clear that the isometries of $\mathbb{S}_\varepsilon^2$ preserve circles and angles, and we have just seen that the same is the case for the stereographic projection.

Exercise 2.5. Show that any generalized circle in $\mathbb{D}_\varepsilon$ is such a projection.

In the next exercise, η_ε is the circle $\{u \in \mathbb{R}^2 \mid u_1^2 + u_2^2 = 1/|\varepsilon|\}$.

Exercise 2.6. Let $C = \pi_\varepsilon(\gamma)$ be a generalized circle.

(a) If $\varepsilon > 0$, γ is a geodesic if and only if C intersects η_ε in two opposite points.
(b) If $\varepsilon < 0$, γ is a geodesic if an only if C intersects η_ε orthogonally.

2.2 Möbius Transformation of $\widehat{\mathbb{R}}^2$

The Poincaré model uses the following type of mappings extensively.

Definition 2.7. A mapping $\widehat{\mathbb{R}}^2 \longrightarrow \widehat{\mathbb{R}}^2$ which preserves angles and generalized circles is a *Möbius transformation of* $\widehat{\mathbb{R}}^2$.
    The group of all Möbius transformations is denoted by $\mathrm{M\ddot{o}b}(\widehat{\mathbb{R}}^2)$.

Example 2.8. By Remark 2.2(iii) and Theorem 2.4, for any isometry $g : \mathbb{S}_1^2 \to \mathbb{S}_1^2$ the conjugate mapping $\pi_1 \circ g \circ \pi_1^{-1} : \widehat{\mathbb{R}}^2 \to \widehat{\mathbb{R}}^2$ is a Möbius transformation.

The Möbius transformations have a powerful notation using the complex numbers. With the identification $\mathbb{R}^2 \ni (x,y) \mapsto x + iy \in \mathbb{C}$ and setting $\widehat{\mathbb{C}} = \mathbb{C} \cup \{\infty\}$, we will write $\mathrm{M\ddot{o}b}(\widehat{\mathbb{R}}^2) = \mathrm{M\ddot{o}b}(\widehat{\mathbb{C}})$.
    The group $\mathrm{M\ddot{o}b}(\widehat{\mathbb{C}})$ obviously contains the following transformations, where $\bar{z}$ is the complex conjugate of z:

$$z \mapsto az, \; a \in \mathbb{C} \smallsetminus \{0\},$$
$$z \mapsto z + b, \; b \in \mathbb{C}, \tag{15}$$
$$z \mapsto \bar{z}.$$

It also contains the transformation

$$z \mapsto \frac{1}{\bar{z}}. \tag{16}$$

To prove that this is a Möbius transformation we simply observe that (16) is the complex notation for the mapping $\pi_1 \circ S_3 \circ \pi_1^{-1}$, where S_3 is as in Example 1.1.

In the next theorem, $\mathrm{GL}(2,\mathbb{C})$ denotes the group of all complex 2×2-matrices with determinant $\neq 0$.

Theorem 2.9. *The Möbius transformations of* $\widehat{\mathbb{C}}$ *are generated by* (15)–(16), *and each of them is of one of the following forms*:

$$z \mapsto \frac{az+b}{cz+d}, \quad \begin{pmatrix} a & b \\ c & d \end{pmatrix} \in \mathrm{GL}(2,\mathbb{C}); \quad z \mapsto \frac{a\bar{z}+b}{c\bar{z}+d}, \quad \begin{pmatrix} a & b \\ c & d \end{pmatrix} \in \mathrm{GL}(2,\mathbb{C}).$$

Proof. Let m be a mapping of this form. If $c = 0$, then either $m(z) = \frac{a}{d}z + \frac{b}{d}$ or $m(z) = \frac{a}{d}\bar{z} + \frac{b}{d}$. If $c \neq 0$, then either $m(z) = \frac{a}{c} + \frac{b-\frac{ad}{c}}{cz+d}$ or $m(z) = \frac{a}{c} + \frac{b-\frac{ad}{c}}{c\bar{z}+d}$. In either case, m is a composition of mappings of types (15)–(16) and thus a Möbius transformation.

To prove that all Möbius transformations are as in the theorem we first note that transformations of this form can map any point to ∞. It suffices therefore to consider the ones which fix ∞. But any Möbius transformation fixing ∞ preserves straight lines and angles, and is thus a similarity transformation $z \mapsto az + b$, or $z \mapsto a\bar{z} + b$. $\square$

2.3 The Poincaré Disc Model

In Section 2.1 we saw that $\pi_{-1} : \mathbb{S}^2_{-1} \setminus \{(0,0,-1)\} \to \mathbb{D}_{-1}$ is one-to-one and onto, and the induced metric on $\mathbb{D}_{-1}$ is given by Exercise 2.1. Restricting π_{-1} to the sheet $\mathbb{S}^{2+}_{-1} = \mathrm{H}^2_{-1}$ (c.f. Definition 1.6) we arrive at the following Riemannian manifold

$$\mathbb{D} = \{z = x + iy \in \mathbb{C} \mid |z| < 1\}, \tag{17}$$

endowed with the metric

$$ds^2 = \frac{4(dx^2 + dy^2)}{(1 - (x^2 + y^2))^2} = \frac{4\,dz\,d\bar{z}}{(1 - |z|^2)^2}. \tag{18}$$

Observe from this definition that $\pi_{-1} : \mathrm{H}^2_{-1} \to \mathbb{D}$ is an isometry.

The unit disc $\mathbb{D}$ together with the metric (18) is called the *Poincaré disc model* of the hyperbolic plane.

Isometries

Conjugating the generators S_2, R_α and N_t (for $\alpha, t \in \mathbb{R}$) of $\mathrm{Isom}(\mathrm{H}^2_{-1})$ via the stereographic projection π_{-1} we obtain generators of $\mathrm{Isom}(\mathbb{D})$.

$$\begin{array}{ll} \widetilde{\sigma} : z \mapsto \bar{z} & (\widetilde{\sigma} = \pi_{-1} \circ S_2 \circ \pi_{-1}^{-1}) \\[2mm] \widetilde{\rho}_\alpha : z \mapsto e^{i\alpha}z & (\widetilde{\rho}_\alpha = \pi_{-1} \circ R_\alpha \circ \pi_{-1}^{-1}) \\[2mm] \widetilde{\nu}_t : z \mapsto \frac{z+\tanh\frac{t}{2}}{1+(\tanh\frac{t}{2})z} & (\widetilde{\nu}_t = \pi_{-1} \circ N_t \circ \pi_{-1}^{-1}) \end{array} \tag{19}$$

This may be checked by direct computation, where the third case is a bit tedious. Observe that $\widetilde{\nu}_t$ fixes points -1 and 1 on the boundary of $\mathbb{D}$ and maps 0 to $\tanh\frac{t}{2} = \frac{\sinh t}{1+\cosh t}$.

Exercise 2.10. The isometries of $\mathbb{D}$ are the Möbius transformations preserving $\mathbb{D}$. These are the mappings $\widehat{\mathbb{C}} \to \widehat{\mathbb{C}}$ of one of the following types:

$$z \mapsto \frac{az+b}{\bar{b}z+\bar{a}}, \qquad \begin{pmatrix} a & b \\ \bar{b} & \bar{a} \end{pmatrix} \in \mathrm{GL}(2,\mathbb{C}),\ \det \begin{pmatrix} a & b \\ \bar{b} & \bar{a} \end{pmatrix} > 0,$$

$$z \mapsto \frac{\frac{a}{\bar{z}}+b}{\frac{b}{\bar{z}}+\bar{a}}, \qquad \begin{pmatrix} a & b \\ \bar{b} & \bar{a} \end{pmatrix} \in \mathrm{GL}(2,\mathbb{C}),\ \det \begin{pmatrix} a & b \\ \bar{b} & \bar{a} \end{pmatrix} < 0.$$

(Hint: write these mappings as products of the ones given in (19).)

Matrix Notation

Let $\mathrm{GL}(2,\mathbb{C})^{\sim}$ be the subgroup of $\mathrm{GL}(2,\mathbb{C})$ formed by all matrices of the form $M = \begin{pmatrix} a & b \\ \bar{b} & \bar{a} \end{pmatrix}$. We will associate to any such M the Möbius transformation of $\mathbb{D}$ (and thus an isometry of $\mathbb{D}$) as in Exercise 2.10, i.e.:

$$z \mapsto M[[z]] = \begin{cases} \dfrac{az+b}{\bar{b}z+\bar{a}}, & \text{if } \det M > 0; \\[2ex] \dfrac{\frac{a}{\bar{z}}+b}{\frac{b}{\bar{z}}+\bar{a}}, & \text{if } \det M < 0; \end{cases} \qquad z \in \mathbb{D}. \tag{20}$$

Exercise 2.11. Check that this correspondence respects the group laws of $\mathrm{GL}(2,\mathbb{C})^{\sim}$ and $\mathrm{Isom}\,\mathbb{D}$, and that its kernel is $\{ \begin{pmatrix} \lambda & 0 \\ 0 & \lambda \end{pmatrix} \mid \lambda \in \mathbb{R} \smallsetminus \{0\} \}$.

Geodesics

In order to see how the geodesics in $\mathbb{D}$ look like, we first note that the image in $\mathbb{D}$ under the projection $\pi_{-1} : \mathrm{H}^2_{-1} \to \mathbb{D}$ of geodesic η in H^2_{-1} (Lemma 1.7) is the segment $]-1,1[$ on the real line. For the general case we then have

Theorem 2.12. *The geodesics of $\mathbb{D}$ are the generalized circles of $\mathbb{D}$ orthogonal to the boundary $\partial\mathbb{D}$.*

Proof. The isometries of $\mathbb{D}$ are the Möbius transformations of $\widehat{\mathbb{C}}$ preserving $\mathbb{D}$ (and $\partial\mathbb{D}$). Therefore, the images of $]-1,1[= \mathbb{D} \cap \mathbb{R}$ under the isometries of $\mathbb{D}$ are intersections with $\mathbb{D}$ of generalized circles of $\mathbb{C}$ orthogonal to $\partial\mathbb{D}$. As $\mathrm{Möb}(\widehat{\mathbb{C}})$ sends any triple of distinct points in $\widehat{\mathbb{C}}$ to any other such triple (and three points determine a circle), there exists, for any given pair of points $u, v \in \partial\mathbb{D}$, a Möbius transformation of $\widehat{\mathbb{C}}$ operating isometrically on $\mathbb{D}$ mapping $\{-1,1\}$ to $\{u,v\}$, and hence the segment $]-1,1[$ onto the circular orthogonal arc through u and v. This shows that all arcs in question are geodesics. That these are the only geodesics is a consequence of Theorem 1.5. $\square$

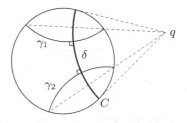

Fig. 4. A Euclidean construction of the common perpendicular of two non-Euclidean lines.

Example 2.13 (Common perpendiculars). If γ_1, γ_2 are non-intersecting geodesics in $\mathbb{D}$ represented by disjoint circles C_1, C_2 in $\mathbb{C}$, then there exists a geodesic δ in $\mathbb{D}$ perpendicular to both of them, called the *common perpendicular* of γ_1 and γ_2.

Here is a construction of δ using elementary Euclidean geometry. Draw straight lines through the intersection points of $\partial\mathbb{D}$ and C_1, respectively C_2. Let q be the intersection point of these lines and draw a circle C with center q such that C intersects $\partial\mathbb{D}$ orthogonally. By the Secant-Segments theorem, C intersects also C_1 and C_2 orthogonally and thus yields the common perpendicular.

In Lecture 4 we will introduce a computational tool to evaluate this perpendicular in a very simple way (see Theorem 4.7).

Exercise 2.14. Let $p, \omega \in \mathbb{D}$ be such that

$$\omega = \frac{2p}{1 + |p|^2}, \qquad p = \frac{\omega}{1 + \sqrt{1 - |\omega|^2}}.$$

(a) The matrix

$$M_p = \frac{i}{\sqrt{1 - |\omega|^2}} \begin{pmatrix} 1 & -\omega \\ \bar{\omega} & -1 \end{pmatrix} = \frac{i}{1 - |p|^2} \begin{pmatrix} 1 + |p|^2 & -2p \\ 2\bar{p} & -(1 + |p|^2) \end{pmatrix}$$

has determinant 1 and represents an isometry $m_p \in \mathrm{Isom}(\mathbb{D})$, $m_p(z) = M_p[[z]]$, with the following properties for $z \in \mathbb{D}$,

$$m_p(z) = z \iff z = p, \qquad m_p^2 = \mathrm{id}. \qquad (21)$$

Such an isometry is called a *half-turn with center p*. Observe that the square of the matrix is $M_p^2 = \begin{pmatrix} -1 & 0 \\ 0 & -1 \end{pmatrix}$.

(b) Assume $z_1, z_2 \in \mathbb{D}$ and let p be the midpoint of the geodesic arc from z_1 to z_2 (so that $m_p(z_1) = z_2$ and $m_p(z_2) = z_1$). Then p and M_p are given in the above form with

$$\omega = \frac{z_1(1 - |z_2|^2) + z_2(1 - |z_1|^2)}{1 - |z_1 z_2|^2}.$$

(c) Assume $q \in \mathbb{C}$, $|q| > 1$. Then there exists a circle C in $\mathbb{C}$ with center q (in the Euclidean sense) orthogonal to the boundary $\partial\mathbb{D}$ of $\mathbb{D}$ defining a geodesic $\gamma = C \cap \mathbb{D}$. In this case the matrix

$$M_\gamma = \frac{i}{\sqrt{|q|^2 - 1}} \begin{pmatrix} 1 & -q \\ \bar{q} & -1 \end{pmatrix}$$

has determinant -1 and represents an isometry $m_\gamma \in \mathrm{Isom}(\mathbb{D})$, $m_\gamma(z) = M_\gamma[[z]]$, with the following properties for $z \in \mathbb{D}$,

$$m_\gamma(z) = z \iff z \in \gamma, \qquad m_\gamma^2 = \mathrm{id}. \tag{22}$$

Such an isometry is called a *symmetry with axis* γ. Observe that here $M_\gamma^2 = \begin{pmatrix} 1 & 0 \\ 0 & 1 \end{pmatrix}$.

2.4 The Upper Half-Plane Model

The Möbius transformation

$$\psi : w \mapsto i\frac{1+w}{1-w} \quad \text{with inverse} \quad \psi^{-1} : z \mapsto \frac{z-i}{z+i}, \tag{23}$$

maps $\mathbb{D}$ in a one-to-one manner onto the so-called *upper half-plane*

$$\mathbb{H} = \{z \in \mathbb{C} \mid \mathrm{Im}\,z > 0\}. \tag{24}$$

The upper half-plane model obtained in this way is the most frequently used model of the hyperbolic plane in the literature. In order to get a description of it we translate the preceding section from $\mathbb{D}$ to $\mathbb{H}$ using ψ.

Exercise 2.15. The induced metric on $\mathbb{H}$ via ψ^{-1} is

$$ds^2 = \frac{dx^2 + dy^2}{y^2} = \frac{dz\,d\bar{z}}{(\mathrm{Im}\,z)^2}.$$

Isometries

Conjugating the generators (19) of $\mathrm{Isom}(\mathbb{D})$ with ψ yields the following generators of $\mathrm{Isom}(\mathbb{H})$,

$$\begin{aligned} \sigma &: z \mapsto -\bar{z} & (\sigma = \psi \circ \tilde{\sigma} \circ \psi^{-1}) \\ \rho_\alpha &: z \mapsto \frac{\cos\frac{\alpha}{2}\,z + \sin\frac{\alpha}{2}}{-\sin\frac{\alpha}{2}\,z + \cos\frac{\alpha}{2}} & (\rho_\alpha = \psi \circ \tilde{\rho}_\alpha \circ \psi^{-1}) \\ \nu_t &: z \mapsto e^t z & (\nu_t = \psi \circ \tilde{\nu}_t \circ \psi^{-1}) \end{aligned} \tag{25}$$

Of course, one may also check directly that: a) the mappings in (25) are isometries; b) the isometries in this list generate $\mathrm{Isom}(\mathbb{H})$. (For the latter one has to use Theorem 1.5.)

The isometry group may again be characterized as follows.

Exercise 2.16. The isometries of $\mathbb{H}$ are the Möbius transformations preserving $\mathbb{H}$. These are the mappings $\widehat{\mathbb{C}} \to \widehat{\mathbb{C}}$ of one of the following types:

$$z \mapsto \frac{az+b}{cz+d}, \qquad \begin{pmatrix} a & b \\ c & d \end{pmatrix} \in \mathrm{GL}(2,\mathbb{R}), \ \det \begin{pmatrix} a & b \\ c & d \end{pmatrix} > 0,$$

$$z \mapsto \frac{a\bar{z}+b}{c\bar{z}+d}, \qquad \begin{pmatrix} a & b \\ c & d \end{pmatrix} \in \mathrm{GL}(2,\mathbb{R}), \ \det \begin{pmatrix} a & b \\ c & d \end{pmatrix} < 0.$$

Matrix Notation

Similarly to the case of the disc model we will associate to $M \in \mathrm{GL}(2,\mathbb{R})$ the Möbius transformation as in Exercise 2.16, where now we write $M[z]$ instead of $M[[z]]$.

$$z \mapsto M[z] = \begin{cases} \dfrac{az+b}{cz+d}, & \text{if } \det M > 0; \\[2mm] \dfrac{a\bar{z}+b}{c\bar{z}+d}, & \text{if } \det M < 0; \end{cases} \qquad z \in \mathbb{H}. \qquad (26)$$

Note that the mapping $z \mapsto \bar{z}$ is orientation reversing. It follows that the isometries given by the top line of (26) are orientation preserving, and the isometries given by the bottom line are orientation reversing.

Exercise 2.17. Check that the correspondence (26) respects the group laws of $\mathrm{GL}(2,\mathbb{R})$ and $\mathrm{Isom}\,\mathbb{H}$, and that its kernel is $\{ \left(\begin{smallmatrix} \lambda & 0 \\ 0 & \lambda \end{smallmatrix} \right) \mid \lambda \in \mathbb{R} \smallsetminus \{0\} \}$.

Two important groups arise here:

$$\begin{aligned} \mathrm{SL}(2,\mathbb{R}) &= \{ M \in \mathrm{GL}(2,\mathbb{R}) \mid \det M = 1 \}, \\ \mathrm{PSL}(2,\mathbb{R}) &= \mathrm{SL}(2,\mathbb{R})/\{\pm\mathbf{1}\}. \end{aligned} \qquad (27)$$

In the second line, $\mathbf{1}$ is the identity matrix, and $/\{\pm\mathbf{1}\}$ (read: "modulo plus minus $\mathbf{1}$") means that two matrices which differ by a multiplication by -1 are considered equivalent.

Exercise 2.17 says, in particular, that the correspondence (26) establishes an isomorphism between $\mathrm{PSL}(2,\mathbb{R})$ and the group $\mathrm{Isom}^+(\mathbb{H})$ of all orientation preserving isometries of $\mathbb{H}$.

Geodesics

To get hold of the geodesics of $\mathbb{H}$ we use the fact that the Möbius transformation $\psi : \mathbb{D} \to \mathbb{H}$ is an isometry (because the Riemannian metric of $\mathbb{H}$ is the one induced by ψ^{-1}). Since isometries preserve geodesics and since ψ preserves angles and generalized circles, we may immediately translate Theorem 2.12 into the following.

Theorem 2.18. *The geodesics of $\mathbb{H}$ are the half-circles centered on the real axis and the vertical straight half-lines.*

Exercise 2.19. Give a direct proof of Theorem 2.18 proceeding in two steps.

(a) For any points p, q on the half imaginary axis $L = \{z \in \mathbb{H} \mid \mathrm{Re}z = 0\}$ the arc on L from p to q is the shortest differentiable curve in $\mathbb{H}$ from p to q.

(b) For any $p, q \in \mathbb{H}$ there exists $\phi \in \mathrm{Isom}(\mathbb{H})$ sending L to a generalized circle through p and q with center on the real axis.

Exercise 2.20. (a) $\mathrm{Isom}^{+}(\mathbb{H})$ is generated by the mappings

$$z \mapsto z + a, \quad z \mapsto \lambda z, \quad z \mapsto -1/z; \quad a \in \mathbb{R}, \lambda > 0.$$

(b) For $p, q \in \mathbb{H}$ the hyperbolic distance $d(p,q)$ from p to q is given by

$$\cosh d(p,q) = 1 + \frac{1}{2\mathrm{Im}(p)\mathrm{Im}(q)}|p - q|^{2}. \tag{28}$$

(c) For $A = \left(\begin{smallmatrix} a & b \\ c & d \end{smallmatrix}\right) \in \mathrm{SL}(2,\mathbb{R})$ the distance from i to $A[\mathrm{i}]$ satisfies

$$2\cosh d(A[\mathrm{i}], \mathrm{i}) = a^{2} + b^{2} + c^{2} + d^{2}.$$

(Hint: for (a), glance at the proof of Theorem 2.9. For (b), show that the formula is invariant under the generators in (a) and reduce the proof to the case where $p, q \in L$ with L as in Exercise 2.19.)

Exercise 2.21.

(a) Let $p = r + \mathrm{i}s \in \mathbb{H}$, $r \in \mathbb{R}$, $s > 0$. The following matrix represents a half-turn $h_p : \mathbb{H} \to \mathbb{H}$ with center p (c.f. Exercise 2.14(a)),

$$H_p = \frac{1}{s}\begin{pmatrix} -r & r^{2} + s^{2} \\ -1 & r \end{pmatrix}.$$

Observe that $h_p^2 = \mathrm{id}$ and $H_p^2 = -1$.

(b) Let $r \in \mathbb{R}$ be the center and $s > 0$ the radius of a half-circle representing a geodesic $\gamma \in \mathbb{H}$. Then the following matrix represents the symmetry $h_\gamma : \mathbb{H} \to \mathbb{H}$ with axis γ (c.f. Exercise 2.14(c)),

$$H_\gamma = \frac{1}{s}\begin{pmatrix} r & -r^{2} + s^{2} \\ 1 & -r \end{pmatrix}.$$

How does H_γ look like if γ is a vertical half-line? Observe that in either case $h_\gamma^2 = \mathrm{id}$ and $H_\gamma^2 = 1$.

(c) Let $H \in \mathrm{GL}(2,\mathbb{R})$, $H \neq \pm 1$ and assume that $H^2 = \pm 1$. Then trace $H = 0$. H represents a half turn if $H^2 = -1$, a symmetry if $H^2 = +1$.

2.5 Types of Orientation Preserving Isometries

We classify the different types of orientation preserving (or "direct") isometries of the hyperbolic plane using the upper half-plane model and give a geometric interpretation of the traces of the corresponding matrices.

Here, and also later on, it is useful to work with *half-traces*. We use the following notation,

$$\mathrm{tr}(A) = \tfrac{1}{2}\mathrm{trace}(A). \qquad (29)$$

The classification hinges on the fixed points and is given by the following exercise.

Exercise 2.22. Let $A \in \mathrm{SL}(2,\mathbb{R})$ be different from the identity matrix and consider its action $z \mapsto A[z]$ on $\widehat{\mathbb{C}}$, respectively $\mathbb{H}$.

— If $|\mathrm{tr}(A)| < 1$, then A has two fixed points in $\mathbb{C}$, p and $\bar{p}$, with one of them, say p in $\mathbb{H}$. In this case, A is said to be of *elliptic type*.
— If $|\mathrm{tr}(A)| > 1$, then A has two distinct fixed points in $\mathbb{R} \cup \infty = \partial\mathbb{H}$. In this case, A is said to be of *hyperbolic type*.
— If $|\mathrm{tr}(A)| = 1$, then A has a unique fixed point, and this point lies on $\mathbb{R} \cup \infty = \partial\mathbb{H}$. In this case, A is said to be of *parabolic type*.

Fig. 5 shows the three cases. In the elliptic case, A represents a rotation with fixed point p (in the sense of the hyperbolic geometry of $\mathbb{H}$). The dotted line is a distance circle of center p. It is invariant under A and orthogonal to the geodesics through p.

The figure in the middle shows the hyperbolic case. The dark circle meets $\partial\mathbb{H}$ at the two fixed points and is the axis of the action of A. Any geodesic orthogonal to the axis is mapped to another such geodesic. The dotted circle is orthogonal to these and is invariant under the action of A.

The parabolic case, on the right-hand side, is a limit case between the first and the second. The center of rotation and the invariant axis are both "collapsed" into a point at infinity. The dotted circle is again invariant under the action of A. Circles of this type are called *horocycles*.

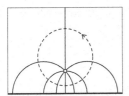 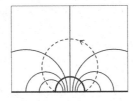 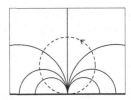

Fig. 5. The three types of orientation preserving isometries of the hyperbolic plane: elliptic, hyperbolic, and parabolic.

Elliptic Type

If $A \in \mathrm{SL}(2, \mathbb{R})$ represents a Möbius transformation with fixed point $z = i$, then a simple calculation shows that there exists $\alpha \in \,]-\pi, \pi]$ such that $A = \pm A_\alpha$, where

$$A_\alpha = \begin{pmatrix} \cos\frac{\alpha}{2} & \sin\frac{\alpha}{2} \\ -\sin\frac{\alpha}{2} & \cos\frac{\alpha}{2} \end{pmatrix}. \tag{30}$$

Hence, A is a matrix representation of the generator ρ_α given in the list (25). This shows that the only elliptic transformations with fixed point i are the ρ_α's.

The corresponding isometry of the unit disc, $\widetilde{\rho}_\alpha = \psi^{-1} \circ \rho_\alpha \circ \psi$ from the list (19) is a Euclidean rotation with angle of rotation α. Since the Euclidean and hyperbolic angular measures coincide, and $\psi : \mathbb{D} \to \mathbb{H}$ preserves angles, it follows that if v is a unit tangent vector at point i, and ρ_α sends v to v', then $\sphericalangle(v, v') = \alpha$.

Now let $z \mapsto \phi(z) = A[z]$ with $A \in \mathrm{SL}(2, \mathbb{R})$ be an arbitrary elliptic transformation with fixed point, say $p = r + is$. The mapping

$$z \mapsto \mu(z) = M[z], \quad z \in \mathbb{H}, \quad M = \frac{1}{\sqrt{s}} \begin{pmatrix} 1 & -r \\ 0 & s \end{pmatrix}, \tag{31}$$

is an isometry sending p to i, and by our discussion there exists $\alpha \in \,]-\pi, \pi]$ such that $\mu \circ \phi \circ \mu^{-1} = \rho_\alpha$, and hence $\phi = \mu^{-1} \circ \rho_\alpha \circ \mu$. Note that this implies $A = \pm M^{-1} A_\alpha M$. We say that ϕ is a *rotation*, p is its *center*, and α its *angle of rotation*.

Using the trace rule $\mathrm{tr}(M^{-1} A_\alpha M) = \mathrm{tr} A_\alpha$ we arrive at the following result: *Any elliptic transformation $z \mapsto A[z]$, $A \in \mathrm{SL}(2, \mathbb{R})$, is a rotation, and its angle of rotation α satisfies*

$$\cos\frac{\alpha}{2} = |\mathrm{tr} A|. \tag{32}$$

Observe that the same holds in $\mathbb{D}$ with $A \in \mathrm{GL}(2, \mathbb{C})^{\sim}$.

Exercise 2.23.

(a) The rotation with center $p \in \mathbb{D}$ and angle α is given as follows, where M_p is as in Exercise 2.14(a)

$$A = \cos\frac{\alpha}{2}\, \mathbf{1} + \sin\frac{\alpha}{2}\, M_p.$$

(b) The rotation with center $p \in \mathbb{H}$ and angle α is given as follows, where H_p is as in Exercise 2.21(a)

$$A = \cos\frac{\alpha}{2}\, \mathbf{1} + \sin\frac{\alpha}{2}\, H_p.$$

Hyperbolic Type

If $A \in \mathrm{SL}(2, \mathbb{R})$ represents a Möbius transformation with fixed points 0 and ∞, then a simple calculation shows that there exists $t \in \mathbb{R}$ such that $A = \pm B_t$, where

$$B_t = \begin{pmatrix} e^{\frac{t}{2}} & 0 \\ 0 & e^{-\frac{t}{2}} \end{pmatrix}. \tag{33}$$

Hence, A is a matrix representation of the generator ν_t from the list in (25). This shows that the only hyperbolic transformations with fixed points 0 and ∞ are the ν_t's.

For any point $p = r + is \in \mathbb{H}$ the distance formula (28) yields

$$\cosh d(p, \nu_t(p)) = 1 + \frac{(e^t - 1)^2 |r + is|^2}{2e^t s^2} \geq 1 + \frac{(e^t - 1)^2}{2e^t} = \cosh(t),$$

with equality if and only if $r = 0$, i.e. if and only if p lies on the axis of ν_t.

Now let $z \mapsto \phi(z) = A[z]$ with $A \in \mathrm{SL}(2, \mathbb{R})$ be an arbitrary hyperbolic transformation with fixed points $u, w \in \partial\mathbb{H}$. Its axis is $\gamma_A = C \cap \mathbb{H}$, where C is the generalized circle intersecting $\partial\mathbb{H}$ orthogonally at u and w. The points u and w are sometimes called the *endpoints at infinity* of γ_A. The mapping

$$z \mapsto \mu(z) = M[z], \quad z \in \mathbb{H}, \quad M = \begin{pmatrix} \frac{w}{w-u} & \frac{-uw}{w-u} \\ \frac{-1}{w} & 1 \end{pmatrix}, \tag{34}$$

sends u to 0 and w to ∞. Since $\mu \circ \phi \circ \mu^{-1}$ fixes 0 and ∞, it follows from the above discussion that there exists a unique $t_A \in \mathbb{R}$ such that $\mu \circ \phi \circ \mu^{-1} = \nu_{t_A}$, and hence $\phi = \mu^{-1} \circ \nu_{t_A} \circ \mu$. As in the preceding paragraph this yields $A = \pm M^{-1} B_{t_A} M$ and $|\mathrm{tr}(A)| = |\mathrm{tr}(B_{t_A})| = \cosh(t_A/2)$.

Observing that $p \in \mathbb{H}$ lies on γ_A if and only if $\mu(p)$ lies on the axis of ν_{t_A}, and using that

$$d(p, \phi(p)) = d(\mu(p), \mu(\phi(p))) = d(\mu(p), \mu \circ \phi \circ \mu^{-1}(\mu(p)))$$
$$= d(\mu(p), \nu_{t_A}(\mu(p))),$$

we arrive at the following result.

Theorem 2.24. *Let $z \mapsto A[z]$ with $A \in \mathrm{SL}(2, \mathbb{R})$ be a hyperbolic transformation and let $\ell_A > 0$ be the real number defined by*

$$\cosh(\ell_A/2) = |\mathrm{tr}(A)|.$$

Then $d(p, A[p]) \geq \ell_A$, for any $p \in \mathbb{H}$, with equality if and only p lies on the axis of A. $\square$

The number ℓ_A is called the *displacement length* of A.

Exercise 2.25.

(a) Let $q \in \mathbb{C}$, $|q| > 1$ be the center of the generalized circle representing a geodesic γ in $\mathbb{D}$ and define M_γ as in Exercise 2.14(c). The hyperbolic isometry with axis γ and displacement length $\ell \geq 0$ is given by

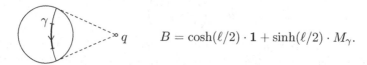

$$B = \cosh(\ell/2) \cdot \mathbf{1} + \sinh(\ell/2) \cdot M_\gamma.$$

Furthermore, the points on the axis are shifted in the counterclockwise sense.

(b) Let $r \in \mathbb{R}$ be the center and $s > 0$ the radius of a generalized circle representing a geodesic $\gamma \in \mathbb{H}$ and define H_γ as in Exercise 2.21(b). The hyperbolic isometry with axis γ and displacement length $\ell \geq 0$ is given by

$$B = \cosh(\ell/2) \cdot \mathbf{1} + \sinh(\ell/2) \cdot H_\gamma.$$

Furthermore, points on γ are shifted away from $r-s$ towards $r+s$.

Parabolic Type

If $A \in \mathrm{SL}(2,\mathbb{R})$ represents a Möbius transformation with unique fixed point ∞, then an easy exercise shows that

$$A = \begin{pmatrix} 1 & b \\ 0 & 1 \end{pmatrix} \tag{35}$$

for some $b \in \mathbb{R}$. Using a transformation as in (34), we see that any parabolic isometry of $\mathbb{H}$ is conjugate to one as given by (35).

Lecture 3:

Fuchsian Groups and Hyperbolic Surfaces

One of the reasons why hyperbolic geometry is studied by physicists is the enormous variety of groups and surfaces built upon it with their large number of interesting and sometimes unexpected properties. We give here a brief introduction to the main concepts: Fuchsian group, fundamental domain, hyperbolic surface, and study a number of examples. Emphasis is made to avoid using techniques from differential geometry.

3.1 Fuchsian Groups

The following notion goes back to Poincaré who introduced it in connection with differential equations of Fuchsian type. (Volume II of the collected works, beginning with [17], provides an exciting view of the development of the concept.)

Definition 3.1. A *Fuchsian group* is a discrete group of orientation preserving isometries of the hyperbolic plane.

In this definition, "discreteness" means the following. Let $\mathcal{H}$ be any model of the hyperbolic plane. Then a subgroup Γ of $\mathrm{Isom}(\mathcal{H})$ is called *discrete* or, more properly, *acts discretely* on $\mathcal{H}$, if for any $p \in \mathcal{H}$ the *orbit*

$$\Gamma(p) = \{\phi(p) \mid \phi \in \Gamma\}$$

is a discrete subset of $\mathcal{H}$, i.e. $\Gamma(p)$ has no accumulation points in $\mathcal{H}$.

Example 3.2. The following simple example plays a fundamental role in the proof of Selberg's trace formula (see Marklof's lectures in this book). For the description we use the upper half-plane model $\mathbb{H}$.

For some fixed $\lambda > 1$ we consider the hyperbolic transformation

$$T_\lambda : z \mapsto \lambda z, \quad z \in \mathbb{H}.$$

The group in question is

$$\Gamma_\lambda = \{T_\lambda^n \mid n \in \mathbb{Z}\}.$$

Using the distance formula (28) we see that for any $z \in \mathbb{H}$ the distance $d(z, T_\lambda^n(z))$ goes to infinity as $n \to \pm\infty$. Hence, the action of Γ_λ is discrete and Γ_λ is a Fuchsian group.

Here is a way to visualize the action of Γ_λ. The figure shows two geodesics γ_0, γ_1 represented by half-circles of center 0 and radius 1, respectively λ. The mapping T_λ sends γ_0 to γ_1, and the strip F_λ between γ_0 and γ_1,

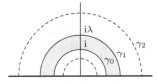

$$F_\lambda = \{z \in \mathbb{H} \mid 1 \le |z| \le \lambda\},$$

is mapped to a similar such strip between γ_1 and $\gamma_2 = T_\lambda(\gamma_1)$. The various strips $T_\lambda^n(F_\lambda)$, $n \in \mathbb{Z}$, cover the entire hyperbolic plane, and two distinct strips are either disjoint, or they intersect each other in a geodesic.

Discrete Subgroups of SL(2, ℝ)

Representing isometries by matrices we come across another type of discreteness. Assume we are working with SL(2, ℝ). Any $A = \left(\begin{smallmatrix} a & b \\ c & d \end{smallmatrix}\right) \in \mathrm{SL}(2, \mathbb{R})$ may be identified with $(a, b, c, d) \in \mathbb{R}^4$ and we call a subgroup $G \in \mathrm{SL}(2, \mathbb{R})$ a *discrete subgroup* if G is discrete as a subset of $\mathbb{R}^4$.

The next theorem shows that the two notions of discreteness are equivalent. (For a more general version of the theorem we refer to [5], section 5.3.)

Theorem 3.3. *Let $G \in \mathrm{SL}(2, \mathbb{R})$ be a subgroup and $\Gamma \subset \mathrm{Isom}(\mathbb{H})$ the group of all transformations $z \mapsto A[z]$, $z \in \mathbb{H}$, with $A \in G$. Then G is a discrete subgroup of $\mathrm{SL}(2, \mathbb{R})$ if and only if Γ is a Fuchsian group.*

Proof. If G contains an infinite convergent sequence of pairwise distinct matrices A_n, then for any $z \in \mathbb{H}$ the images $A_n[z]$ converge, and so Γ cannot be a Fuchsian group.

Conversely, if Γ is not Fuchsian, then for some $p \in \mathbb{H}$ the orbit $\Gamma(p)$ has an accumulation point. This implies that there exists an infinite sequence of pairwise distinct matrices $A_n = \left(\begin{smallmatrix} a_n & b_n \\ c_n & d_n \end{smallmatrix}\right) \in G$ such that $A_n[p]$ converges. From this we conclude that the distances from i to $A_n[\mathrm{i}]$ are bounded by a constant. By the formula in Exercise 2.20(c), these distances satisfy

$$2 \cosh d(A_n[\mathrm{i}], \mathrm{i}) = a_n^2 + b_n^2 + c_n^2 + d_n^2.$$

Hence, all A_n are contained in a ball of finite radius in $\mathbb{R}^4$, and so G is not a discrete subset of $\mathbb{R}^4$. □

Exercise 3.4. (a) Show that the following is a discrete subgroup of $\mathrm{SL}(2, \mathbb{R})$

$$\mathrm{SL}(2, \mathbb{Z}) = \left\{ \begin{pmatrix} a & b \\ c & d \end{pmatrix} \in \mathrm{SL}(2, \mathbb{R}) \;\middle|\; a, b, c, d \in \mathbb{Z} \right\}.$$

(b) Show that for any integer $N \geq 1$, the set

$$G_N = \left\{ \begin{pmatrix} a & b \\ c & d \end{pmatrix} \in \mathrm{SL}(2, \mathbb{Z}) \;\middle|\; \begin{pmatrix} a & b \\ c & d \end{pmatrix} \equiv \begin{pmatrix} 1 & 0 \\ 0 & 1 \end{pmatrix} \,(\mathrm{mod}\ N) \right\}$$

is a discrete subgroup of $\mathrm{SL}(2, \mathbb{Z})$.

The symbol $\equiv \ldots (\mathrm{mod}\ N)$ means that the corresponding matrix elements are congruent modulo N, i.e. $a = kN + 1$, for some $k \in \mathbb{Z}$; $b = jN$, for some $j \in \mathbb{Z}$, etc.

3.2 Fundamental Domains

The visualization in Example 3.2 brings us to the next concept, where for the sake of generality we briefly replace the hyperbolic plane by an arbitrary Riemannian manifold.

Definition 3.5. Let X be a Riemannian manifold and Γ a group of isometries acting on X. A subset $F \subset X$ is called a *fundamental domain* for this action if the following holds.

(i) For any $x \in X$ there exists $g \in \Gamma$ sending x to $g(x) \in F$.

(ii) If $x, y \in F$ and if $y = g(x)$ for some $g \in \Gamma$, $g \neq$ id, then x and y lie on the boundary of F.

A different way of phrasing this is to say that the images $g(F)$, $g \in \Gamma$ *cover* X *without overlapping*.

Example 3.6. In Example 3.2, F_λ is a fundamental domain for the action of $\Gamma = \{T_\lambda^n \mid n \in \mathbb{Z}\}$. Here is another one:

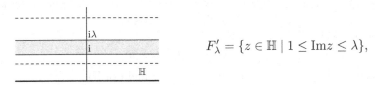

$$F_\lambda' = \{z \in \mathbb{H} \mid 1 \leq \mathrm{Im}\,z \leq \lambda\},$$

In the following we deal with generated subgroups, where we will use the following notation. If G is a group and $g_1, \ldots, g_n \in G$, then

$$H = \langle g_1, \ldots, g_n \rangle_G \tag{36}$$

denotes the subgroup of G *generated* by $g_1, \ldots, g_n$, i.e. the group formed by all elements that are products of the g_i's and their inverses, such as $g_2 g_3 g_5^{-1} g_5^{-1} g_1$, etc.

Example 3.7. In this example we *begin* with a domain F and *then* construct a Fuchsian group having F as fundamental domain. The description is in the disc model.

Consider a triple $\{c_1, c_2, c_3\}$ of pairwise non-intersecting geodesics in $\mathbb{D}$ as shown on the left-hand side in Fig. 6. Each of them separates $\mathbb{D}$ into two so-called *geodesic half-planes*, and we assume the relative positions of the geodesics to be such that any two of them lie in the same half-plane determined

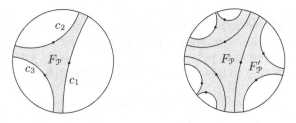

Fig. 6. A fundamental domain for a Fuchsian group generated by three half-turns. The right diagram shows a beginning of the corresponding tiling of the unit disc.

by the third. The intersection of the three half-planes in question is then a domain $F_{\mathcal{P}}$ bounded by c_1, c_2, c_3 (shaded area on the left-hand side in Fig. 6).

Choose points $p_i \in c_i$, and let $m_{p_i} : \mathbb{D} \to \mathbb{D}$ be the half-turn with center p_i, $i = 1, 2, 3$ (cf. Exercise 2.14). Half-turn m_{p_1}, e.g. exchanges the two half-planes determined by c_1 and maps $F_{\mathcal{P}}$ onto a similar such domain $F'_{\mathcal{P}}$ adjacent to $F_{\mathcal{P}}$ along c_1.

In the same way we have images of $F_{\mathcal{P}}$ along c_2 and c_3. This gives us a domain F^1 formed by four copies of $F_{\mathcal{P}}$ with six boundary geodesics $m_{p_i}(c_j)$, $i, j = 1, 2, 3$, $i \neq j$. The half-turns with centers $m_{p_i}(p_j)$ are the mappings $m_{p_i} m_{p_j} m_{p_i}^{-1}$, and applying them to the adjacent copies of $F_{\mathcal{P}}$ we obtain a domain F^2 with twelve boundary geodesics and images of p_1, p_2, p_3 on them. We may then apply the same procedure to F^2 to get F^3, and so on. In this way we obtain a *tiling* of $\mathbb{D}$, i.e. a covering with copies of $F_{\mathcal{P}}$, where any two distinct copies are either disjoint or share a geodesic.

We claim that

$$\Gamma_{\mathcal{P}} = \langle m_{p_1}, m_{p_2}, m_{p_3} \rangle_{\mathrm{Isom}(\mathbb{D})}$$

is a Fuchsian group and $F_{\mathcal{P}}$ is a fundamental domain of $\Gamma_{\mathcal{P}}$.

We already know that the images of $F_{\mathcal{P}}$ cover $\mathbb{D}$. Hence, we will be done if we can show that if $g, h \in \Gamma_{\mathcal{P}}$ and $g \neq h$, then $g(F_{\mathcal{P}}) \cap h(F_{\mathcal{P}})$ is either empty or a boundary geodesic. Applying h^{-1} to both domains we see that it suffices to prove this for the case $h = \mathrm{id}$. But, for $i = 1, 2, 3$, we have $m_{p_i}(F_{\mathcal{P}}) \subset D_i$, where D_i is the geodesic half-plane adjacent to $F_{\mathcal{P}}$ along c_i. For $j \neq i$, m_{p_j} maps D_i into D_j, and for $k \neq j$, m_{p_k} maps D_j into D_k, and so on. It follows that if $g \in \Gamma_{\mathcal{P}}$ is different from id, m_{p_1}, m_{p_2} and m_{p_3}, then $g(F_{\mathcal{P}}) \cap F_{\mathcal{P}} = \emptyset$. This concludes the proof of the claim.

Exercise 3.8. Let $\Gamma'_{\mathcal{P}}$ be the subgroup of $\Gamma_{\mathcal{P}}$ of all fixed point free isometries of $\Gamma_{\mathcal{P}}$. Then

(a) $\Gamma'_{\mathcal{P}}$ is generated by the hyperbolic elements $\tau_1 = m_{p_2} m_{p_3}$, $\tau_2 = m_{p_3} m_{p_1}$, $\tau_3 = m_{p_1} m_{p_2}$, and we have $\tau_1 \tau_2 \tau_3 = \mathrm{id}$.
(b) $F_{\mathcal{P}} \cup F'_{\mathcal{P}}$ is a fundamental domain for the action of $\Gamma'_{\mathcal{P}}$.

Example 3.9. This is a counterpart to Example 3.7, where instead of half-turns we take reflections. For later use we will, however, formulate it a bit differently.

Let T_1, T_2, T_3 be hyperbolic transformations with axes a_1, a_2, a_3, respectively, whose relative positions are as shown in Fig. 7 such that we have pairwise disjoint common perpendiculars b_1, b_2, b_3 with b_i opposite of a_i, $i = 1, 2, 3$. In addition to this we make the assumption that

$$T_1 T_2 T_3 = \mathrm{id}.$$

How does the group $\Gamma = \Gamma_{\mathcal{T}} = \langle T_1, T_2 \rangle_{\mathrm{Isom}(\mathbb{D})}$ look like?

For an answer we introduce the reflections S_1, S_2, S_3 with respect to b_1, b_2, b_3, respectively and first claim that

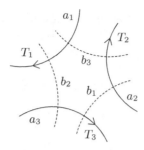

Fig. 7. Generators of a Fuchsian group.

$$T_1 = S_2 S_3, \quad T_2 = S_3 S_1, \quad T_3 = S_1 S_2.$$

For the proof we let b_2' be the perpendicular to a_1 half way between b_3 and $T_1(b_3)$ so that for the reflection S_2' at b_2' we get $S_2' S_3 = T_1$. In a similar way we let S_1' with axis b_1' orthogonal to a_2 be the reflection such that $T_2 = S_3 S_1'$. With this setting we have $T_3^{-1} = T_1 T_2 = S_2' S_3 S_3 S_1' = S_2' S_1'$. In particular, the axis of T_3 coincides with the axis of $S_2' S_1'$. Hence, the axis of T_3 is perpendicular to b_2', b_1', and therefore b_2', b_1' are identical with b_2, b_1, the unique perpendiculars between a_3 and a_1, a_2. The claim follows.

Now consider the group $\tilde{\Gamma} = \langle S_1, S_2, S_3 \rangle_{\mathrm{Isom}(\mathbb{D})}$. With the same argument as in Example 3.7 we see that the domain $\tilde{F} \subset \mathbb{D}$ bounded by b_1, b_2, b_3 is a fundamental domain for the action of $\tilde{\Gamma}$. In particular, $\tilde{\Gamma}$ is a discrete group. Since T_1, T_2 are products of S_1, S_2, S_3, it follows that Γ is a subgroup of $\tilde{\Gamma}$ and therefore discrete as well.

We leave it as an exercise to show that $F = \tilde{F} \cup S_1(\tilde{F})$ is a fundamental domain of Γ. From the shape of this domain it follows that there exists a constant $c > 0$ such that for any $T \in \Gamma \smallsetminus \{\mathrm{id}\}$ and any $p \in \mathbb{D}$ we have $d(p, T(p)) \geq c$. Altogether we have the following properties: $\langle T_1, T_2 \rangle_{\mathrm{Isom}(\mathbb{D})}$ *is a Fuchsian group and all its elements are hyperbolic.*

3.3 Hyperbolic Surfaces

The examples in the preceding section lead us to the notion of a hyperbolic surface. Roughly speaking, these are surfaces on which small regions are the same as small regions of the hyperbolic plane.

In the following definition, a *circular disc in* $\mathbb{H}$ is a subset Ω of type

$$\Omega = B_\delta(p) = \{z \in \mathbb{H} \mid d(z, p) < \delta\}.$$

Definition 3.10. A *hyperbolic surface* is a point set $\mathcal{S}$ together with a family $\mathcal{A}$ of mappings with the following properties.

(i) Each $\alpha \in \mathcal{A}$ is a one-to-one mapping $\alpha : W_\alpha \to \Omega_\alpha$, where W_α is a subset of $\mathcal{S}$ and Ω_α is a circular disc in $\mathbb{H}$.

(ii) For any $x \in \mathcal{S}$ there exists $\alpha \in \mathcal{A}$ such that $x \in W_\alpha$.

(iii) If $\alpha, \beta \in \mathcal{A}$ and the intersection $D = W_\alpha \cap W_\beta$ is not empty, then there exists an isometry $g \in \mathrm{Isom}(\mathbb{H})$ such that

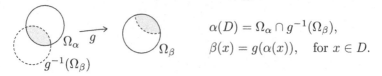

$$\alpha(D) = \Omega_\alpha \cap g^{-1}(\Omega_\beta),$$
$$\beta(x) = g(\alpha(x)), \quad \text{for } x \in D.$$

Example 3.11. We consider again the domain $F_\lambda = \{z \in \mathbb{H} \mid 1 \le |z| \le \lambda\}$ as in Example 3.2. Assuming, for a second, that F_λ is made of fabric, we may fold F_λ up and paste the two boundary lines together to obtain a cylindrical surface $\mathcal{S}_\lambda$ as shown on the right-hand side in Fig. 8.

Mathematically, $\mathcal{S}_\lambda$ is constructed as follows. Recall that the mapping $z \mapsto T_\lambda(z) = \lambda z$ is an isometry sending the lower boundary geodesic γ_0 of F_λ to the upper boundary geodesic γ_1.

For any $z \in \gamma_0$ we let $\mathbf{z}$ be the unordered pair $\mathbf{z} = \{z, T_\lambda(z)\}$, and for any z in the interior of F_λ ($1 < |z| < \lambda$) we set $\mathbf{z} = \{z\}$. The set $\mathcal{S}_\lambda$ is then defined as

$$\mathcal{S}_\lambda = \{\mathbf{z} \mid z \in F_\lambda\}.$$

We will say that $\mathcal{S}_\lambda$ has been obtained from F_λ by *pasting* together γ_0 and γ_1 via T_λ. It becomes a hyperbolic surface in a natural way via the following family $\mathcal{A}$ of mappings.

For p in the interior of F we let δ_p be a quarter of the distance from p to the boundary and define $W_{\alpha_p} = \{\mathbf{z} \mid d(z, p) < \delta_p\}$, $\alpha_p(\mathbf{z}) = z$, for $\mathbf{z} \in W_{\alpha_p}$. For $p \in \gamma_0$ we set $\delta_p = \lambda/4$ and define

$$W_{\alpha_p} = \{\mathbf{z} \mid z \in F_\lambda, \ d(z, p) < \delta_p \text{ or } d(z, T_\lambda(p)) < \delta_p\},$$

$$\alpha_p(\mathbf{z}) = \begin{cases} z, & \text{if } d(z, p) < \delta_p; \\ T_\lambda^{-1}(z), & \text{if } d(z, T_\lambda(p)) < \delta_p; \end{cases} \quad \mathbf{z} \in W_{\alpha_p}.$$

Conditions (i)–(iii) of Definition 3.10 are readily checked, and so $\mathcal{S}_\lambda$ with $\mathcal{A}$ is a hyperbolic surface.

Fig. 8. Pasting γ_0 to γ_1 yields $\mathcal{S}_\lambda$. The figure on the right-hand side shows the part of $\mathcal{S}_\lambda$ corresponding to the domain in F_λ delimited by the two radial lines.

On a hyperbolic surface S one can "do geometry" by identifying any W_α with the corresponding $\Omega_\alpha \subset \mathbb{H}$. As an illustration we define the distance between two points in S.

Definition 3.12. Let $p, q \in S$. A *geodesic trail* from p to q is a mapping $c : [a, b] \to S$ from an interval $[a, b]$ to S such that

(i) $c(a) = p$, $c(b) = q$.
(ii) There exists a subdivision $t_0 = a \leq t_1 \leq \cdots \leq t_N = b$ such that each $c([t_i, t_{i+1}])$ is contained in some W_{α_i} and $\alpha_i(c([t_i, t_{i+1}]))$ is a geodesic arc in $\Omega_i = \alpha_i(W_{\alpha_i})$, $i = 0, \ldots, N - 1$.

The *length* $\ell(c)$ of the trail is defined as the sum of the lengths of the arcs $\alpha_i(c([t_i, t_{i+1}]))$. (By property (iii) in Definition 3.10, the definition of a trail and its length do not depend on the choice of the subdivision.)

Let $\mathcal{C}_{pq}$ be the set of all trails from p to q. If this set is empty then we say that p and q have infinite distance: $d(p, q) = \infty$. If $\mathcal{C}_{pq} \neq \emptyset$, the distance from p to q is defined as

$$d(p, q) = \inf_{c \in \mathcal{C}_{pq}} \ell(c). \tag{37}$$

To simplify the next exercise we prove a technical lemma.

Lemma 3.13. *If $p \in W_\alpha$ and $q \notin W_\alpha$ then $d(p, q)$ is greater than or equal to the distance in $\mathbb{H}$ from $\alpha(p)$ to the boundary of $\Omega_\alpha = \alpha(W_\alpha)$.*

Proof. Consider a geodesic trail $c : [a, b] \to S$ with $c(a) = p$, $c(b) = q$, and let $a = t_0 \leq t_1 \leq \cdots \leq t_N = b$ be a subdivision such that each $c([t_i, t_{i+1}])$ is contained in W_{α_i} for some $\alpha_i \in \mathcal{A}$. Take k to be the index such that $c([t_i, t_{i+1}]) \subset W_\alpha$ for all $i < k$ whilst $c([t_k, t_{k+1}])$ itself is no longer entirely contained in W_α. Then $D = W_\alpha \cap W_{\alpha_k}$ is not empty, and we let $g \in \mathrm{Isom}(\mathbb{H})$ be the isometry for which $\alpha(D) = \Omega_\alpha \cap g^{-1}(\Omega_{\alpha_k})$. Then $\alpha_k(D) = g(\Omega_\alpha) \cap \Omega_{\alpha_k}$, and there exists $t^* \in [t_k, t_{k+1}]$ such that $\alpha_k(c([t_k, t^*])) \subset \alpha_k(D)$ and $\alpha_k(c(t^*))$ lies on the boundary of $\alpha_k(D)$ in Ω_{α_k}. It follows that $\alpha(c([a, t^*[))$ goes from $\alpha(p)$ to the boundary of Ω_α, and hence the length of c is greater than or equal to the distance from $\alpha(p)$ to the boundary. $\square$

The exercise is to show that S together with d is a metric space.

Exercise 3.14. For any $p, q, r \in S$ the following hold.

(a) $d(p, q) \geq 0$, with equality if and only if $p = q$.
(b) $d(p, q) = d(q, p)$.
(c) $d(p, q) + d(q, r) \geq d(p, r)$.

An *isometry* between hyperbolic surfaces will always be understood to be an isometry with respect to the distance function (37).

Let us finally introduce the geodesics.

Definition 3.15. A *geodesic* in S is a mapping $c : I \to S$ from an interval I to S such that whenever $c([t_1, t_2]) \subset W_\alpha$ for some $t_1 \leq t_2 \in I$ and $\alpha \in \mathcal{A}$, then the curve $t \mapsto \alpha(c(t))$, $t \in [t_1, t_2]$ is a geodesic arc in $\mathbb{H}$ parametrized with constant speed.

We will say that c is a *unit speed geodesic* if this constant is 1, i.e. if in the representations $t \mapsto \alpha(c(t))$ we have $d(\alpha(c(t)), \alpha(c(t'))) = |t - t'|$.

Remark 3.16. If I is a closed interval, say $I = [a, b]$, then c is called a *geodesic arc*. By compactness there exists in this case a subdivision $t_0 = a \leq t_1 \leq \cdots \leq t_N = b$, such that each sub-arc $c([t_i, t_{i+1}])$ is contained in some W_α, $\alpha \in \mathcal{A}$. Hence, geodesic arcs are special cases of geodesic trails.

Theorem 3.17. *Let $c, c' : I \to S$ be geodesics. If $c(t) = c'(t)$ for all t in some interval $(t_0, t_1) \subset I$, then $c(t) = c'(t)$ for all $t \in I$.*

Proof. Otherwise we find $t_2 \in I$ where the two curves bifurcate in the sense that, for arbitrarily small $\varepsilon > 0$ we have $c(t) = c'(t)$ for all $t \in \,]t_2 - \varepsilon, t_2[$ (respectively, $]t_2, t_2 + \varepsilon[\,$), but $c(t) \neq c'(t)$ for some $t \in [t_2, t_2 + \varepsilon[$ (respectively, $]t_2 - \varepsilon, t_2]$). Taking $\alpha \in \mathcal{A}$ with $c(t_2) \in W_\alpha$ and ε sufficiently small, we then have $c(]t_2 - \varepsilon, t_2 + \varepsilon[), c'(]t_2 - \varepsilon, t_2 + \varepsilon[) \subset W_\alpha$ (by (iii) in Definition 3.10 we may take the same α for both curves). We now have bifurcating geodesic arcs $\alpha(c(]t_2 - \varepsilon, t_2 + \varepsilon[))$, $\alpha(c'(]t_2 - \varepsilon, t_2 + \varepsilon[))$ in $\mathbb{H}$, which is impossible. $\square$

Exercise 3.18. If $\varphi : S \to S'$ is an isometry between hyperbolic surfaces and $c : I \to S$ is a geodesic then $\varphi \circ c : I \to S'$ is a geodesic.

3.4 Quotient Surfaces

In this section we construct hyperbolic surfaces out of Fuchsian groups. Throughout the section, Γ denotes a Fuchsian group operating on $\mathbb{H}$.

The construction is based on the following two observations. The first observation says that orbits are equivalence classes.

Lemma 3.19. *Let $p, q \in \mathbb{H}$. If $\Gamma(p) \neq \Gamma(q)$ then $\Gamma(p) \cap \Gamma(q) = \emptyset$.*

Proof. If $z \in \Gamma(p)$, then $z = R(p)$ for some $R \in \Gamma$ and

$$\Gamma(z) = \{T(z) \mid T \in \Gamma\} = \{TR(p) \mid R \in \Gamma\} = \{T'(p) \mid T' \in \Gamma\} = \Gamma(p).$$

Hence, if $z \in \Gamma(p) \cap \Gamma(q)$ then $\Gamma(p) = \Gamma(z) = \Gamma(q)$. $\square$

Lemma 3.20. *For any $p \in \mathbb{H}$ there exists $\delta > 0$, such that the discs $B_\delta(p')$, $p' \in \Gamma(p)$, are pairwise disjoint.*

Proof. Since the points of the orbit $\Gamma(p)$ do not accumulate at p there exists $\delta > 0$ such that $B_{2\delta}(p) \cap \Gamma(p) = \{p\}$. For any $R, T \in \Gamma$ with $R(p) \neq T(p)$ we have therefore $d(R(p), T(p)) = d(p, R^{-1}T(p)) \geq 2\delta$, and so the discs of radius δ around $R(p)$ and $T(p)$ are disjoint. $\square$

From now on we restrict our considerations to groups *acting without fixed points*, that is, for any $p \in \mathbb{H}$ and any $T \in \Gamma$ different from the identity we assume that $T(p) \neq p$.

Definition 3.21. Let $\Gamma : \mathbb{H} \to \mathbb{H}$ be a Fuchsian group acting without fixed points. The space $\mathbb{H}/\Gamma$ (read: "$\mathbb{H}$ mod Γ") whose points are the orbits:

$$\mathbb{H}/\Gamma = \{ z = \Gamma(z) \mid z \in \mathbb{H} \},$$

is called the *quotient surface* of Γ.

$\mathbb{H}/\Gamma$ has the structure of a hyperbolic surface in a natural way: For any $p \in \mathbb{H}$ we let δ_p be the largest number for which the discs of radius $3\delta_p$ around the points in the orbit $\Gamma(p)$ are pairwise disjoint. (The factor 3 is for technical reasons.) By Lemma 3.20, $\delta_p > 0$. In the disc $\Omega_{\alpha_p} = \{ z \in \mathbb{H} \mid d(z,p) < \delta_p \}$ the points have pairwise disjoint orbits, and so we obtain a one-to-one mapping $\alpha_p : W_{\alpha_p} \to \Omega_{\alpha_p}$ by setting

$$W_{\alpha_p} = \{ z \in \mathbb{H}/\Gamma \mid d(z,p) < \delta_p \}, \quad \alpha_p(z) = z, \text{ for } z \in W_{\alpha_p}.$$

Lemma 3.22. $\mathbb{H}/\Gamma$ *together with the set* $\mathcal{A}$ *of all mappings* α_p, $p \in \mathbb{H}$, *is a hyperbolic surface.*

Proof. We already know that the mappings α_p satisfy conditions (i) and (ii) of Definition 3.10. To prove condition (iii), let us assume that $W_{\alpha_p} \cap W_{\alpha_q} \neq \emptyset$. Then there exists $S \in \Gamma$ such that $\Omega_{\alpha_p} \cap S(\Omega_{\alpha_q}) \neq \emptyset$. This S plays the role of g^{-1} in condition (iii), and it only remains to show that $\Omega_{\alpha_p} \cap T(\Omega_{\alpha_q}) = \emptyset$ for all $T \in \Gamma$ with $T \neq S$.

Suppose this were not the case. Then we would have the following situation.

$$6\delta_q \leq d(Sq, Tq) < 2(\delta_p + \delta_q),$$

$$6\delta_p \leq d(p, TS^{-1}p) < 2(\delta_p + \delta_q) + d(Sq, Tq).$$

(In the last inequality we have used that TS^{-1} sends Sq to Tq and thus $d(p, TS^{-1}p) \leq 2d(p, Sq) + d(Sq, Tq)$.) From the first line we obtain $2\delta_q < \delta_p$, and from both lines together, $\delta_p < 2\delta_q$, a contradiction. $\square$

Exercise 3.23. Let $\Gamma_\lambda = \{ T_\lambda^n \mid n \in \mathbb{Z} \}$ with fundamental domain F_λ be the Fuchsian group as in Example 3.2, and $\mathcal{S}_\lambda$ the hyperbolic surface obtained by pasting together the boundary geodesics of F_λ as in Example 3.11. Show that $\mathcal{S}_\lambda$ and $\mathbb{H}/\Gamma_\lambda$ are isometric.

Exercise 3.24. For $p = \Gamma(p), q = \Gamma(q) \in \mathbb{H}/\Gamma$ the distance is

$$d(p, q) = \min\{ d(p, q') \mid q' \in \Gamma(q) \}$$
$$= \min\{ d(p', q') \mid p' \in \Gamma(p), \ q' \in \Gamma(q) \}.$$

Exercise 3.25. Let $\phi \in \mathrm{Isom}(\mathbb{H})$ and define

$$\Gamma' = \phi\Gamma\phi^{-1} \stackrel{\mathrm{def}}{=} \{\phi T \phi^{-1} \mid T \in \Gamma\}. \tag{38}$$

Then $\mathbb{H}/\Gamma$ and $\mathbb{H}/\Gamma'$ are isometric.

3.5 Closed Geodesics and Conjugacy Classes

For the study of chaotic systems the distribution of the lengths of the closed geodesics on a surface $\mathbb{H}/\Gamma$ is a point of interest. We show how these lengths are related to the members of Γ.

Definition 3.26. Let $\ell > 0$. A *closed* or *periodic* geodesic of period ℓ on a hyperbolic surface S is a unit speed geodesic $c : \mathbb{R} \to S$ satisfying

$$c(t) = c(t + \ell), \quad t \in \mathbb{R}.$$

If we take any $a \in \mathbb{R}$, then the mapping $c : [a, a + \ell] \to S$ is a closed geodesic arc of length ℓ. One calls therefore ℓ also the *length* of c.

Example 3.27. The following example is given in $\mathbb{D}$. Let c_1, c_1', c_2, c_2' be pairwise disjoint geodesics in $\mathbb{D}$ with the configuration as shown in Fig. 9, where c_i and c_i' are symmetric with respect to the imaginary axis, $i = 1, 2$. We also assume that the corresponding circles do not meet on the boundary of $\mathbb{D}$, so that for any pair c_i, c_j' there exists a *common perpendicular*, i.e. a geodesic orthogonal to both. Let, in particular, a_i be the common perpendicular of the pair c_i, c_i', and denote by T_i the hyperbolic isometry with axis a_i sending c_i to c_i', $i = 1, 2$. We have already seen in Example 3.9 that

$$\Gamma_{\mathcal{C}} = \langle T_1, T_2 \rangle_{\mathrm{Isom}(\mathbb{D})}$$

is a Fuchsian group. ($\mathcal{C}$ stands for $\{c_1, c_1', c_2, c_2'\}$.)

A fundamental domain for $\Gamma_{\mathcal{C}}$ is given by the region $F_{\mathcal{C}}$ in $\mathbb{D}$ delimited by c_1, c_1', c_2, c_2'. The shaded area in the figure is the part of $F_{\mathcal{C}}$ bounded by the common perpendiculars.

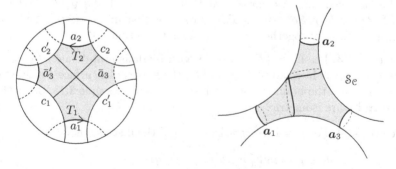

Fig. 9. A fundamental domain for the group $\Gamma_{\mathcal{C}}$ and the quotient surface $S_{\mathcal{C}} = \mathbb{H}/\Gamma_{\mathcal{C}}$.

The surface $S_C = \mathbb{H}/\Gamma_C$ is the same as the one obtained from F_C by pasting c_i to c_i' via T_i, $i = 1, 2$. The figure on the right-hand side aims at giving an intuitive view of this surface.

Here are a few closed geodesics on S_C: consider the arc $\bar{a}_i$ on a_i from c_i to c_i'. Pasting c_i to c_i' via T_i we get a geodesic arc $\boldsymbol{a}_i$ on S_C with matching ends. Since a_i crosses c_i and c_i' under the same angle the two ends match smoothly, and so $\boldsymbol{a}_i$ is a closed geodesic.

For the next example we let $\bar{a}_3$ be the arc from c_1' to c_2 on the common perpendicular of these geodesics, and $\bar{a}_3'$ the analogous arc from c_2' to c_1. The image $T_2(\bar{a}_3)$ matches smoothly with $\bar{a}_3'$, and the image $T_1(\bar{a}_3')$ matches smoothly with $\bar{a}_3$. Hence $\bar{a}_3$ and $\bar{a}_3'$ together yield a closed geodesic $\boldsymbol{a}_3$ on S_C.

Our last example is the 'figure eight" geodesic on S_C shown on the right-hand side in the figure. This geodesic is obtained by pasting together the perpendicular arcs from c_1 to c_2 and from c_2' to c_1 via T_1 and T_2, where we leave it as an exercise to show that T_1 and T_2 bring their endpoints smoothly together.

The preceding example may give the impression that constructing closed geodesics is a painful procedure. We will now show that for surfaces of the form $\mathbb{H}/\Gamma$ we get them rather easily.

So let again $\Gamma : \mathbb{H} \to \mathbb{H}$ be a Fuchsian group acting without fixed points. We denote by

$$\pi : \mathbb{H} \to \mathbb{H}/\Gamma \tag{39}$$

the *natural projection*, i.e. the mapping sending $z \in \mathbb{H}$ to its orbit $\pi(z) := \boldsymbol{z} = \Gamma(z)$. This mapping has, of course, the property that

$$\pi(T(z)) = \pi(z), \quad \text{for any } T \in \Gamma. \tag{40}$$

Furthermore, π is a *local isometry*, i.e. for any $p \in \mathbb{H}$ there exists a neighborhood $U_p \subset \mathbb{H}$ of p such that the restriction $\pi : U_p \to \pi(U_p)$ is an isometry. In fact, such a neighborhood is Ω_{α_p}, and $\pi : \Omega_{\alpha_p} \to \pi(\Omega_{\alpha_p})$ is the inverse of $\alpha_p : W_{\alpha_p} \to \Omega_p$, and by Exercise 3.24 α_p is an isometry.

Since $\pi : \mathbb{H} \to \mathbb{H}/\Gamma$ is a local isometry, any geodesic $c : \mathbb{R} \to \mathbb{H}$ is sent to a geodesic $\pi \circ c : \mathbb{R} \to \mathbb{H}/\Gamma$.

The following simple observations are crucial.

Observation 3.28. If $T \in \Gamma$ is a hyperbolic transformation of displacement length ℓ, and $a_T : \mathbb{R} \to \mathbb{H}$ is its axis parametrized with unit speed such that $T(a_T(t)) = a_T(t + \ell)$, $t \in \mathbb{R}$, then $\pi \circ a_T$ is a closed geodesic of period ℓ on $\mathbb{H}/\Gamma$.

Proof. This follows from the fact that $\pi \circ T(z) = \pi(z)$ for all $z \in \mathbb{H}$, and therefore $\pi(a_T(t + \ell)) = \pi(T(a_T(t))) = \pi(a_T(t))$. $\square$

Observation 3.29. If $T \in \mathrm{Isom}(\mathbb{H})$ is a hyperbolic isometry with axis a_T, and $\phi \in \mathrm{Isom}(\mathbb{H})$, then $\phi T \phi^{-1}$ is a hyperbolic isometry with axis $\phi(a_T)$.

Proof. $\phi T \phi^{-1}(\phi(a_T)) = \phi(T(a_T)) = \phi(a_T)$. $\square$

The two observations together imply that if $T \in \Gamma$, then all T' of the form $T' = STS^{-1}$ with $S \in \Gamma$ yield the same geodesic in $\mathbb{H}/\Gamma$.

Elements g, g' in a group G with the relationship that $g' = hgh^{-1}$ for some $h \in G$, are said to be *conjugate in G*. This relationship is an equivalence relation, and the equivalence class

$$[g]_G = \{hgh^{-1} \mid h \in G\}$$

is called the *conjugacy class of g in G*. Thus, the above implication is that the entire conjugacy class $[T]_\Gamma$ gives us *one* geodesic in $\mathbb{H}/\Gamma$.

We will now show that the conjugacy classes and the closed geodesics are in one-to-one correspondence. For this we have to add a complement to our definition of a closed geodesic.

First, if $c, c' : \mathbb{R} \to \mathcal{S}$ are parametrized closed geodesics with unit speed, and if c and c' only differ by a parameter change of type $c'(t) = c(t+\omega)$, $t \in \mathbb{R}$, where ω is a constant, then we will consider c and c' equivalent. Second, if ℓ is a period, then for any positive integer n, the multiple $n\ell$ is also a period. We have therefore to mention somewhere which period is wanted. For this reason we define a (non-parametrized) *closed geodesic of period ℓ* formally as a pair (c, ℓ), where ℓ is a positive real number and c is the equivalence class of a closed unit speed geodesic of period ℓ.

Theorem 3.30. *The mapping $T \mapsto \pi \circ a_T$ establishes a one-to-one correspondence between the conjugacy classes of Γ and the closed geodesics in $\mathbb{H}/\Gamma$.*

Proof. We have already seen that conjugate elements yield the same geodesic, so it makes sense to define $\rho([T]_\Gamma) = (a_T, \ell)$, where a_T is the equivalence class of $\pi \circ a_T$ and ℓ is the displacement length of T.

ρ is onto: Let $c : \mathbb{R} \to \mathbb{H}/\Gamma$ be a closed unit speed geodesic of period ℓ. We have to find $T \in \Gamma$ with displacement length ℓ such that $\pi \circ a_T = c$.

For this we choose $p \in \mathbb{H}$ with $\pi(p) = c(0)$, and let $a^1 : [0, \delta_p] \to B_{\delta_p} = \Omega_{\alpha_p} \subset \mathbb{H}$ with $a^1(0) = p$ be the arc satisfying $\pi \circ a^1(t) = c(t)$, $t \in [0, \delta_p]$. This arc extends to a unit speed geodesic $a : \mathbb{R} \to \mathbb{H}$, and by Theorem 3.17 we have $\pi \circ a(t) = c(t)$, for all $t \in \mathbb{R}$.

Now $\pi(a(t+\ell)) = c(t+\ell) = c(t) = \pi(a(t))$, for any t, i.e. $a(t+\ell)$ and $a(t)$ lie in the same Γ-orbit. This holds in particular for $t \in [0, \delta_p]$. Since the images of $B_{\delta_p}(p) = B_{\delta_p}(a(0))$ under Γ are pairwise disjoint, there exists $T \in \Gamma$ such that $a(t+\ell) = T(a(t))$ for all $t \in [0, \delta_p]$. It follows that $a = a_T$, and we are done.

ρ is one-to-one: Here we use a similar argument. If $\pi \circ a_{T'} = c$, then there exists $S \in \Gamma$ such that $S \circ a_T(t) = a_{T'}(t)$, for $t \in [0, \delta_p]$ (again because

$a_T(t)$ and $a_{T'}(t)$ lie in the same Γ-orbit, and there exists only one $S \in \Gamma$ such that $S(B_{\delta_p}(a_T(0)))$ intersects $B_{\delta_p}(a_{T'}(0)))$. It follows that $S(a_T) = a_{T'}$. By Observation 3.29, STS^{-1} is a hyperbolic isometry with axis $a_{T'}$. Since STS^{-1} also has the same displacement length and the same direction as T' it follows that $STS^{-1} = T'$. $\square$

In the next two exercises we consider the closed geodesics for some simple examples.

Exercise 3.31. Let $\Gamma_\lambda = \{T_\lambda^n \mid n \in \mathbb{Z}\}$ be the group already studied in Examples 3.2 and 3.11. Its quotient surface $S_\lambda = \mathbb{H}/\Gamma_\lambda$ is a cylinder as depicted in Fig. 8.

(a) The lengths of the closed geodesics on S_λ are precisely the numbers $n \log \lambda$, $n = 1, 2, \ldots$.
(b) Two cylinders S_λ and $S_{\lambda'}$ are isometric if and only if $\lambda = \lambda'$.

The example in this exercise is atypical inasmuch as the group is cyclic and, in particular commutative. The situation changes drastically as soon as we have more than one generator, as seen in the next exercise.

Exercise 3.32. Let $\Gamma_e = \langle T_1, T_2 \rangle_{\mathrm{Isom}(\mathbb{D})}$ be the two-generator group studied in Example 3.27.

Let us call a *word* any expression $w = {n_1 \atop i_1}{n_2 \atop i_2} \ldots {n_k \atop i_k}$, where the i_κ are alternately chosen from $\{1, 2\}$ (i.e. $i_1 i_2 i_3 \ldots$ is either $121\ldots$ or $212\ldots$), and all n_κ are nonzero integers. For any such word we have the element

$$T^w = T_{i_1}^{n_1} T_{i_2}^{n_2} \ldots T_{i_k}^{n_k} \in \Gamma_e.$$

(a) Repeat the argument given in Example 3.7 applying it to the fundamental domain F_e to show that for any such word w one has $T^w \neq \mathrm{id}$.
(b) If $w \neq w'$ then $T^w \neq T^{w'}$.

For any word $w = {n_1 \atop i_1}{n_2 \atop i_2} \ldots {n_k \atop i_k}$, the *word length* $\|w\|$ is defined as

$$\|w\| = |n_1| + \cdots + |n_k|. \tag{41}$$

(c) For any integer $n \geq 1$ there are exactly $4 \cdot 3^{n-1}$ different words w of length $\|w\| = n$.
(d) There are more than $3^n/n$ different conjugacy classes $[T^w]_{\Gamma_e}$ with a word w of length $\|w\| = n$.

For the next question we choose a test point $p \in \mathbb{H}$ and let δ be the larger of the two distances $d(p, T_1(p))$, $d(p, T_2(p))$.

(e) $d(p, T^w(p)) \leq \delta \|w\|$.
(f) On $S_e = \mathbb{H}/\Gamma_e$ there are more than $3^n/n$ different closed geodesics of lengths $\leq \delta n$.

In Marklof's lecture it is shown that on a compact hyperbolic surface the number $N(\ell)$ of closed geodesics of length $\leq \ell$ is asymptotically equal to e^ℓ/ℓ as $\ell \to \infty$.

3.6 Side Pairings

In this section we describe a construction of Fuchsian groups due to Poincaré, generally known as Poincaré's *polygon theorem*. In order to convey the idea in a not too complicated way and without having to resort to more sophisticated techniques we will carry this out in a restricted setting and refer the reader to the article [9] for a thorough treatment of the subject.

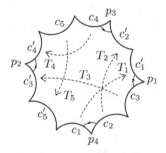

Fig. 10. The cycle of p_1 under the side pairing $T_1, \ldots, T_5$ is $p_1 p_2 p_3 p_4 (p_1)$.

Thus, let us consider a *compact convex geodesic polygon* F in the hyperbolic plane $\mathbb{H}$. By this we mean that F is a compact subset of $\mathbb{H}$ with non-empty interior which can be obtained as the intersection of a finite number of geodesic half-planes. Fig. 10 shows an example.

We say that F has a *side pairing* $T_1, \ldots, T_n$, if F has $2n$ sides (each side being a geodesic arc) which form pairs c_1, c_1'; c_2, c_2'; $\ldots$; c_n, c_n', and if $T_1, \ldots, T_n \in \mathrm{Isom}^+(\mathbb{H})$ are orientation preserving isometries such that

$$T_i(c_i) = c_i', \quad T_i(F) \cap F = c_i', \quad i = 1, \ldots, n. \tag{42}$$

In other words, each T_i moves F to a neighboring domain adjacent to F along side c_i'. Of course, this necessitates that $\ell(c_i) = \ell(c_i')$, $i = 1, \ldots, n$.

We group the boundary points of F into equivalence classes as follows. If x lies on a side c_i but is not a vertex of F, then we will say that x and $T_i(x)$ are equivalent to each other and form an equivalence class. If x is one of the vertices p of F, then the equivalence class is a *cycle of vertices* determined by the following procedure (we take cycle $p_1 p_2 p_3 p_4 (p_1)$ in Fig. 10 as a guiding example):

Set $p_1 = p$. Draw a circular arc of some small radius $\varepsilon > 0$ in F centered at p_1 connecting the two sides of F at p_1 with each other in the counterclockwise sense (in Fig. 10 this arc goes from c_1' to c_3). Its endpoint is equivalent to a boundary point at distance ε from a second vertex, p_2, and we draw an arc of radius ε as before, now centered at p_2 (from c_3' to c_4'). Its endpoint is equivalent to a boundary point at distance ε of a third vertex, p_3, around which we draw again an arc, and so on. After a number k of steps we are back at the initial point of the first arc and we stop. The sequence of vertices obtained by this procedure is the *cycle* of p. We will denoted it by $p_1 \ldots p_k(p_1)$ and write $p_{k+1} := p_1$.

Exercise 3.33. Determine the cycles in each of the following cases.

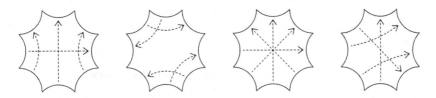

Fig. 11. Some side pairings of an octagon.

For any vertex p of F we denote by $\vartheta(p) \in]0, \pi]$ the measure of the interior angle of F at p.

Definition 3.34. A compact convex geodesic polygon F in $\mathbb{H}$ with side pairing $T_1, \ldots, T_n$ is *of type S*, if for any cycle of vertices $p_1 p_2 \ldots p_k(p_1)$ the following conditions are satisfied.

(i) $\vartheta(p_1) + \cdots + \vartheta(p_k) = 2\pi$.
(ii) $\vartheta(p_i) + \vartheta(p_{i+1}) \leq \pi$, $i = 1, \ldots, k$.

Condition (i) is essential for the subsequent construction. Condition (ii) is not necessary mathematically, but simplifies the subsequent exposition considerably. For the general case we refer to [9]. A large family of examples satisfying (i) and (ii) is provided by the symmetric octagons studied in Lecture 6.

Exercise 3.35. If F is as in Definition 3.34, then $n \geq 4$.

(Hint: apply the Gauss-Bonnet formula, Theorem 1.22).

Theorem 3.36 (Poincaré). *Let F be a compact convex geodesic polygon in $\mathbb{H}$ with side pairing $T_1, \ldots, T_n$, and assume F is of type S. Then $\Gamma_{\mathcal{T}} = \langle T_1, \ldots, T_n \rangle_{\mathrm{Isom}(\mathbb{H})}$ is a Fuchsian group with fundamental domain F. Moreover, all elements of $\Gamma_{\mathcal{T}} \smallsetminus \{\mathrm{id}\}$ are hyperbolic.*

Proof. The first step is to show that there exists a *tiling* of $\mathbb{H}$ with copies of F, i.e. a family of domains $F^0, F^1, F^2, \ldots$, where each F^i is isometric to F and

(i) $\bigcup_{i \geq 0} F^i = \mathbb{H}$.
(ii) If $i \neq j$, then $F^i \cap F^j$ is either empty, or a common vertex, or a common side of F^i and Fj.

In order to describe an explicit procedure to obtain such a tiling we color the sides of F using n different colors such that the sides in each pair c_i, c'_i have the same color. Furthermore we draw all c'_i dotted and the c_i undotted. Whenever we will take a copy $\phi(F)$, $\phi \in \text{Isom}(\mathbb{H})$, we color and dot the sides of $\phi(F)$ in the same way so that ϕ is color and dot preserving.

We will say that two such copies $\phi_1(F)$, $\phi_2(F)$ are a *matching pair* if ϕ_1 and ϕ_2 are orientation preserving isometries, and if $\phi_1(F)$ is adjacent to $\phi_2(F)$ along a common side, where the matching sides must have the same color and opposite dotting. For example, for each T_i, $i = 1, \ldots, n$, the couple F, $T_i(F)$ is a matching pair.

Clearly, if $\phi \in \text{Isom}^+(\mathbb{H})$, and s is any side of $\phi(F)$, then there exists a unique $\psi \in \text{Isom}^+(\mathbb{H})$ such that $\psi(F)$ is adjacent to $\phi(F)$ along s, and such that the two copies are a matching pair.

Let us investigate how matching pairs pile up around a vertex p of F. Let $p_1 p_2 \ldots p_k (p_1)$ with $p_1 = p$ be the cycle of p. Then there are uniquely determined isometries

$$\phi_1^p = \text{id}, \phi_2^p, \ldots, \phi_k^p \in \text{Isom}^+(\mathbb{H})$$

such that

$$\phi_i^p(p_i) = p, \quad i = 1, \ldots, k,$$
$$\phi_i(F) \text{ and } \phi_{i+1}(F) \text{ are a matching pair}, \quad i = 1, \ldots, k - 1. \tag{43}$$

By condition (i) of Definition 3.34, the interior angles of the domains $\phi_i^p(F)$ at p sum up to 2π, and it follows that also $\phi_k^p(F)$ and F are a matching pair. Furthermore, since all domains are convex, the intersection $\phi_i^p(F) \cap F_j^p(F)$, for $i \neq j$, is always either a matching side or the single point p.

As an example, the isometries for the cycle $p_1 p_2 p_3 p_4 (p_1)$ in Fig. 10 for $p = p_1$ are

$$\phi_1^p = \text{id}, \quad \phi_2^p = T_3^{-1}, \quad \phi_3^p = T_3^{-1}T_4, \quad \phi_4^p = T_3^{-1}T_4T_2.$$

We will obtain the tiling by the below algorithm and first finish the proof of the theorem.

Each T_i sends F to a matching neighbor, and if F', F'' is any matching pair, then $T_i(F')$, $T_i(F'')$ is a matching pair. These two properties together imply that T_i preserves the entire tiling, and it follows that the same is true for any $\phi \in \Gamma_\mathfrak{I} = \langle T_1, \ldots, T_n \rangle_{\text{Isom}(\mathbb{H})}$. Moreover these ϕ also preserve the coloring and the dotting of the sides.

Now if F^* is any copy of F in the tiling, then there exists a sequence of copies

$$F_0 = F, \ F_1, F_2, \ldots, F_{N-1}, \ F_N = F^*$$

in the tiling such that any couple F_j, F_{j+1} is a matching pair. From this we conclude that there exists $\phi \in \Gamma_\mathfrak{I}$ such that $\phi(F) = F^*$. Indeed, for each

$j = 0, \ldots, N-1$ the common side of the pair F_j, F_{j+1} has the color of one of the sides $c_{i(j)}$ of F, and so

$$F^* = T^{\varepsilon_0}_{i(0)} T^{\varepsilon_1}_{i(1)} \cdots T^{\varepsilon_{N-1}}_{i(N-1)}(F)$$

for suitable exponents $\varepsilon_j = \pm 1$, $j = 0, \ldots, N-1$.

We have now seen that the images of F under $\Gamma_{\mathcal{T}}$ cover $\mathbb{H}$. Since the only color and dot preserving isometry of F onto itself is the identity, this means that F is a fundamental domain of $\Gamma_{\mathcal{T}}$, and $\Gamma_{\mathcal{T}}$ is discrete.

For the statement about the hyperbolicity of the elements in $\Gamma_{\mathcal{T}} \setminus \{\mathrm{id}\}$ we first note that from the fashion in which F is sent to its adjacent neighbors it follows that there is some positive constant δ (depending on the shape of F) such that

$$d(x, g(x)) \geq \delta$$

for any $x \in F$ and any $g \in \Gamma_{\mathcal{T}}$, $g \neq \mathrm{id}$. We next observe that this inequality holds, in fact, for *all* $x \in \mathbb{H}$. Namely, any such x lies in the image $h(F)$ for some $h \in \Gamma_{\mathcal{T}}$, and if $g \in \Gamma_{\mathcal{T}}$ is any element different from the identity, then

$$d(x, g(x)) = d(h^{-1}(x), h^{-1}g(x)) = d(h^{-1}(x), h^{-1}gh(h^{-1}(x)) \geq \delta,$$

because $h^{-1}(x) \in F$, and $h^{-1}gh$ is just another element of $\Gamma_{\mathcal{T}}$. Now for elliptic and parabolic elements $\phi \in \mathrm{Isom}(\mathbb{H})$ the infimum over all distances $d(x, \phi(x))$, $x \in \mathbb{H}$, is zero. Hence the $g \in \Gamma_{\mathcal{T}} \setminus \{\mathrm{id}\}$ must be hyperbolic. $\square$

Algorithm 3.37. Let $F \subset \mathbb{H}$ be a polygon with side pairing as in Definition 3.34. Then the following infinite procedure yields a tiling of $\mathbb{H}$.

1 Set $D := F$.
2 For each side s of D, draw the matching copy of F across s.
3 Fill-in the missing copies for each vertex p of D using (43).
4 Set D to be the union of all copies of F obtained at this step.
5 Continue with 2.

Proof. By condition (ii) of Definition 3.34, a matching pair of copies of F always forms a convex polygon. This, together with condition (i) implies that if instruction 4 is completed for the first time, then D is a compact geodesic polygon tiled with copies of F. Furthermore, each vertex p of D belongs to at most two copies of F, and so, by condition (ii), the interior angle of D at p is $\leq \pi$. This implies that D is again convex, and the algorithm can go into the second round.

Now assume that we are at instruction 2 of round j and that D is tiled with copies of F such that each vertex p of D belongs to at most two copies of F. Then our argument used for round 1 remains valid, and so also the domain D at the beginning of round $j+1$ has these properties. Hence, the algorithm produces an ever growing tiling, and we easily see that if we let it run indefinitely, then eventually any point of $\mathbb{H}$ will be covered. $\square$

We finish this lecture with a guided exercise in which we generalize the Gauss-Bonnet formula for triangles (Theorem 1.22) to hyperbolic surfaces obtained in the above way.

We first remark that if $P \subset \mathbb{H}$ is a convex geodesic polygon domain with interior angles $\alpha_1, \ldots, \alpha_N$, then

$$\text{area} P = (N-2)\pi - (\alpha_1 + \cdots + \alpha_N). \tag{44}$$

For $N = 3$ this is Theorem 1.22 in the case of the hyperbolic plane. For larger N one obtains the formula by decomposing P into $N-2$ triangles.

Exercise 3.38 (Gauss-Bonnet formula for polygon surfaces). Let F with side pairing $T_1, \ldots, T_n$ and Fuchsian group $\Gamma_{\mathcal{T}}$ be as in Theorem 3.36 and let n_c be the number of vertex cycles. We denote by $\mathcal{S} = \mathbb{H}/\Gamma_{\mathcal{T}}$ the quotient surface as defined in Section 3.4,

A *triangle* in $\mathcal{S}$ is a subset $D \subset \mathcal{S}$ which is isometric to a filled geodesic triangle in $\mathbb{H}$. The area of D, its sides and interior angles, etc., are defined as for the corresponding triangle in $\mathbb{H}$. A collection $\mathfrak{D} = \{D_1, \ldots D_f\}$ of such triangles is called a *triangulation* of $\mathcal{S}$ if it satisfies the following conditions.

(i) $D_1 \cup \cdots \cup D_f = \mathcal{S}$.
(ii) If $i \neq j$, then $D_i \cap D_j$ is either empty or a common vertex or a common side of D_i and D_j.

A point of $\mathcal{S}$ is a *vertex of* $\mathfrak{D}$ if it is the vertex of one of the triangles. A geodesic arc in $\mathcal{S}$ is an *edge* of $\mathfrak{D}$ if it is the side of one of the triangles. We denote by $n_0(\mathfrak{D})$ the number of vertices, by $n_1(\mathfrak{D})$ the number of edges, and set $n_2(\mathfrak{D}) = f$, the number of *faces*. The *Euler characteristic of the triangulation* is defined as

$$\chi_{\mathfrak{D}} = n_0(\mathfrak{D}) - n_1(\mathfrak{D}) + n_2(\mathfrak{D}). \tag{45}$$

Show that

(a) $\text{area} F = \text{area} D_1 + \cdots + \text{area} D_f$.
(b) $3n_2(\mathfrak{D}) = 2n_1(\mathfrak{D})$.
(c) $\text{area} F = -2\pi\chi_{\mathfrak{D}}$.
(d) $\chi_{\mathfrak{D}} = n_c - n + 1$.
(e) $n_c - n + 1$ is an even number.

By (a), the sum of the areas of the triangles in any triangulation of $\mathcal{S}$ is the same, and one may understand this value as an elementary definition of $\text{area}\,\mathcal{S}$, the *area* of the surface. Of course, this coincides with the area as defined in Riemannian geometry.

By (c), the Euler characteristic is also the same for any triangulation, and we call this value the *Euler characteristic* of $\mathcal{S}$, denoted by $\chi_{\mathcal{S}}$. This coincides with the Euler characteristic as defined in combinatorial surface topology.

With these notations, the result of exercise (c) is

$$\boxed{\text{area}\,\mathcal{S} = -2\pi\chi_{\mathcal{S}}.}\tag{46}$$

This is the Gauss-Bonnet formula for $\mathcal{S}$. Instead of the Euler characteristic one also uses the *genus* $g_{\mathcal{S}} = 1 - \chi_{\mathcal{S}}/2$, and the formula then is

$$\text{area}\,\mathcal{S} = 4\pi(g_{\mathcal{S}} - 1).\tag{47}$$

For more on the topology of surfaces we refer to [16]. For further reading about groups of Möbius transformations and the geometry of hyperbolic surfaces we refer to [5], [6], [19].

Lecture 4:

The Matrix Model

When dealing with computational matters concerning Fuchsian groups, such as to find an "optimal" set of generators, or to enumerate lengths of closed geodesics etc., one is frequently confronted with elementary questions from hyperbolic geometry. A typical example: can one decide on a computational level, rather than by looking at a print out, whether a set of geodesics has a given configuration such as, for example, the triple in Fig. 6 or the eight geodesics on the left in Fig. 9?

In this lecture we propose a model of the hyperbolic plane, the *matrix model*, which provides a calculus that is rather easy to deal with and suitable for this type of questions. The new feature being that points, geodesics and isometries are treated on the same level, all of them represented by matrices. Moreover, the resulting calculus is such that most of the time the matrices need not be written out explicitly. The model was introduced by Semmler in [18].

The following exposition is self contained and written up in a way which allows it *to be read independently of the earlier lectures*. In the exercises, however, we will sometimes give the link to the corresponding approach in the other models.

We will work with $\mathrm{M}(2,\mathbb{R})$, the set of all 2×2 real matrices, and frequently deal with the following subsets (some of them are groups, some are not):

$$\mathrm{GL}(2,\mathbb{R}),\ \mathrm{GL}^{+}(2,\mathbb{R}),\ \mathrm{GL}^{-}(2,\mathbb{R}),\ \mathrm{SL}(2,\mathbb{R}),\ \mathrm{SL}^{-}(2,\mathbb{R}),\ \mathrm{T0}(2,\mathbb{R}),$$

which, respectively, consist of all invertible, $\det > 0$, $\det < 0$, $\det = 1$, $\det = -1$, and $\mathrm{tr} = 0$ matrices. We recall that $\mathrm{tr}A = \frac{1}{2}\mathrm{trace}A$, and observe that

$$A \in \mathrm{T0}(2,\mathbb{R}) \implies \det A = -\mathrm{tr}A^{2}.\tag{48}$$

For typographical reasons, the matrix product is sometimes denoted by a dot: $AB = A \cdot B$.

4.1 Basic Ingredients

We begin with the basic objects of geometry: points, lines, rigid motions. To motivate the definitions given here we occasionally resort to the upper half-plane model.

Points

The set of points in the matrix model will be denoted by P_M. It is defined as follows, where the motivation comes from the fact that in the upper half-plane points can be identified with the half-turns around them, and these in turn can be represented by real 2×2 matrices with trace 0.

$$P_M = \{p \in \mathrm{SL}(2,\mathbb{R}) \mid \mathrm{tr}\, p = 0,\ p_{21} < 0\}. \tag{49}$$

Here, $p = \left(\begin{smallmatrix} p_{11} & p_{12} \\ p_{21} & p_{22} \end{smallmatrix}\right)$, and the condition $p_{21} < 0$ is motivated by the fact that p and $-p$ describe the same half-turn. Sometimes it is convenient to work with any matrix $v \in \mathrm{GL}^+(2,\mathbb{R})$ and normalize it only if necessary:

$$p = \mathrm{sgn}(-v_{21}) \frac{1}{\sqrt{\det v}}\, v \quad \text{for any} \quad v \in \mathrm{GL}^+(2,\mathbb{R}).$$

Observe that we can write any point $p \in P_M$ as

$$p = \frac{1}{s}\begin{pmatrix} -r & r^2 + s^2 \\ -1 & r \end{pmatrix}. \tag{50}$$

with uniquely determined $r \in \mathbb{R}$ and $s > 0$. Comparing this with Exercise 2.21(a) we see that p represents the Möbius transformation fixing point $r+is$ in the upper half-plane (see Section 2.4). Hence, (50) gives us the link between P_M and the upper half-plane model. In order to pass to the hyperboloïd model (Section 1.1) it suffices to express p in the basis I, J, K:

$$p = xI + yJ + zK$$

with

$$I = \begin{pmatrix} 1 & 0 \\ 0 & -1 \end{pmatrix}, \qquad J = \begin{pmatrix} 0 & 1 \\ 1 & 0 \end{pmatrix}, \qquad K = \begin{pmatrix} 0 & 1 \\ -1 & 0 \end{pmatrix} \tag{51}$$

and map point p to point $(x, y, z) \in \mathrm{H}^2_{-1} = \{(x_1, x_2, x_3) \in \mathbb{R}^3 \mid -x_1^2 - x_2^2 + x_3^2 = 1,\ x_3 > 0\}$ (Definition 1.6 and (2)).

Note that

$$I^2 = J^2 = 1, \quad K^2 = -1$$
$$IJ = -JI = K, \quad IK = -KI = J, \quad JK = -KJ = -I, \tag{52}$$

where $\mathbf{1}$ is the unit matrix.

Exercise 4.1. Let $\pi_{-1} : H^2_{-1} \to \mathbb{D}$ be the stereographic projection from the hyperboloïd model to the disc model defined in (13), and $\psi : \mathbb{D} \to \mathbb{H}$ the isometry defined in (23). Finally, let $\sigma : \mathbb{D} \to \mathbb{D}$ be the (orientation reversing) isometry defined by $\sigma(z) = i\bar{z}$.

If $x = (x_1, x_2, x_3) \in H^2_{-1}$ and $r + is = \psi(\sigma(\pi(x)))$, then

$$\frac{1}{s} \begin{pmatrix} -r & r^2 + s^2 \\ -1 & r \end{pmatrix} = x_1 I + x_2 J + x_3 K.$$

Distances

The *distance* $d(p,q)$ between points $p, q \in P_M$ is defined by the formula

$$\cosh d(p,q) = -\mathrm{tr}(p \cdot q). \tag{53}$$

Writing p and q in the form

$$p = \frac{1}{s} \begin{pmatrix} -r & r^2 + s^2 \\ -1 & r \end{pmatrix}, \quad q = \frac{1}{\sigma} \begin{pmatrix} -\rho & \rho^2 + \sigma^2 \\ -1 & \rho \end{pmatrix},$$

we get

$$-\mathrm{tr}(p \cdot q) = 1 + \frac{1}{2s\sigma}((r-\rho)^2 + (s-\sigma)^2). \tag{54}$$

This shows that $d(p,q) \geq 0$, with equality if and only if $p = q$. Clearly, one also has $d(p,q) = d(q,p)$. In Theorem 4.14 we will show, using tools arising from the matrix model, that d also satisfies the triangle inequality and, hence, P_M with d is a metric space. Comparing (54) with Exercise 2.20(b) we see that the mapping $p \mapsto r + is$ given by (50) is an isometry between the matrix model and the upper half-plane model.

For non-normalized points formula (53) reads

$$\cosh d(p,q) = \frac{|\mathrm{tr}(pq)|}{\sqrt{\mathrm{tr}(p^2)\mathrm{tr}(q^2)}}. \tag{55}$$

Isometries

Any $\alpha \in \mathrm{GL}^+(2,\mathbb{R})$ acts on $M(2,\mathbb{R})$ by conjugation and, restricted to P_M, this is an isometric action because traces and determinants are invariant. One observes that elements in $\mathrm{GL}^-(2,\mathbb{R})$, acting on $p \in P_M$ by conjugation change the sign of p_{21} and the result has to be multiplied by -1 in this case. Altogether, an isometric action of $\mathrm{GL}(2,\mathbb{R})$ on P_M is given in the following way:

$$p \mapsto \begin{cases} \alpha p \alpha^{-1} & \text{if } \alpha \in \mathrm{GL}^+(2,\mathbb{R}), \\ -\alpha p \alpha^{-1} & \text{if } \alpha \in \mathrm{GL}^-(2,\mathbb{R}). \end{cases} \tag{56}$$

We point out that another way to deal with the signs would be to attach a charge to every point, corresponding to the sign of $-p_{21}$, and to double the points in this way. Then a "positive" point becomes "negative" when acted upon by conjugation with a matrix of negative determinant. However, this leads to cumbersome sign rules, and so we have opted for the model where only "positive" points are admitted.

Exercise 4.2. Let $\alpha = \left(\begin{smallmatrix} a & b \\ c & d \end{smallmatrix} \right) \in \mathrm{GL}(2, \mathbb{R})$, and $p = \frac{1}{s} \left(\begin{smallmatrix} -r & r^2+s^2 \\ -1 & r \end{smallmatrix} \right)$. If $r' + is' = \frac{a(r+is)+b}{c(r+is)+d}$, then $\alpha \cdot p \cdot \alpha^{-1} = \frac{1}{s'} \left(\begin{smallmatrix} -r' & r'^2+s'^2 \\ -1 & r' \end{smallmatrix} \right)$.

Geodesics

While one would expect geodesics to be defined as sets of points, we will take another approach which will be the source of the computational strength of the matrix model.

Consider $B \in \mathrm{GL}^-(2, \mathbb{R})$ with $\mathrm{tr}B = 0$. As shown in Exercise 4.4, the set of points

$$G_B = \{p \in P_M \mid \mathrm{tr}(p \cdot B) = 0\}$$

corresponds to a geodesic in $\mathbb{H}$. Our intention is to understand B itself as the geodesic rather than G_B. Since $G_B = G_{\lambda \cdot B}$ for all nonzero $\lambda \in \mathbb{R}$, we will normalize B by dividing it by the square root of $-\det B$. This leads to the following definition: The set of lines in the matrix model is

$$L_M := \mathrm{SL}^-(2, \mathbb{R}) \cap \mathrm{T0}(2, \mathbb{R}) = \{B \in \mathrm{M}(2, \mathbb{R}) \mid \mathrm{tr}B = 0, \ \det B = -1\}. \quad (57)$$

Observe that with $B \in L_M$ we also have $-B \in L_M$. The members of L_M are called the *oriented lines* or *oriented geodesics*.

The orientation of the lines will be a helpful tool and allows us among other things to orient the triangles, and to introduce an orientation on the model itself.

It is reasonable to define the action of $\alpha \in \mathrm{GL}(2, \mathbb{R})$ on L_M by

$$B \mapsto \alpha B \alpha^{-1}, \quad B \in L_M. \quad (58)$$

In contrast to the action on points (see (56)), no sign change is made here. So (58) is really an operation on oriented lines.

Exercise 4.3. Let $B \in \mathrm{GL}^-(2, \mathbb{R})$ and assume that $\mathrm{tr}B = 0$. Then, for any $p \in P_M$ one has $-BpB^{-1} = p \iff \mathrm{tr}(p \cdot B) = 0$.

Exercise 4.4. To give an interpretation of the preceding exercise in the upper half-plane we consider a point $p = r + is \in \mathbb{H}$ and a geodesic $\gamma \subset \mathbb{H}$ defined by the Euclidean circle with center $\rho \in \mathbb{R}$ and radius $\sigma > 0$. The half-turn around p and the symmetry with axis γ are given, respectively, by

$$H_p = \frac{1}{s} \left(\begin{matrix} -r & r^2+s^2 \\ -1 & r \end{matrix} \right), \quad H_\gamma = \frac{1}{\sigma} \left(\begin{matrix} \rho & -\rho^2+\sigma^2 \\ 1 & -\rho \end{matrix} \right)$$

(Exercise 2.21). If we write $p = \rho + \lambda e^{it}$, $t \in {]0, \pi[}$, then p lies on γ if and only if $\lambda = \sigma$. Show by direct calculation that

$$p \in \gamma \iff \mathrm{tr}(H_p \cdot H_\gamma) = 0.$$

Fixed Objects

If $B \in M(2, \mathbb{R})$ with $\mathrm{tr}B = 0$, then B will be fixed by any element of the form $\alpha = \lambda\mathbf{1} + \mu B$ with real numbers λ, μ and nonzero determinant, $\mathbf{1}$ being the identity matrix. In fact,

$$(\lambda\mathbf{1} + \mu B)B = \lambda B + \mu B^2 = B(\lambda\mathbf{1} + \mu B)$$

and hence

$$(\lambda\mathbf{1} + \mu B)B(\lambda\mathbf{1} + \mu B)^{-1} = B. \tag{59}$$

For given B the isometries defined by the (non-normalized) $\lambda\mathbf{1} + \mu B$ form a one-parameter family. If B represents a point in P_M we call these isometries *rotations*, and the point is their *center of rotation*; if B represents a line in L_M we call them *translations* and the line is their *axis*.

4.2 Incidence and Perpendicularity

Two objects $u, v \in T0(2, \mathbb{R})$ are called *orthogonal* iff $\mathrm{tr}(uv) = 0$. This notion does not need normalization, and the property of being orthogonal is invariant under the action of $\alpha \in GL(2, \mathbb{R})$:

$$\mathrm{tr}((\alpha u \alpha^{-1}) \cdot (\alpha v \alpha^{-1})) = \mathrm{tr}((\alpha u v \alpha^{-1})) = \mathrm{tr}(uv).$$

We also note the following simple fact.

If $u, v \in T0(2, \mathbb{R})$ and $uv \in T0(2, \mathbb{R})$ then $uv = -vu$. (60)

By (54), two points are never orthogonal. By Exercise 4.3, a point p is orthogonal to the line described by $B \in GL^-(2, \mathbb{R})$ if and only if it is fixed by the action of B. In this case we will also say that the point and the line are *incident*.

If $A, B \in L_M$ are lines with $\mathrm{tr}(AB) = 0$, we may let act one upon the other:

$$ABA^{-1} = -B,$$

(see (60)). This changes the oriented line B into its opposite $-B$. At the same time any point $p \in P_M$ incident with A remains fixed:

Fig. 12. A diagram of perpendicular lines.

48   Aline Aigon-Dupuy, Peter Buser, and Klaus-Dieter Semmler

$$-ApA^{-1} = p,$$

and any point incident with B is mapped to a point incident with B. From this we see that orthogonality in L_M corresponds to perpendicularity of geodesics in the other models (see also Exercise 4.4), and we will also say that A and B are *perpendicular*.

The above considerations lead us to the following operation:

Definition 4.5. For $X, Y \in T0(2, \mathbb{R})$ we define

$$X \wedge Y := \frac{1}{2}(XY - YX) = XY - \operatorname{tr}(XY) \cdot 1.$$

Clearly, the product $\wedge : T0(2, \mathbb{R}) \times T0(2, \mathbb{R}) \to T0(2, \mathbb{R})$ defined in this way is bilinear, and the notation suggests that it has similar properties as the classical exterior product.

Theorem 4.6. *For $X, Y, \ldots \in T0(2, \mathbb{R})$ the following identities hold, where the left-hand side in* (v) *is the determinant of a 3×3 matrix.*

$$
\begin{align}
X \wedge Y &= -Y \wedge X; \qquad X \wedge X = 0 \tag{i} \\
\operatorname{tr}(X \cdot (Y \wedge Z)) &= \operatorname{tr}((X \wedge Y) \cdot Z); \qquad \operatorname{tr}(X \cdot (X \wedge Y)) = 0 \tag{ii} \\
X \wedge (Y \wedge Z) &= \operatorname{tr}(XY)Z - \operatorname{tr}(XZ)Y \tag{iii} \\
\operatorname{tr}((X \wedge Y) \cdot (Z \wedge W)) &= \operatorname{tr}(XW) \cdot \operatorname{tr}(YZ) - \operatorname{tr}(XZ) \cdot \operatorname{tr}(YW) \tag{iv} \\
-\det(\operatorname{tr}(A_i \cdot B_j)) &= \operatorname{tr}((A_1 \wedge A_2) \cdot A_3)\operatorname{tr}((B_1 \wedge B_2) \cdot B_3) \tag{v}
\end{align}
$$

Proof. As all expressions are multilinear (except for the corollaries in (i) and (ii)), it suffices to check the identities for the basis elements I, J, K in (51). Using

$$\operatorname{tr}I^2 = \operatorname{tr}J^2 = -\operatorname{tr}K^2 = 1, \quad I \wedge J = K, \ I \wedge K = J, \ J \wedge K = -I, \tag{61}$$

this becomes a simple exercise. $\square$

In the next theorem we use common geometric language: if point p is incident with line A we say that A *goes through* p; if for two lines A, B there exists a point incident with both we say that A and B *intersect*, etc.

Theorem 4.7. *Let $p, q, A, B \in T0(2, \mathbb{R})$. Then the following hold.*

(i) *If p, q are distinct points, then $p \wedge q$ represents the line through p and q.*

(ii) *If p is a point and A a line, then $p \wedge A$ represents the line through p perpendicular to A.*

(iii) *If A, B are lines and $\det(A \wedge B) > 0$, then $A \wedge B$ represents the point incident with A and B.*

(iv) *If A, B are lines and $\det(A \wedge B) < 0$, then $A \wedge B$ represents the line perpendicular to A and B.*

Proof. Calculate using Theorem 4.6 together with (48) and interpret. $\square$

Remark 4.8. It can occur that A and B are distinct lines and $\det(A \wedge B) = 0$. This corresponds to the case of geodesics in the upper half-plane model given by circles which meet on the real line.

One may include the determinant zero elements into the matrix model by treating them as "points at infinity", but we will not do this here.

For points and lines there exist various types of distances defined as follows, where the elements do not have to be normalized. For points p, q the distance is given by formula (55) and is repeated here for convenience,

$$\cosh d(p,q) = \frac{|\mathrm{tr}(pq)|}{\sqrt{\mathrm{tr}(p^2)\mathrm{tr}(q^2)}}. \tag{(55)}$$

If A and B represent non-intersecting lines we define their *distance* by

$$\cosh d(A,B) = \frac{|\mathrm{tr}(AB)|}{\sqrt{\mathrm{tr}(A^2)\mathrm{tr}(B^2)}}, \tag{62}$$

while for intersecting lines we define an *angle* by

$$\cos \sphericalangle(A,B) = \frac{\mathrm{tr}(AB)}{\sqrt{\mathrm{tr}(A^2)\mathrm{tr}(B^2)}}. \tag{63}$$

This must be understood as an oriented angle between oriented lines. Finally, the *oriented distance* of a point p to an oriented line A is defined by

$$\sinh d(A,p) = \frac{\mathrm{tr}(A \cdot p)}{\sqrt{-\mathrm{tr}(A^2)\mathrm{tr}(p^2)}}. \tag{64}$$

In (62) and (63) one has, of course, to verify that the values on the right-hand side lie in the proper intervals. This can be done using the following calculation (with (48)),

$$\det(A \wedge B) = -\mathrm{tr}((A \wedge B) \cdot (A \wedge B)) = -\mathrm{tr}^2(AB) + \mathrm{tr}A^2\mathrm{tr}B^2.$$

Example 4.9. To illustrate the coherence of the preceding definitions we solve the following exercise: Let $A, B \in L_M$ be lines with common perpendicular $C = A \wedge B$, and let p, q be the intersection points of C and A, B respectively. Then

$$d(A,B) = d(p,q).$$

Solution: A and B are normalized so that $\mathrm{tr}A^2 = \mathrm{tr}B^2 = 1$ (see (48)). Using Theorem 4.7 and Theorem 4.6(iii) we get

$$p = A \wedge C = A \wedge (A \wedge B) = B - \mathrm{tr}(AB) \cdot A,$$
$$q = B \wedge C = B \wedge (A \wedge B) = \mathrm{tr}(AB) \cdot B - A.$$

Calculating $p^2 = B^2 - \mathrm{tr}(AB) \cdot BA - \mathrm{tr}(AB) \cdot AB + \mathrm{tr}^2(AB) \cdot A^2$, and using that tr is an additive function (with respect to addition of matrices) we get the following expressions for $\operatorname{tr} p^2$ and, with a similar calculation, for $\operatorname{tr} q^2$ and $\operatorname{tr}(pq)$

$$\operatorname{tr} p^2 = 1 - \operatorname{tr}^2(AB), \quad \operatorname{tr} q^2 = 1 - \operatorname{tr}^2(AB), \quad \operatorname{tr}(pq) = \operatorname{tr}(AB) \cdot (1 - \operatorname{tr}^2(AB)).$$

Hence,

$$\frac{\operatorname{tr}^2(pq)}{\operatorname{tr} p^2 \cdot \operatorname{tr} q^2} = \operatorname{tr}^2(AB),$$

which proves the claim. $\square$

Example 4.10 (To mark off a point on a line). The simplest step in a geometric construction is to mark off a point on a line at some given distance from another point on that line. In the matrix model this is achieved in the following way.

Take $p \in P_M$, $A \in L_M$ such that p lies on A, i.e. $\operatorname{tr}(p^2) = -1$, $\operatorname{tr}(A^2) = 1$, $\operatorname{tr}(pA) = 0$. For any given number $t \in \mathbb{R}$ we have

$$\sqrt{1+t^2}\, 1 + tA \in \operatorname{SL}(2,\mathbb{R}).$$

Thus, conjugation with this matrix is an isometry with axis A. A direct calculation yields

$$(\sqrt{1+t^2}\, 1 + tA) \cdot p \cdot (\sqrt{1+t^2}\, 1 + tA)^{-1} = (2t^2 + 1)p + 2t\sqrt{1+t^2}\, A \wedge p.$$

Writing $2t\sqrt{1+t^2} = s$ we obtain as image the following point $p_s \in P_M$ lying on A,

$$p_s = \sqrt{1+s^2}\, p + sA \wedge p.$$

The distance between p and p_s satisfies $\cosh d(p, p_s) = -\operatorname{tr}(pp_s) = \sqrt{s^2 + 1}$, and parameter s may be interpreted as the hyperbolic sine of the oriented distance from p to p_s. Note that the product of the half-turns around p and p_s is as follows (where we use that p lies on A so that $(A \wedge p)p = Ap^2$),

$$-p_s p = \sqrt{1+s^2}\, 1 + sA.$$

4.3 Triangles

In the following we want to emphasize the *linearity* in the approach to geometric problems via the matrix model. As an example we consider the classical properties of triangles well known in Euclidean geometry. We point out that no explicit representation by matrices is needed and no reference to a particular basis such as I, J, K is made.

We begin with an exercise that can be solved in a similar way as Example 4.9.

Exercise 4.11. Assume either $v, w \in P_M$ or $v, w \in L_M$. Set

$$m = u + v, \qquad m' = u - v.$$

(Here u, v have to be normalized, for otherwise the expressions $u+v$ and $u-v$ are geometrically not meaningful.) Then m and m' are incident with $u \wedge v$, and

$$\text{tr}^2(m \cdot u) = \text{tr}^2(m \cdot v), \qquad \text{tr}^2(m' \cdot u) = \text{tr}^2(m' \cdot v).$$

When v, w are points, then m can be interpreted as the midpoint and m' the perpendicular bisector of the two points. What are the interpretations of m, m' when v, w are lines?
(Hint: consider the isometric actions of m and m' on u and v.)

Exercise 4.12. Let $A, B, A', B' \in L_M$ be such that $\text{tr}(AB) = 0$, $\text{tr}(A'B') = 0$. Then there exists $g \in \text{GL}(2, \mathbb{R})$ such that $gAg^{-1} = A'$, $gBg^{-1} = B'$.
(Hint: show first that for $g_1 = A + A'$ one gets $g_1 A g_1^{-1} = A'$, and then continue.)

In the next theorem, three objects $X, Y, Z \in \text{T0}(2, \mathbb{R})$ are called *concurrent* if there exists $W \in \text{T0}(2, \mathbb{R})$ orthogonal to all three of them: $\text{tr}(XW) = \text{tr}(YW) = \text{tr}(ZW) = 0$. There are various cases for this situation, a triple of lines going through the same point, a triple of lines having a common perpendicular, a triple of points lying on the same line, etc.

Theorem 4.13. *In a triangle the three perpendicular bisectors are concurrent, the three altitudes are concurrent, the three medians are concurrent and the three angular bisectors are concurrent.*

Proof. Let $a, b, c \in P_M$ be the vertices of the triangle. We assume that a, b and c are in normalized form. All further elements will be written in non-normalized form.

By Exercise 4.11, the perpendicular bisectors are the lines $a-b$, $b-c$, $c-a$, perpendicular to the sides $a \wedge b$, $b \wedge c$, $c \wedge a$ and passing through the midpoints $a + b$, $b + c$, $c + a$, respectively. Introducing p_m as the element incident with the first two we get

$$p_m = (a - b) \wedge (b - c) = a \wedge b + b \wedge c + c \wedge a.$$

The expression on the right-hand side is invariant under cyclic permutation of the letters, and we conclude that p_m is incident with all three lines. Observe that p_m may be a point or a line or an element with determinant zero ("point at infinity").

The altitudes of the triangle are

$$H_a = a \wedge (b \wedge c), \quad H_b = b \wedge (c \wedge a), \quad H_c = c \wedge (a \wedge b).$$

Introducing $p_h = H_a \wedge H_b$ as the element incident with the first two we get

$$p_h = (\operatorname{tr}(ab)\,c - \operatorname{tr}(ac)\,b) \wedge (\operatorname{tr}(bc)\,a - \operatorname{tr}(ba)\,c)$$
$$= \operatorname{tr}(ab)\operatorname{tr}(bc)\,c \wedge a + \operatorname{tr}(bc)\operatorname{tr}(ca)\,a \wedge b + \operatorname{tr}(ca)\operatorname{tr}(ab)\,b \wedge c.$$

Again this last expression is invariant under cyclic permutation of the letters, and p_h is incident with all altitudes.

The midpoint of a and b is $a+b$, and the median passing through $a+b$ and c becomes $(a+b)\wedge c$; idem for the other two sides of the triangle. Introducing p_g as the element incident with the medians through c and a we obtain

$$p_g = ((a+b)\wedge c) \wedge ((b+c)\wedge a)$$
$$= (\operatorname{tr}(((a+b)\wedge c)(b+c)))\,a - (\operatorname{tr}(((a+b)\wedge c)\,a))\,(b+c)$$
$$= (\operatorname{tr}((a\wedge c)\,b))a - (\operatorname{tr}((b\wedge c)a))\,(b+c)$$
$$= -(\operatorname{tr}((a\wedge b)c))\,(a+b+c).$$

Here we have, surprisingly, that if a, b, c are points written in normalized form then, as in the Euclidean case, the intersection of the three medians is represented by $a+b+c$.

The proof for the angular bisectors is left as an exercise. □

The formulas occurring in the preceding proof as well as many others can be given a more concise form if we introduce the following notation (still for $a, b, c \in P_M$ in normalized form),

$$x = \cosh d(a,b) = -\operatorname{tr}(ab),$$
$$y = \cosh d(b,c) = -\operatorname{tr}(bc), \tag{65}$$
$$z = \cosh d(c,a) = -\operatorname{tr}(ca).$$

With this notation the expression $\operatorname{tr}((a\wedge b)c)$, for example, can be rewritten in the following symmetric way,

$$\operatorname{tr}^2((a\wedge b)c) = 1 - x^2 - y^2 - z^2 + 2xyz = \det H(a,b,c), \tag{66}$$

where

$$H(a,b,c) = \begin{pmatrix} 1 & x & z \\ x & 1 & y \\ z & y & 1 \end{pmatrix}. \tag{67}$$

The right-hand side in (66) shows that the left-hand side is invariant under cyclic permutation, and the left-hand side shows that the right-hand side is nonnegative. It is interesting to observe that the latter property is equivalent to the triangle inequality:

Theorem 4.14 (Triangle inequality). *Let* $S = S(a,b,c)$, $M = M(a,b,c)$, $L = L(a,b,c)$ *be the lengths of, respectively, the small, the medium and the large side of triangle* a,b,c, *and define* $H(a,b,c)$ *as in (65), (67). Then* $\det H(a,b,c) \geq 0$, *and*

$$L(a,b,c) \leq S(a,b,c) + M(a,b,c), \quad \text{with equality iff } \det H(a,b,c) = 0.$$

Proof. A simple calculation using the addition theorem for the cosh-function yields

$$(\cosh(S+M) - \cosh L)(\cosh L - \cosh(M-S)) = 1 - x^2 - y^2 - z^2 + 2xyz,$$

where by (66) the right-hand side is nonnegative. Observe that the condition $\det H(a,b,c) = 0$ means that a, b, c lie on the same line. □

The reader may have observed, that the preceding arguments can be extended to triples which do not exclusively consist of points. As an example we propose the following exercise in which "$q \in A$" stands for "q is incident with A."

Exercise 4.15.

(i) If $p \in P_M$ and $A \in L_M$ then $d(p, A) = \min\{d(p, q) \mid q \in A\}$.
(ii) If $A \in L_M$ and p, $q \in P_M$ then $|d(p, A) - d(q, A)| \le d(p, q)$.

The determinant $H(a,b,c)$, (again for normalized $a, b, c \in P_M$) contains further information. While the classical "length of base times length of altitude = twice the area" has no analogue in the hyperbolic case (see (44)) it is still true that "sinh(base) times sinh(altitude)" is an invariant which does not depend of the choice of the base: let h_a, h_b, h_c be the lengths of the altitudes, then

$$\sinh^2 d(a,b) \sinh^2 h_c = \sinh^2 d(a,b) \sinh^2 d(a \wedge b, c) = \frac{(x^2-1)\mathrm{tr}^2((a \wedge b)c)}{\mathrm{tr}^2(ab) - \mathrm{tr}(a^2)\mathrm{tr}(b^2)}.$$

Here the right-hand side is equal to the determinant of $H(a,b,c)$ which is invariant under cyclic permutation, and therefore

$$\sinh d(a,b) \sinh h_c = \sinh d(b,c) \sinh h_a = \sinh d(c,a) \sinh h_b$$
$$= \sqrt{\det H(a,b,c)}. \quad (68)$$

Here, as before, we have used Theorem 4.6. The various interpretations of X, Y, Z, W in the identity (iv) of the theorem lead to a variety of trigonometric formulae. We finish this lecture with two examples.

The cosine of the angle γ between the oriented lines $c \wedge a$, $c \wedge b$ is given by (63), and using identity (iv) we get the cosine law:

$$\cos \gamma = \frac{\mathrm{tr}((c \wedge a)(c \wedge b))}{\sqrt{\mathrm{tr}((c \wedge a)^2)\mathrm{tr}((c \wedge b)^2)}} = \frac{yz - x}{\sqrt{y^2 - 1}\sqrt{z^2 - 1}}. \quad (69)$$

From this we obtain

$$\frac{\sin^2 \gamma}{\sinh^2 d(a,b)} = \frac{1 - \cos^2 \gamma}{\cosh^2 d(a,b) - 1} = \frac{(y^2-1)(z^2-1) - (yz-x)^2}{(x^2-1)(y^2-1)(z^2-1)},$$

where the numerator equals $1 - x^2 - y^2 - z^2 + 2xyz$. Therefore, the right-hand side is, once again, invariant under cyclic permutation and we get the sine law for the interior angles α, β, γ at a, b, c, respectively:

$$\frac{\sin \gamma}{\sinh d(a,b)} = \frac{\sin \alpha}{\sinh d(b,c)} = \frac{\sin \beta}{\sinh d(c,a)} = \frac{\sqrt{\det H(a,b,c)}}{\sqrt{(x^2-1)(y^2-1)(z^2-1)}}. \quad (70)$$

For an extensive set of further formulas of this kind we refer to [5] or [6] or the classical book of Fenchel [11].

Lecture 5:

Constellations and Helling Matrices

In this lecture we introduce a technique allowing one to determine the geometric configuration of a set of objects from $\mathrm{T0}(2, \mathbb{R})$ in a symbolic computational way.

5.1 The Helling Matrix of a Constellation

By a *geometric constellation* we mean an ordered set of points and oriented lines.

$$\mathcal{C} = (v_1, \dots, v_n)$$

(all normalized), where two such constellations are considered equal if they differ only by an isometry. Hence, a constellation captures geometric and combinatorial data of an aggregate, but not its position.

A good way of extracting the geometry of a configuration is the so called *Helling-matrix* (used intrinsically in [13, 14]):

Definition 5.1. To any constellation $\mathcal{C} = (v_1, \dots, v_n)$ of n objects, we assign the $n \times n$- matrix $H = H(\mathcal{C})$ defined by

$$H_{ij} = H(\mathcal{C})_{ij} = -\mathrm{tr}(v_i v_j), \quad i, j = 1 \dots, n.$$

The matrix is symmetric and has 1's and -1's on the diagonal telling us which v_i is a point, and which is a line. Whenever v_i, v_j are two points, then $H_{ij} \geq 1$.

Of course, if we apply an isometry with positive determinant to the whole constellation, the Helling matrix will remain invariant. If the determinant of the isometry is negative then some of the matrix entries may change sign.

Remark 5.2. Technically speaking, $H(\mathcal{C})$ is a Gram matrix with respect to the bilinear form $x, y \mapsto -\mathrm{tr}(xy)$, $x, y \in \mathrm{T0}(2, \mathbb{R})$. Since $\mathrm{T0}(2, \mathbb{R})$ is a three-dimensional vector space it follows that the rank of $H(\mathcal{C})$ is at most 3.

We now study various elementary configurations.

Example 5.3 (Two points, one line). Our first example is given by an oriented line $A \in L_M$ and two points $p, q \in P_M$, all written in normal form so that A has determinant -1, and p, q have determinant 1, and $p_{21} < 0$, $q_{21} < 0$.

Fig. 13 shows two different constellations, one with p, q lying on the same side of A, and one where p and q are separated by A. How can one characterize these cases computationally?

$$O(p, q, A) < 0 \qquad\qquad\qquad , \qquad\qquad\qquad O(p, q, A) > 0$$

Fig. 13. Two constellations: one non-separating, the other separating.

The Helling matrix of the constellation is

$$H(p, q, A) = \begin{pmatrix} 1 & -\mathrm{tr}(pq) & -\mathrm{tr}(pA) \\ -\mathrm{tr}(pq) & 1 & -\mathrm{tr}(qA) \\ -\mathrm{tr}(pA) & -\mathrm{tr}(qA) & -1 \end{pmatrix}.$$

By (64), $\mathrm{tr}(pA)$ and $\mathrm{tr}(qA)$ are the hyperbolic sines of the oriented distances from the points to the line. If we now extract the following product from the Helling matrix,

$$O(p, q, A) = \mathrm{tr}(pA)\mathrm{tr}(qA)\mathrm{tr}(pq), \tag{71}$$

then we have a quantity whose sign remains constant if we move around p and q in a continuous way but without crossing A, while the sign changes whenever p or q move across A. Since A appears twice in the expression of $O(p, q, A)$, the sign is independent of the orientation of A. Finally, the trick with the factor $\mathrm{tr}(pq)$ is that in this way also p and q appear twice, and hence the sign will not change if we abandon the normalizations and multiply A, p, q by nonzero factors.

Since in the normalized case, $\mathrm{tr}(p, q) < 0$, we get the following picture: $O(p, q, A) < 0$ if p, q lie on the same side of A; $O(p, q, A) > 0$ if the points lie on opposite sides; and $O(p, q, A) = 0$ if at least one of the points lies on A.

Example 5.4 (Three lines). Let $A, B, C \in L_M$ be lines with pairwise positive distances. Fig. 14 shows the two possible configurations schematically. In the case shown on the right, line B intersects the common perpendicular $A \wedge C$ of A and C and separates A, C. In the case shown on the left none of the lines separates the other two.

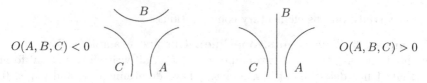

$O(A, B, C) < 0$ $O(A, B, C) > 0$

Fig. 14. Two constellations: one non-separating, the other separating. The first diagram is the "infinite triangle"-, the second diagram the "three parallels" constellation.

In order to characterize the two cases computationally we look at the same expression as in Example 5.3,

$$O(A, B, C) = \operatorname{tr}(AB)\operatorname{tr}(BC)\operatorname{tr}(CA), \tag{72}$$

which is a product of entries of the Helling matrix $H(A, B, C)$. As all lines appear twice in this expression, the sign does not depend on the orientations of the lines.

Since the distances are positive all factors of $O(A, B, C)$ have absolute value > 1. Assuming without loss of generality that $|\operatorname{tr}(AB)| \leq |\operatorname{tr}(BC)| \leq |\operatorname{tr}(CA)|$ and using that

$$\operatorname{tr}((B \wedge (A \wedge C))^2) = \operatorname{tr}((\operatorname{tr}(AB)C - \operatorname{tr}(BC)A)^2)$$
$$= \operatorname{tr}((\operatorname{tr}^2(AB))C^2 - \operatorname{tr}(AB)\operatorname{tr}(BC)(CA + AC) + \operatorname{tr}^2(BC)A^2)$$
$$= \operatorname{tr}^2(AB) + \operatorname{tr}^2(BC) - 2\operatorname{tr}(AB)\operatorname{tr}(BC)\operatorname{tr}(CA)$$

we easily prove that lines with the "infinite triangle" constellation satisfy $O(A, B, C) < 0$, while for lines with the "three parallels" constellation we have $O(A, B, C) > 0$, as indicated in Fig. 14.

Example 5.5 (Three isometries). Helling matrices also show up in connection with isometries. In this example we consider a triple of isometries

$$\alpha = a\mathbf{1} + A, \ \beta = b\mathbf{1} + B, \ \gamma = c\mathbf{1} + C \in \operatorname{SL}(2, \mathbb{R}),$$
$$\text{with } A, B, C \in \operatorname{T0}(2, \mathbb{R}),$$

where $\mathbf{1}$ is the identity matrix, and we recall that the action on P_M and L_M is by conjugation.

Caution: in the current example the normalized elements are α, β, γ, and *not* A, B, C. Observing that $1 = \det(a\mathbf{1} + A) = a^2 - \operatorname{tr}(A^2)$, etc., we have

$$\operatorname{tr}(A^2) = a^2 - 1, \quad \operatorname{tr}(B^2) = b^2 - 1, \quad \operatorname{tr}(C^2) = c^2 - 1.$$

For such a triple we consider the Helling matrix $H = H(A, B, C) = (H_{ij})$ of the non-normalized constellation A, B, C. It contains various informations about the isometries. For example, the signs of the diagonal elements H_{ii} tell us whether the corresponding isometries are hyperbolic, elliptic, or parabolic.

We are now particularly interested in the situation where A, B, C are lines and the triple is such that $\alpha\beta\gamma = -1$ (the choice of the sign is for practical reasons, recall that for $g \in \mathrm{SL}(2,\mathbb{R})$ both g and $-g$ represent the same isometry). This situation occurs in connection with the question under what conditions a group $\Gamma = \langle \alpha, \beta \rangle$ generated by two hyperbolic elements is *hyperbolic*, i.e. *all* its elements are hyperbolic. A well-known criterion is the following: *If A, B, C, are pairwise disjoint lines with the "infinite triangle" constellation, then Γ is hyperbolic* (cf. Example 3.9).

Our aim is to show that the hypotheses of this criterion can be read off from $H(A, B, C)$.

Since $\alpha\beta\gamma = -1$, and $(c\mathbf{1} + C)^{-1} = (c\mathbf{1} - C)$, we have $(a\mathbf{1} + A)(b\mathbf{1} + B) = -c\mathbf{1} + C$, and therefore $-c = ab + \mathrm{tr}(AB)$. By cyclic permutation the same holds for $-a$, $-b$, and $H = H(A, B, C)$ becomes

$$H = -\begin{pmatrix} \mathrm{tr}(AA) & \mathrm{tr}(AB) & \mathrm{tr}(AC) \\ \mathrm{tr}(AB) & \mathrm{tr}(BB) & \mathrm{tr}(BC) \\ \mathrm{tr}(AC) & \mathrm{tr}(BC) & \mathrm{tr}(CC) \end{pmatrix} = \begin{pmatrix} 1 - a^2 & ab + c & ac + b \\ ab + c & 1 - b^2 & bc + a \\ ac + b & bc + a & 1 - c^2 \end{pmatrix}.$$

Now A, B, C are lines if and only if $H_{ii} < 0$, $i = 1, 2, 3$. Furthermore,

$$\mathrm{tr}((A \wedge B)^2) = \mathrm{tr}^2(AB) - \mathrm{tr}(A^2)\mathrm{tr}(B^2) = a^2 + b^2 + c^2 + 2abc - 1.$$

The expression on the right is invariant under cyclic permutation, and we obtain the following,

$$A \wedge B \text{ is a line} \iff B \wedge C \text{ is a line} \iff C \wedge A \text{ is a line}$$
$$\iff H_{12}^2 - H_{11}H_{22} > 0.$$

Finally, by Example 5.4, A, B, C have the "infinite triangle" constellation if and only if

$$O(A, B, C) = \mathrm{tr}(AB)\mathrm{tr}(BC)\mathrm{tr}(CA) = -H_{12}H_{23}H_{31} < 0.$$

In Lecture 6 we will use the following existence theorem for two-generator hyperbolic groups.

Theorem 5.6. *For any three numbers $a, b, c > 1$ there exist, up to isometry, uniquely determined lines $A, B, C \in \mathrm{T0}(2, \mathbb{R})$ such that $a\mathbf{1} + A$, $b\mathbf{1} + B$, $c\mathbf{1} + C \in \mathrm{SL}(2,\mathbb{R})$ and*

$$(a\mathbf{1} + A)(b\mathbf{1} + B)(c\mathbf{1} + C) = -\mathbf{1}.$$

These lines are pairwise disjoint, have the "infinite triangle" constellation, and $\Gamma = \langle a\mathbf{1} + A,\ b\mathbf{1} + B \rangle$ is a discrete hyperbolic group.

Proof. Observe that $\det(a\mathbf{1} + A) = a^2 - \mathrm{tr}(A^2)$ and $\mathrm{tr}(A^2) = a^2 - 1 > 0$, so that A and similarly B, C are indeed lines.

We begin with the uniqueness, which as often in such cases, will then give a hint for the definition of A, B, C. So assume that A, B, C with $(a\mathbf{1} +$

A), $(b\mathbf{1} + B)$, $(c\mathbf{1} + C) \in \mathrm{SL}(2, \mathbb{R})$ and $(a\mathbf{1} + A)(b\mathbf{1} + B)(c\mathbf{1} + C) = -\mathbf{1}$ are given. Then as in Example 5.5,

$$\mathrm{tr}(A^2) = a^2 - 1, \qquad \mathrm{tr}(B^2) = b^2 - 1, \qquad \mathrm{tr}(C^2) = c^2 - 1,$$
$$\mathrm{tr}(AB) = -(ab + c), \quad \mathrm{tr}(BC) = -(bc + a), \quad \mathrm{tr}(CA) = -(ca + b).$$

Since

$$\mathrm{tr}((A \wedge B)^2) = \mathrm{tr}^2(AB) - \mathrm{tr}(A^2)\mathrm{tr}(B^2) = a^2 + b^2 + c^2 + 2abc - 1 > 0,$$

$A \wedge B$ is a line perpendicular to A and B. This proves that A and B have positive distance. By cyclic permutation the same holds for B, C and C, A. Finally,

$$O(A, B, C) = \mathrm{tr}(AB)\mathrm{tr}(BC)\mathrm{tr}(CA) = (-ab - c)(-bc - a)(-ca - b) < 0,$$

and so we have shown that A, B, C have the "infinite triangle" constellation.

The perpendicularity of A and $A \wedge B$ allows us to conjugate all elements with a suitable $g \in \mathrm{GL}(2, \mathbb{R})$ (cf. Exercise 4.12) such that afterwards $A = x_1 I$ and $A \wedge B = \lambda J$ for suitable $x_1, \lambda \in \mathbb{R}$. Line B is then of the form $B = y_1 I + y_3 K$. Conjugation of all elements with J, if necessary, yields $y_3 > 0$ (because $J^{-1}KJ = JKJ = -K$), and with an additional conjugation by K, if necessary, we achieve that also $x_1 > 0$. We have thus shown that, up to isometry,

$$A = x_1 I, \quad B = y_1 I + y_3 K, \quad \text{with } x_1 > 0, \; y_3 > 0.$$

From the fact that $\mathrm{tr}(A^2) = a^2 - 1$, $\mathrm{tr}(AB) = -(ab + c)$ and $\mathrm{tr}(B^2) = b^2 - 1$ we obtain explicit expressions for x_1, y_1, y_3, and using that $-c + C = (a\mathbf{1} + A)(b\mathbf{1} + B)$ we also obtain C explicitly. The result is

$$A = a'I,$$
$$B = -\frac{ab + c}{a'}I + \frac{s}{a'}K, \qquad (73)$$
$$C = -\frac{ac + b}{a'}I + sJ + \frac{as}{a'}K,$$

with

$$a' = \sqrt{a^2 - 1}, \quad s = \sqrt{a^2 + b^2 + c^2 + 2abc - 1}.$$

Hence the claimed uniqueness, up to isometry.

For the existence, we *define* A, B, C by (73). A simple calculation then shows that $\det(a\mathbf{1} + A) = \det(b\mathbf{1} + B) = \det(c\mathbf{1} + C) = 1$, and $(a\mathbf{1} + A)(b\mathbf{1} + B) = -c\mathbf{1} + C$, as required. The fact that the group is discrete and hyperbolic was shown in Example 3.9 □

Exercise 5.7. Show the following fact in a similar way: for any three numbers $x, y, z > 1$ satisfying $\det \begin{pmatrix} 1 & x & z \\ x & 1 & y \\ z & y & 1 \end{pmatrix} \geq 0$ (cf. (67)) there exist, up to isometry, uniquely determined points $p_1, p_2, p_3 \in P_M$ such that $H(p_1, p_2, p_3) = \begin{pmatrix} 1 & x & z \\ x & 1 & y \\ z & y & 1 \end{pmatrix}$. (Hint: the solution may be found in the proof of Theorem 5.16.)

Since for $g \in \mathrm{SL}(2, \mathbb{R})$ the isometric action of g and $-g$ is the same, the sign in a condition such as $\alpha\beta\gamma = \pm 1$ seems to have no importance. It *does*, however contain geometric information as, for example, in the following corollary to Theorem 5.6.

Theorem 5.8. *Let* $\alpha = a\mathbf{1} + A$, $\beta = b\mathbf{1} + B$, $\gamma = c\mathbf{1} + C \in \mathrm{SL}(2, \mathbb{R})$, *with* $A, B, C \in \mathrm{T0}(2, \mathbb{R})$ *and* $a, b, c > 1$ *be hyperbolic transformations satisfying* $\alpha\beta\gamma = \pm 1$.

(i) *If* $\alpha\beta\gamma = -1$, *then* A, B, C *are pairwise disjoint and have the "infinite triangle" constellation.*

(ii) *If* $\alpha\beta\gamma = +1$, *then either* A, B, C *have the "three parallels" constellation or* A, B, C *mutually intersect.*

Proof. If the product is -1, then Theorem 5.6 implies that the three lines have pairwise positive distances and form the "infinite triangle" constellation. Now assume that $\alpha\beta\gamma = +1$. Then $(a\mathbf{1} + A)(b\mathbf{1} + B) = c\mathbf{1} - C$, and therefore $ab + \mathrm{tr}(AB) = c$. By cyclic permutation, $\mathrm{tr}(AB) = c - ab$, $\mathrm{tr}(BC) = a - bc$, $\mathrm{tr}(CA) = b - ca$. Furthermore,

$$\mathrm{tr}((A \wedge B)^2) = (c - ab)^2 - (a^2 - 1)(b^2 - 1) = a^2 + b^2 + c^2 - 2abc - 1.$$

By cyclic permutation, $\mathrm{tr}((A \wedge B)^2) = \mathrm{tr}((B \wedge C)^2) = \mathrm{tr}((C \wedge A)^2)$. Hence the trichotomy: either the three lines intersect each other, or they are disjoint but with pairwise zero distance, or they have pairwise positive distances. Leaving the intermediate case (a limit situation of the third case) as an exercise, we now assume that the distances are positive, so that $A \wedge B$ is a line and $a^2 + b^2 + c^2 - 2abc - 1 > 0$.

Without loss of generality we may further assume that $1 < a \leq b \leq c$, and thus $a < bc$, $b < ac$. An elementary argument shows that these inequalities together with $a^2 + b^2 + c^2 - 2abc - 1 > 0$ imply that $c > ab$. This yields $\mathrm{tr}(AB)\mathrm{tr}(BC)\mathrm{tr}(CA) = (c - ab)(a - bc)(b - ca) > 0$, and so A, B, C have the "three parallels" constellation (cf. Example 5.4). $\square$

5.2 Quadrilaterals

A *quadrilateral* is a set of four points $p_1, p_2, p_3, p_4 \in P_M$ written in normalized form. The Helling matrix $H = H(p_1, p_2, p_3, p_4)$ is a symmetric matrix with 1's in the diagonal, and all matrix entries are ≥ 1. It encodes the lengths of the sides as well as the lengths of the diagonals of the quadrilateral.

There is a redundancy in this encoding, given by the fact that

$$\det H(p_1, p_2, p_3, p_4) = 0. \tag{74}$$

(By Remark 5.2, all Helling matrices have rank ≤ 3.) The equation $\det H = 0$ is a polynomial equation in which each matrix entry occurs with degree two. This means that, given any five of the six essential entries, there are at most

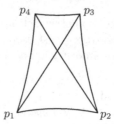

Fig. 15. A quadrilateral with diagonals

two solutions for the remaining one. In Euclidean geometry the corresponding relation between the sides and the diagonals of a Euclidean quadrilateral is known as *Heron's formula*. Equation (74) is the hyperbolic version of it.

In Lecture 6 we will deal with quadrilaterals of a particular shape which we call strongly convex. As in Euclidean geometry a quadrilateral is *convex* if the vertices are distinct, and if we can relabel them such that two consecutive vertices p_i, p_{i+1} (with $p_{4+1} := p_1$) always lie on the same side of the line through the two other vertices. We now say that the quadrilateral is *strongly convex* if, in addition, for each pair of opposite sides the lines containing these sides are disjoint.

Let us denote the (non-normalized) line carrying p_i and p_j by A_{ij}: $A_{ij} = p_i \wedge p_j$. Then A_{12} intersects A_{34} if and only if

$$\text{tr}((A_{12} \wedge A_{34})^2) = \text{tr}^2(A_{12}A_{34}) - \text{tr}(A_{12}^2)\text{tr}(A_{34}^2) < 0.$$

Using rule (iv) of Theorem 4.6 several times we obtain the right-hand side in terms of the Helling matrix,

$$\text{tr}^2((p_1 \wedge p_2)(p_3 \wedge p_4)) - \text{tr}((p_1 \wedge p_2)^2)\text{tr}((p_3 \wedge p_4)^2)$$
$$= (H_{14}H_{23} - H_{13}H_{24})^2 - (H_{12}^2 - 1)(H_{34}^2 - 1).$$

Abbreviating

$$sc(i, j, k, l) := (H_{il}H_{jk} - H_{ik}H_{jl})^2 - (H_{ij}^2 - 1)(H_{kl}^2 - 1),$$

we get the following criterion: if after a suitable permutation of the indices

$$sc(1, 2, 3, 4) \geq 0 \quad \text{and} \quad sc(4, 1, 2, 3) \geq 0,$$

then the quadrilateral is strongly convex.

Frequently the following condition is sufficient, which avoids looking at all permutations of the indices. Set

$$g(i, j, k, l) := H_{ij}H_{kl} - H_{ik}H_{jl} + H_{il}H_{jk},$$
$$gm(i, j, k, l) := \min\{g(i, j, k, l), g(i, k, j, l), g(i, j, l, k)\}. \tag{75}$$

Lemma 5.9. *If $gm(1, 2, 3, 4) < 0$, then the quadrilateral is strongly convex.*

Proof. It is easily checked that $gm(i,j,k,l) = \min_\sigma g(\sigma(i), \sigma(j), \sigma(k), \sigma(l))$, where σ runs through *all* permutations of the letters i, j, k, l. Therefore, $gm(i,j,k,l)$ is invariant under permutations, and we may assume that the three numbers $x = H_{12}H_{34}$, $y = H_{14}H_{23}$, $z = H_{13}H_{24}$ are given in increasing order: $1 \leq x \leq y \leq z$. (By (54) all entries of H are ≥ 1.)

If now $gm(1,2,3,4)$ is negative then $z > x + y$, which implies $z - y > x$ and $z - x > y$, and hence $(z-y)^2 > x^2$ and $(z-x)^2 > y^2$. We obtain

$$0 < (z-y)^2 - x^2 < sc(1,2,3,4) \quad \text{and} \quad 0 < (z-x)^2 - y^2 < sc(4,1,2,3).$$

□

5.3 Relative Orientations of Triangles

In an up-to-isometry setting, it does not make sense to talk about the orientation of a triangle. However, it is reasonable to ask whether two triangles p_i, p_j, p_k and p_l, p_m, p_n in a given set of points have the *same* orientation. To extract this property from the Helling matrix of these points we use the following notation.

For any matrix H we denote by $H_{i_1..i_m, j_1..j_n}$ the $m \times n$-matrix formed by the common part of rows $i_1 \ldots, i_m$ and columns $j_1, \ldots, j_n$:

$$H_{i_1..i_m, j_1..j_n} = (g_{st}) \quad \text{with} \quad g_{st} = h_{i_s j_t}. \tag{76}$$

The main case will be

$$H_{ijk,lmn} = \begin{pmatrix} h_{il} & h_{im} & h_{in} \\ h_{jl} & h_{jm} & h_{jn} \\ h_{kl} & h_{km} & h_{kn} \end{pmatrix}.$$

Theorem 5.10. *If $H = H(p_1, \ldots p_N)$, then for all $i,j,k,l,m,n = 1, \ldots, N$,*

$$\mathrm{tr}(p_i p_j p_k)\mathrm{tr}(p_l p_m p_n) = \det(H_{ijk,lmn}).$$

Proof. This is formula (v) of Theorem 4.6 together with the fact that

$$\mathrm{tr}((u \wedge v)w) = \mathrm{tr}(uvw), \quad \text{for all} \quad u, v, w \in \mathrm{T0}(2, \mathbb{R}).$$

□

Combining the theorem with what we saw in Example 5.3 we get a criterion for the relative position of four points:

Lemma 5.11. *Let $p_1, p_2, p_3, p_4 \in P_M$ be distinct points (normalized or not), and let $H = H(p_1, p_2, p_3, p_4)$. Then p_3, p_4 lie on the same side of the line through p_1, p_2 if and only if $\det(H_{123,124}) > 0$.* □

If we understand triangles as ordered triples, then by analogy to the Euclidean case one may say that p_1, p_2, p_3 and p_1, p_2, p_4 have the same orientation if p_3, p_4 lie on the same side of the line through p_1, p_2, and opposite orientations otherwise.

More generally, if in a set of points with corresponding Helling matrix H two ordered triples, say p_i, p_j, p_k and p_l, p_m, p_n are considered, then, based on Lemma 5.11 we will say that they have the *same orientation* if $\det(H_{ijk,lmn}) > 0$, and *opposite orientations* otherwise.

The following simple exercise checks that the usual features of "orientation" are satisfied. In item (c), $\alpha(p)$ denotes the image of p under the isometry $\alpha \in GL(2, \mathbb{R})$, as defined in (56).

Exercise 5.12. (a) To have the same orientation is an equivalence relation.
(b) If $p_1, p_2, p_3 \in P_M$ and σ is a permutation of three indices, then p_1, p_2, p_3 and $p_{\sigma(1)}, p_{\sigma(2)}, p_{\sigma(3)}$ have the same orientation if σ is an even permutation, opposite orientations if σ is odd.
(c) If $p, q, r \in P_M$ and $\alpha \in GL(2, \mathbb{R})$ is an isometry, then triangles p, q, r and $\alpha(p), \alpha(q), \alpha(r)$ have the same orientation if $\alpha \in GL^+(2, \mathbb{R})$, opposite orientations if $\alpha \in GL^-(2, \mathbb{R})$.

5.4 How to Construct the Constellation out of Matrix Data

This section addresses the following question: given an $N \times N$ matrix H that looks like a Helling matrix, is there a constellation $\mathcal{C}$ such that $H = H(\mathcal{C})$, and if so, how can one find it? For simplicity we restrict our considerations to the case corresponding to points in normal form.

Example 5.13. Consider a triple of points $p_1, p_2, p_3 \in P_M$. Then the distances satisfy the triangle inequality. In Theorem 4.14 we saw that this is equivalent to the inequality $\det H(p_1, p_2, p_3) \geq 0$. Hence, a necessary condition for a 3×3-matrix to belong to a triple of points is that its determinant be nonnegative.

We now make the following definition, where we use the notation introduced in (76).

Definition 5.14. An $N \times N$-matrix H is a *point-type matrix* if it is of rank 3, and if for any choice of distinct indices $i, j, k \in \{1, \ldots, N\}$,

$$H_{ij} = H_{ji} \geq 1, \quad H_{ii} = 1, \quad \det H_{ijk,ijk} > 0.$$

If $p_1, \ldots, p_N \in P_M$ are points *in general position*, i.e. if no three of them lie on the same line, then $H(p_1, \ldots, p_N)$ satisfies the requirements of this definition, because under this hypothesis all triangle inequalities are strict (see Theorem 4.14). Constellations in which aligned triples occur, however, are excluded. We have made this choice in order to avoid extra considerations of the limit cases.

Here are some useful linear algebra facts for $N = 4$:

Lemma 5.15. *Let H be a 4×4 point-type matrix and let M be the so-called adjoint, i.e. the 4×4 matrix $M = (M_{ij})$ whose entries are*

$$M_{ij} = (-1)^{i+j} \det H_{jkl,ikl} \quad with \quad \{i, j, k, l\} = \{1, 2, 3, 4\}.$$

($H_{jkl,ikl}$ is meant to be the matrix obtained by erasing line i and column j in H.) Then

$$H \cdot M = 0.$$

Furthermore, there exist real numbers T_1, T_2, T_3, T_4 such that

$$M_{ij} = (-1)^{i+j} T_i T_j, \quad i, j = 1, \ldots, 4.$$

These numbers are uniquely determined by H up to a common sign change.

Proof. The determinant formula yields

$$H \cdot M = \det(H) \, \mathbb{I},$$

where $\mathbb{I}$ is the 4×4 unit matrix. (This is Cramer's rule written in a form which holds also for singular matrices.) As H has rank 3 we get $H \cdot M = 0$. Furthermore, the 0-eigenspace of H is of dimension 1 (since $\det(H_{123,123}) > 0$, H has precise rank 3) and so the columns of M are of the form

$$(-1)^i \lambda_i \begin{pmatrix} -T_1 \\ T_2 \\ -T_3 \\ T_4 \end{pmatrix}, \quad i = 1, \ldots, 4,$$

for suitable λ_i, T_j. None of the T_j is zero because $\lambda_j T_j = M_{jj} = \det(H_{ikl,ikl}) > 0$. We may therefore assume wo.l.o.g. that $T_4 = \sqrt{\det(H_{123,123})}$ and then

$$\lambda_4 T_4 = M_{44} = \det(H_{123,123}) = T_4^2.$$

Now M is symmetric and

$$\lambda_i T_4 = (-1)^i M_{i4} = (-1)^i M_{4i} = \lambda_4 T_i = T_4 T_i.$$

It follows that $\lambda_i = T_i$, $i = 1, 2, 3, 4$, and thus $(-1)^{i+j} T_i T_j = M_{ij}$, as claimed. The uniqueness up to a common multiplication by -1 is clear. $\square$

We come to the main result of this section.

Theorem 5.16. *For any $N \times N$ point-type matrix H there exists a constellation of points $\mathcal{C} = \{p_1, \ldots, p_N\}$ such that $H = H(p_1, \ldots, p_N)$. This constellation is uniquely determined up to isometry.*

Proof. We may assume that $N \geq 3$. For the first three points existence and uniqueness was proposed in Exercise 5.7. The solution to this exercise is as follows, where we use the basis I, J, K (see (51) and (61)). The first two points may be put anywhere, for instance,

$$p_1 = K, \quad p_2 = \sqrt{H_{12}^2 - 1}\, J + H_{12} K. \qquad \text{(i)}$$

These points are in normal form and $-\text{tr}(p_1 p_2) = H_{12}$. Since p_1, p_2 and $p_1 \wedge p_2$ are linearly independent, the third point can be written in the form $p_3 = x_1 p_1 + x_2 p_2 + x_3\, p_1 \wedge p_2$. The coefficients x_1, x_2, x_3 can easily be determined from the conditions $-\text{tr}(p_1 p_3) = H_{13}$, $-\text{tr}(p_2 p_3) = H_{23}$, $-\text{tr}(p_3^2) = 1$, and it comes out that there are only two possibilities for p_3, namely

$$p_3 = \frac{1}{H_{12}^2 - 1} \Big((H_{12} H_{23} - H_{13}) p_1 + (H_{12} H_{13} - H_{23}) p_2$$

$$\pm \sqrt{\det(H_{123,123})}\, p_1 \wedge p_2 \Big). \qquad \text{(ii)}$$

For either sign the triple p_1, p_2, p_3 has Helling matrix $H(p_1, p_2, p_3) = H_{123,123}$, and the two solutions differ by the isometry $p \mapsto -IpI^{-1}$ (see (56)). We let p_3 from now on be the one with the plus sign. If we set

$$
\begin{aligned}
t_{123} &:= \sqrt{\det(H_{123,123})}, \\
t_{12i} &:= \frac{\det(H_{12i,123})}{t_{123}}, \quad i = 4, \ldots, N,
\end{aligned}
\qquad \text{(iii)}
$$

then by Lemma 5.15 (with T_4 in the role of t_{123} and T_3 in the role of t_{12i}), $|t_{12i}| = \sqrt{\det(H_{12i,12i})}$. Setting

$$p_i = \frac{1}{H_{12}^2 - 1} \Big((H_{12} H_{2i} - H_{1i}) p_1 + (H_{12} H_{1i} - H_{2i}) p_2 + t_{12i}\, p_1 \wedge p_2 \Big), \qquad \text{(iv)}$$

we have therefore in the same way as for p_3 that $-\text{tr}(p_1 p_i) = H_{1i}$, $-\text{tr}(p_2 p_i) = H_{2i}$, $-\text{tr}(p_i^2) = 1$. Furthermore, using Lemma 5.15 again, we compute that $-\text{tr}(p_3 p_i) = H_{3i}$. If we replace t_{12i} by $-t_{12i}$, this last equality is lost. Hence, p_i is the unique point satisfying $H(p_1, p_2, p_3, p_i) = H_{123i,123i}$.

This accomplishes the uniqueness proof and describes a set of points for which the first three columns of $H(p_1, \ldots, p_N)$ coincide with the first three columns of H. That the rest coincides as well is an immediate consequence of the next lemma. $\square$

Lemma 5.17. *A point type matrix is determined by any three of its columns (or rows).*

Proof. We may assume that the considered columns are the first three. Transposing them gives the first three rows; also, all diagonal elements are equal to 1. For any choice of different $i, j = 4, \ldots, n$, the equation $\det(H_{123i,123j}) = 0$ is linear in H_{ij}, and the coefficient of H_{ij} in this equation is $\det(H_{123,123}) \neq 0$. Hence all elements are uniquely determined. $\square$

It is useful to extend the notation introduced in (iii) above and set

$$t_{ijk} := \frac{\det(H_{ijk,123})}{\sqrt{\det(H_{123,123})}}. \tag{77}$$

Theorem 5.10 then yields

$$t_{ijk} = \text{tr}(p_i p_j p_k), \tag{78}$$

$$t_{ijk} t_{lmn} = \det(H_{ijk,lmn}). \tag{79}$$

A further property of the t_{ijk} is a special case of the so-called triple trace theorem (see, for example, Helling [13, 14], where this and related identities are applied).

Theorem 5.18. *Let ω be a word in terms of a set of M points: $\omega = p_{i_1} \cdots p_{i_M}$. Then $\text{tr}\,\omega$ can be expressed as a polynomial in the entries of the corresponding Helling matrix H if M is even, and in the entries of H and the triple traces t_{ijk} if M is odd.*

Proof. The idea of the proof is to play with two identities for traces, $\text{tr}(XY) = \text{tr}(YX)$ and

$$\text{tr}(XY) + \text{tr}(XY^{-1}) = 2\text{tr}(X)\text{tr}(Y), \quad X, Y \in \text{SL}(2, \mathbb{R}). \tag{80}$$

(To check rule (80), write $X = a\mathbf{1} + A$, $Y = b\mathbf{1} + B$ with $A, B \in \text{T0}(2, \mathbb{R})$. Then $Y^{-1} = b\mathbf{1} - B$ and $\text{tr}(XY) = ab + \text{tr}(AB)$, $\text{tr}(XY^{-1}) = ab - \text{tr}(AB)$, and the relation follows.) We restrict our considerations to $M = 4, 5$; this will be enough to show how it works. For $M = 4$:

$$\begin{aligned}
\text{tr}\,p_i p_j p_k p_l &= 2\text{tr}\,p_i p_j \text{tr}\,p_l p_k - \text{tr}\,p_i p_j p_l p_k \\
&= 2H_{ij}H_{lk} - \text{tr}\,p_k p_i p_j p_l \\
&= 2H_{ij}H_{lk} - (2\text{tr}\,p_k p_i \text{tr}\,p_j p_l - \text{tr}\,p_k p_i p_l p_j) \\
&= 2H_{ij}H_{lk} - 2H_{ki}H_{jl} + \text{tr}\,p_j p_k p_i p_l \\
&= 2H_{ij}H_{lk} - 2H_{ki}H_{jl} + (2\text{tr}\,p_j p_k \text{tr}\,p_i p_l - \text{tr}\,p_j p_k p_l p_i) \\
&= 2H_{ij}H_{lk} - 2H_{ki}H_{jl} + 2H_{jk}H_{il} - \text{tr}\,p_i p_j p_k p_l,
\end{aligned}$$

and hence

$$\text{tr}\,p_i p_j p_k p_l = H_{ij}H_{kl} - H_{ik}H_{jl} + H_{il}H_{jk}. \tag{81}$$

For $N = 5$ the same trick:

$$\begin{aligned}
\text{tr}\,p_i p_j p_k p_l p_m &= \text{tr}\,p_j p_k p_l p_m p_i = -2\text{tr}\,p_j p_k p_l H_{mi} - \text{tr}\,p_j p_k p_l p_i p_m \\
&= -2t_{jkl}H_{mi} - (2\text{tr}\,p_j p_k \text{tr}\,p_l p_i p_m - \text{tr}\,p_k p_j p_l p_i p_m) \\
&= -2H_{mi}t_{jkl} + 2H_{jk}t_{lim} + \text{tr}\,p_l p_i p_m p_k p_j \\
&= -2H_{mi}t_{jkl} - 2H_{jk}t_{lmi} - 2H_{li}t_{mkj} - \text{tr}\,p_l p_m p_k p_j p_i \\
&= -2H_{mi}t_{jkl} - 2H_{jk}t_{lmi} + 2H_{li}t_{jkm} + 2H_{lm}t_{kji} + \tau,
\end{aligned}$$

where $\tau = \operatorname{tr} p_m p_l p_k p_j p_i = -\operatorname{tr} p_i p_j p_k p_l p_m$. Hence

$$\operatorname{tr} p_i p_j p_k p_l p_m = -H_{im} t_{jkl} - H_{jk} t_{ilm} + H_{il} t_{jkm} - H_{lm} t_{ijk}. \qquad (82)$$

Lecture 6:

Constellations of Points and Fuchsian Groups in Genus 2

We round off these lectures with an application to the Fuchsian groups that correspond to hyperbolic surfaces of genus two. We begin with a constellation of points which will give rise to discrete groups and tilings of the hyperbolic plane by geodesic hexagons and octagons. These constellations will be characterized by their Helling matrices. At the end of the lecture we will see that, among other things, the constructions given here cover all hyperbolic surfaces of genus 2. For the major part of this lecture we work with the matrix model.

6.1 The Closing Property

As a prologue we begin with an exercise in Euclidean geometry which requires a sheet of paper, a ruler and a pencil.

Exercise 6.1. Let p, q, r, s, be a parallelogram in the Euclidean plane as shown in Fig. 16, where pr and qs are the diagonals. Take any point x_0 in the plane and perform the following construction: draw the straight line through x_0, s and mark off the point x_1 on it such that s becomes the midpoint of segment $x_0 x_1$. Repeat this with x_1 and r to obtain another point x_2, then the same with x_2 and q to get x_3, and finally with x_3 and p to end up with a point x_4. What do you observe?

The mystery why this construction closes the cycle $x_0, x_1, \ldots$, is solved if we think of half-turns: let h_p, h_q, etc., be the half-turns around the vertices of the parallelogram. Then $x_1 = h_s(x_0)$, $x_2 = h_r(x_1)$, etc. The products

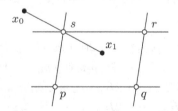

Fig. 16. Draw the polygon $x_0 x_1 x_2 x_3 x_4$.

$h_p h_q$ and $h_r h_s$ are translations, and since the four points are the vertices of a parallelogram, the two translations cancel out.

Coming back to the matrix model of the hyperbolic plane we are led to the following definition.

Definition 6.2. A sequence of points $p_1, \ldots, p_n \in P_M$ has the *closing property*, or simply is *closing* if $p_1 p_2 \cdots p_n = \pm 1$. The sequence is *pre-closing* if $\operatorname{tr}(p_1 p_2 \cdots p_n) = 0$.

Thus, "closing" means that if we understand the p_i as half-turns (operating by $x \mapsto p_i x p_i^{-1} = -p_i x p_i$, $x \in P_M$, see (56)), then the product is the identity.

From a pre-closing polygon $p_1, \ldots, p_n$ we obtain a closing one by putting $p_{n+1} = \pm p_1 \cdots p_n$. Conversely, a closing polygon yields a pre-closing one if we drop the last point.

Remark 6.3. The closing and pre-closing properties are preserved if we perform a cyclic permutation of the points. This allows an interesting operation on such polygons: start with a pre-closing $p_1, \ldots, p_n$, add the last point to get a closing $p_1, p_2, \ldots, p_{n+1}$, then "roll" to get the closing $p_2, \ldots, p_{n+1}, p_1$, drop the last point to get the pre-closing $p_2, \ldots, p_{n+1}$, then roll again and continue.

Exercise 6.4. (a) A triple p_1, p_2, p_3, is never closing.
(b) A quadruple p_1, p_2, p_3, p_4, can only be closing if the points lie on the same line.

6.2 Closing Hexagons

For the description of the Fuchsian groups of genus 2 — a goal of this lecture — we will need a number of properties of closing hexagons and pre-closing pentagons.

We begin by constructing a comprehensive family of examples, where we imitate the so-called Fenchel-Nielsen parameters from hyperbolic surface theory ([6], [12]).

Example 6.5. First we choose three numbers $a, b, c > 1$ to obtain hyperbolic isometries $\alpha = a\mathbf{1} + A$, $\beta = b\mathbf{1} + B$, $\gamma = c\mathbf{1} + C \in \operatorname{SL}(2, \mathbb{R})$ satisfying $\alpha\beta\gamma = -1$ whose axes A, B, C have the infinite triangle constellation (Theorem 5.6). This is done *before* choosing the vertices on A, B, C. (For the reader who is familiar with Fenchel-Nielsen parameters: this step corresponds to choosing the lengths of three non-separating simple closed geodesics for a hyperbolic surface of genus 2.)

If we choose a (normalized) point p_1 on A arbitrarily we can then mark off exactly one (normalized) point p_2 on A such that α becomes $-p_1 p_2$. This is done as in Example 4.10, and the point is

$$p_2 = ap_1 + p_1 \wedge A.$$

(Our A here plays the role of $-sA$ in Example 4.10.) We may check again that indeed $\operatorname{tr}(p_2^2) = -1$ and $-p_1 p_2 = a\mathbf{1} + A$. Similarly, we have free choices of p_3 on B, p_5 on C and then uniquely determined p_4 on B, p_6 on C such that altogether

$$\alpha = -p_1 p_2, \quad \beta = -p_3 p_4, \quad \gamma = -p_5 p_6.$$

Since $\alpha\beta\gamma = -1$ we have

$$p_1 p_2 p_3 p_4 p_5 p_6 = \mathbf{1}.$$

The closing hexagons constructed in this way form a 6-dimensional family. We may parametrize this family by introducing three additional parameters τ_α, τ_β, τ_γ for the positions of p_1, p_3, p_5 on A, B, C. For this we define the "zero positions" to be the intersection points of the perpendiculars $C \wedge A$ with A, respectively $A \wedge B$ with B, respectively $B \wedge C$ with C, and then understand the τ_α etc., as the hyperbolic sines of the oriented distances to these zero positions as in Example 4.10. Observing that $\operatorname{tr}(A^2) = a^2 - 1$, $\operatorname{tr}((C \wedge A)^2) = \operatorname{tr}^2(CA) - \operatorname{tr}(C^2)\operatorname{tr}(A^2) = (ac+b)^2 - (c^2-1)(a^2-1)$, etc., we obtain

$$p_1 = \tau_\alpha \frac{C \wedge A}{\sqrt{S(a,b,c)}} + \sqrt{\tau_\alpha^2 + 1}\, \frac{(C \wedge A) \wedge A}{\sqrt{S(a,b,c)}\sqrt{a^2-1}},$$

$$p_2 = a\,p_1 + p_1 \wedge A,$$

$$p_3 = \tau_\beta \frac{A \wedge B}{\sqrt{S(a,b,c)}} + \sqrt{\tau_\beta^2 + 1}\, \frac{(A \wedge B) \wedge B}{\sqrt{S(a,b,c)}\sqrt{b^2-1}}, \tag{83}$$

$$p_4 = b\,p_3 + p_3 \wedge B,$$

$$p_5 = \tau_\gamma \frac{B \wedge C}{\sqrt{S(a,b,c)}} + \sqrt{\tau_\gamma^2 + 1}\, \frac{(B \wedge C) \wedge C}{\sqrt{S(a,b,c)}\sqrt{c^2-1}},$$

$$p_6 = c\,p_5 + p_5 \wedge C,$$

where

$$S(a,b,c) = a^2 + b^2 + c^2 + 2abc - 1.$$

Exercise 6.6. Given α, β, γ as in Example 6.5, verify by direct computation that the points defined in (83) belong to P_M and form a closing hexagon.

The next theorem gives an intrinsic characterization of the hexagons constructed in the preceding examples. Notice the role of the sign in the product of $p_1, \ldots, p_6$.

Theorem 6.7. Let $p_1, \ldots, p_6 \in P_M$ be such that $p_1 \cdots p_6 = \pm 1$. Then the following statements are equivalent.

(i) $p_1 \cdots p_6 = +1$.
(ii) Lines $p_1 \wedge p_2$, $p_3 \wedge p_4$, $p_5 \wedge p_6$ have the "infinite triangle" constellation.
(iii) For each cyclic choice of $i, j, k, l \in (1, \ldots, 6)$ one has $g(i, j, k, l) < -1$.
(iv) $g(1, 2, 3, 4) < 0$.

By *cyclic choice* we mean that for some cyclic permutation σ of four letters one has $1 \leq \sigma(i) < \sigma(j) < \sigma(k) < \sigma(l) \leq 6$. The symbol $g(i,j,k,l)$ already used in (75) is defined as

$$g(i,j,k,l) = \mathrm{tr}(p_i p_j)\mathrm{tr}(p_k p_l) - \mathrm{tr}(p_i p_k)\mathrm{tr}(p_j p_l) + \mathrm{tr}(p_i p_l)\mathrm{tr}(p_j p_k)$$
$$= \mathrm{tr}(p_i p_j p_k p_l), \tag{84}$$

(see also (81)).

Proof. The equivalence of (i) and (ii) is merely a translation of Theorem 5.8 to $\alpha = -p_1 p_2$, $\beta = -p_3 p_4$, $\gamma = -p_5 p_6$.

To infer (iii) from (i) it suffices to consider the choices 1234, 1235, 1245; all other choices are obtained from these by cyclic permutation. Now

$$p_1 p_2 p_3 p_4 = p_6 p_5,$$
$$p_1 p_2 p_3 p_5 = -p_1 p_2 p_3 p_4 p_5 p_6 p_6 p_5 p_4 p_5 = p_6 \cdot p_5 p_4 p_5^{-1},$$
$$p_1 p_2 p_4 p_5 = p_1 p_2 p_3 p_4 p_5 p_6 p_6 p_5 p_4 p_3 p_4 p_5 = p_6 \cdot (p_5 p_4) p_3 (p_5 p_4)^{-1}.$$

On the right-hand side we have products of two members of P_M (see (56)). Hence the traces are < -1.

To infer (i) from (iv) we write $p_1 \cdots p_6 = \varepsilon\mathbf{1}$, where $\varepsilon = \pm 1$. Then $p_1 p_2 p_3 p_4 = \varepsilon p_6 p_5$ and $g(1,2,3,4) = \mathrm{tr}(p_1 p_2 p_3 p_4) = \varepsilon \mathrm{tr}(p_6 p_5)$. As $g(1,2,3,4)$ and $\mathrm{tr}(p_6 p_5)$ are negative we must have $\varepsilon = 1$. This concludes the proof of the theorem. □

Definition 6.8. Let $p_1, \ldots, p_5 \in P_M$ be normalized points forming a pre-closing pentagon $C_5 = (p_1, p_2, p_3, p_4, p_5)$, and let $p_6 \in P_M$ be the point satisfying
$$p_1 \cdots p_6 = \varepsilon\mathbf{1}, \quad \text{with } \varepsilon \in \{-1, 1\}.$$
Then C_5 respectively, the hexagon $C_6 = (p_1, \ldots, p_6)$ is called *good* if $\varepsilon = +1$.

By Theorem 6.7, a pre-closing pentagon is good as soon as $g(1,2,3,4) < 0$, and we then automatically have $g(i,j,k,l) < -1$, for any cyclic choice of $i,j,k,l \in \{1, \ldots, 5\}$.

In Section 6.4 we will use such pentagons to produce tilings of the hyperbolic plane with octagons and quotient surfaces of genus 2. But let us first consider the corresponding Helling matrices.

6.3 Closing and Pre-Closing Point-Type Matrices

A useful fact for computations is that closing, pre-closing and other geometric properties are encoded in the Helling matrix. To mark the point we start this paragraph with the following definition, where we restrict ourselves to the case that has to do with genus 2.

We use again the notation t_{ijk} introduced in (77).

Definition 6.9.

(i) A 5×5 point-type matrix H^5 is called *pre-closing* iff

$$H_{15}t_{234} + H_{23}t_{451} + H_{45}t_{123} - H_{14}t_{235} = 0. \tag{85}$$

(ii) A 5×5 point-type matrix H^5 is called *good* iff it is pre-closing and $g(1, 2, 3, 4) < 0$.

(iii) A 6×6 point-type matrix H is called *good* iff its last column satisfies

$$
\begin{aligned}
H_{16} &= -g(2, 3, 4, 5) \\
H_{26} &= g(1, 3, 4, 5) - 2H_{12}g(2, 3, 4, 5) \\
H_{36} &= -g(5, 4, 1, 2) + 2 \det H_{123,345} \\
H_{46} &= g(1, 2, 3, 5) - 2H_{45}g(1, 2, 3, 4) \\
H_{56} &= -g(1, 2, 3, 4).
\end{aligned}
\tag{86}
$$

The next exercise explains this definition. The result of the exercise will also be needed in the proof of Theorem 6.12.

Exercise 6.10. The Helling matrix of a pre-closing pentagon is pre-closing. The Helling matrix of a good pentagon or hexagon is good.

For the part concerning the hexagon, show the more general result that, if $p_1, \ldots, p_5 \in P_M$ are points written in normal form, and v_6 is the product $v_6 = -p_5 p_4 p_3 p_2 p_1$, then the quantities $-\mathrm{tr}(p_i v_6)$, $i = 1, \ldots, 6$, satisfy the identities (86).

(Hint: compute with (78) – (82) and (84).)

Example 6.11. Here is an interesting example of a good matrix. It is obtained as the Helling matrix of a pentagon whose so-called outer pentagon (see the next section) is half of the regular octagon with center p_5 and interior angles $\pi/4$. This octagon was used in [3, 4] as an early test case for quantum chaos on hyperbolic surfaces.

$$
H_{\mathrm{reg}} = \begin{pmatrix}
1 & 1 + \sqrt{2} & 3 + 2\sqrt{2} & 5 + 3\sqrt{2} & 1 + \sqrt{2} \\
1 + \sqrt{2} & 1 & 1 + \sqrt{2} & 3 + 2\sqrt{2} & 1 + \sqrt{2} \\
3 + 2\sqrt{2} & 1 + \sqrt{2} & 1 & 1 + \sqrt{2} & 1 + \sqrt{2} \\
5 + 3\sqrt{2} & 3 + 2\sqrt{2} & 1 + \sqrt{2} & 1 & 1 + \sqrt{2} \\
1 + \sqrt{2} & 1 + \sqrt{2} & 1 + \sqrt{2} & 1 + \sqrt{2} & 1
\end{pmatrix}.
$$

Theorem 6.12. *A good 5×5 matrix H^5 comes from a good pentagon. A good 6×6 matrix H comes from a good hexagon.*

Proof. By Theorem 5.16 there exist $p_1, \ldots, p_5 \in P_M$ such that $H(p_1, \ldots, p_5) = H^5$. By (82) and (78), $C_5 = (p_1, \ldots, p_5)$ is pre-closing, and by the remark at the end of the preceding section, the fact that $g(1, 2, 3, 4) < 0$ implies that C_5 is good.

Now take $C_6 = (p_1, \ldots, p_6)$ with $H(C_6) = H$ and set $v_6 = -p_5p_4p_3p_2p_1$. We have to show that $v_6 = p_6$. By Exercise 6.10, $\operatorname{tr}(v_6p_i) = \operatorname{tr}(p_6p_i)$, $i = 1, \ldots, 5$. Because p_1, p_2, p_3 are linearly independent elements of $\mathrm{T0}(2, \mathbb{R})$, and $\mathrm{T0}(2, \mathbb{R})$ is a three-dimensional vector space with basis I, J, K, we conclude that $\operatorname{tr}((v_6 - p_6)I) = \operatorname{tr}((v_6 - p_6)J) = \operatorname{tr}((v_6 - p_6)K) = 0$. This implies that $v_6 = p_6 + \lambda\mathbf{1}$ for some $\lambda \in \mathbb{R}$. Now $v_6, p_6 \in \mathrm{SL}(2, \mathbb{R})$ and so $\det(v_6) = \det(p_6)$. This is only possible for $\lambda = 0$. Thus, C_6 is closing. As $g(1, 2, 3, 4) = -H_{56} < 0$, Theorem 6.7 tells us that C_6 is good. $\square$

Here is an immediate consequence.

Corollary 6.13. *By dropping the last line and column of a good 6×6 matrix H we get a good 5×5 matrix H^5. By adding a sixth line and column to a good 5×5 matrix H^5 following the rules of (86) we get a good 6×6 matrix H.* $\square$

Of course, a proof of this should also be possible solely on the matrix level, and we sketch one here for completeness. It hinges on the following identity. Assume for a moment that H is *any* symmetric 6×6 matrix with 1's in the diagonal and such that $H_{16}, \ldots, H_{56}$ satisfy (86). Denote the columns of H by $\boldsymbol{H}_1, \ldots, \boldsymbol{H}_6$ and abbreviate $\det H_{123,ijk} =: \delta_{ijk}$. Then

$$\delta_{236}\boldsymbol{H}_1 - \delta_{136}\boldsymbol{H}_2 + \delta_{126}\boldsymbol{H}_3 - \delta_{123}\boldsymbol{H}_6 = \boldsymbol{V}, \tag{87}$$

where

$$
\begin{aligned}
&V_1 = V_2 = V_3 = 0, \\
&V_4 = g(1, 2, 3, 4) \det H_{1234,1235} - g(1, 2, 3, 5) \det H_{1234,1234}, \\
&V_5 = g(1, 2, 3, 4) \det H_{1235,1235} - g(1, 2, 3, 5) \det H_{1234,1235}, \\
&V_6 = -(\delta_{234}H_{51} + \delta_{451}H_{23} + \delta_{123}H_{45} - \delta_{235}H_{14})^2 \\
&\qquad + \det H^5 - \det H_{1234,1234} - \det H_{1235,1235} + 2H_{45}\det H_{1234,1235}.
\end{aligned}
\tag{88}
$$

This may be checked, for example, using symbolic computation.

If now H^5 is as in the corollary, and we augment it into H via (86), then $\boldsymbol{V} = 0$, and the sixth column of H becomes a linear combination of the first three. The sixth line having the analog property it follows that H has rank 3. Conversely, if H is given in the first place, then all "$\det H_{\ldots}$" in (88) are zero, and so H^5 satisfies (85). The remaining parts of the proof are then easily filled in.

Example 6.14. In Remark 6.3 we mentioned a way to go from one pre-closing constellation to another. Corollary 6.13 allows a nice description of this procedure in terms of the matrices: take a good H^5, augment it via (86) and replace the upper 5×5 block by the lower one as shown in the following diagram.

$$\begin{pmatrix} 1 & H_{12} & H_{13} & H_{14} & H_{15} & H_{16} \\ H_{12} & 1 & H_{23} & H_{24} & H_{25} & H_{26} \\ H_{13} & H_{23} & 1 & H_{34} & H_{35} & H_{36} \\ H_{14} & H_{24} & H_{34} & 1 & H_{45} & H_{46} \\ H_{15} & H_{25} & H_{35} & H_{45} & 1 & H_{56} \\ H_{16} & H_{26} & H_{36} & H_{46} & H_{56} & 1 \end{pmatrix}$$

Then apply a cyclic permutation of the rows and columns of the new 5×5 matrix and continue in the same way.

The procedure yields infinitely many different matrices, respectively pentagons. In [1] it is used to describe the so-called mapping class group. In [2] and [7] it is used in connection with the uniformization of algebraic curves.

Example 6.15. In Example 6.5 we constructed a 6-dimensional family of good hexagons. Computing the corresponding Helling matrices we arrive at a parametrization of good 6×6 matrices. Here we describe this parametrization without resorting to the underlying points, i.e. simply as a recipe which yields a large family of good matrices.

First we choose real numbers $a, b, c > 1$ and make the "Ansatz"

$$H = \begin{pmatrix} 1 & a & H_{13} & H_{14} & H_{15} & H_{16} \\ a & 1 & H_{23} & H_{24} & H_{25} & H_{26} \\ H_{13} & H_{23} & 1 & b & H_{35} & H_{36} \\ H_{14} & H_{24} & b & 1 & H_{45} & H_{46} \\ H_{15} & H_{25} & H_{35} & H_{45} & 1 & c \\ H_{16} & H_{26} & H_{36} & H_{46} & c & 1 \end{pmatrix}.$$

Abbreviating

$$d = \frac{ab+c}{\sqrt{(a^2-1)(b^2-1)}}, \quad e = \frac{bc+a}{\sqrt{(b^2-1)(c^2-1)}}, \quad f = \frac{ca+b}{\sqrt{(c^2-1)(a^2-1)}},$$

we then take any real numbers $\tau_\alpha, \tau_\beta, \tau_\gamma$ and put

$$H_{23} = \tau_\alpha \tau_\beta + d\sqrt{\tau_\alpha^2 + 1}\sqrt{\tau_\beta^2 + 1},$$
$$H_{13} = -\sqrt{a^2 - 1}(d\tau_\alpha\sqrt{\tau_\beta^2 + 1} + \tau_\beta\sqrt{\tau_\alpha^2 + 1}) + aH_{23},$$
$$H_{24} = \sqrt{b^2 - 1}(\tau_\alpha\sqrt{\tau_\beta^2 + 1} + d\tau_\beta\sqrt{\tau_\alpha^2 + 1}) + bH_{23},$$
$$H_{14} = (abd - \sqrt{a^2 - 1}\sqrt{b^2 - 1})\sqrt{\tau_\alpha^2 + 1}\sqrt{\tau_\beta^2 + 1} - \tau_\alpha\tau_\beta c$$
$$+ bH_{13} + aH_{24} - 2abH_{23},$$

continuing with the remaining entries H_{45}, H_{35}, etc., by obvious cyclic transposition and completing the matrix using (86). One may now check in a direct way that this yields good matrices.

We obtain, in fact, *all of them* in this manner. For a proof we take any good matrix H, put $a = H_{12}$, $b = H_{34}$, $c = H_{56}$ and define d as before. It is then straight forward (albeit tedious) to check that H arises in the above way with

$$\tau_\alpha := \frac{-t_{134} + a\,t_{234}}{\sqrt{a^2 - 1}\sqrt{b^2 - 1}\sqrt{d^2 - 1}} = \frac{-a\,t_{561} + t_{562}}{\sqrt{a^2 - 1}\sqrt{b^2 - 1}\sqrt{d^2 - 1}},$$

$$\tau_\beta := \frac{-b\,t_{123} + t_{124}}{\sqrt{a^2 - 1}\sqrt{b^2 - 1}\sqrt{d^2 - 1}},$$

$$\tau_\gamma := \frac{-c\,t_{345} + t_{346}}{\sqrt{a^2 - 1}\sqrt{b^2 - 1}\sqrt{d^2 - 1}}.$$

6.4 Symmetric Octagons

The considerations of the preceding paragraphs allow us to construct Fuchsian groups and tilings of the hyperbolic plane as, for instance, the octagon tiling shown in Fig. 19. The idea is as follows.

Take a pre-closing pentagon $C_5 = (p_1, p_2, p_3, p_4, p_5)$ and build the *outer pentagon* $D_5 = (q_1, q_2, q_3, q_4, q_5)$ for which the old points p_i become the midpoints of the sides:

$$\begin{aligned}
q_1 &= & p_6 &= -p_1 p_2 p_3 p_4 p_5, \\
q_2 &= p_1^{-1} q_1 p_1 &&= -p_2 p_3 p_4 p_5 p_1, \\
q_3 &= p_2^{-1} q_2 p_2 &&= -p_3 p_4 p_5 p_1 p_2, \\
q_4 &= p_3^{-1} q_3 p_3 &&= -p_4 p_5 p_1 p_2 p_3, \\
q_5 &= p_4^{-1} q_4 p_4 &&= -p_5 p_1 p_2 p_3 p_4.
\end{aligned}$$

Now take the half-turns around the midpoints of this new pentagon, then the half-turns around the midpoints of its images, and so on. If the shape of the pentagon satisfies the necessary "good" conditions we get a tiling.

Here is one of the necessary properties.

Lemma 6.16. *Good pentagons yield convex outer pentagons.*

Proof. From $p_1 \cdots p_6 = 1$ we get $p_5 p_6 p_1 p_2 = p_4 p_3$. Therefore $g(5, 6, 1, 2) = \operatorname{tr}(p_5 p_6 p_1 p_2) = \operatorname{tr}(p_4 p_3) < 0$. This shows that the quadrilateral p_5, q_1, p_1, p_2,

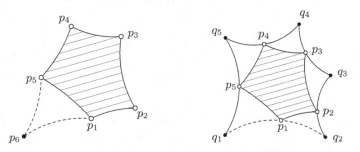

Fig. 17. Successive reflection of p_6 at the vertices of the inner pentagon yields the outer pentagon.

is strongly convex. Hence, q_1 lies on the side of line $p_5 \wedge p_1$ which is opposite to the inner pentagon. As the same holds for q_2 with respect to line $p_1 \wedge p_2$, and for q_4 with respect to line $p_3 \wedge p_4$, we conclude that all points p_2, q_3, p_3, q_4, p_4, q_5, p_5, lie on the same side of line $q_1 \wedge q_2$. The convexity of the outer pentagon follows now by cyclic permutation. $\square$

The next theorem brings us to the tilings.

Theorem 6.17. *Let Q with vertices $q_1, \ldots, q_5$ be a convex pentagon and let $p_1, \ldots, p_5$ be the midpoints of the sides (p_1 the midpoint of side q_1, q_2, etc.). Under this hypothesis the following statements are equivalent.*

(i) $\mathrm{tr}(p_1 \cdots p_5) = 0$.
(ii) *The sum of the interior angles of Q is π.*
(iii) $p_1, \ldots, p_5$ *is a good pentagon.*

Proof. This is shown by examining how the images of Q under the half-turns

$$x \mapsto h_i(x) = p_i x p_i^{-1}$$

pile up at a vertex. Fig. 18 shows the successive images $h_4(Q)$ adjacent to Q along side q_5, q_4, then the images $h_3(Q)$, $h_3 h_4(Q)$ of the couple Q, $h_4(Q)$ adjacent to Q along side q_4, q_3, and so on. At vertex q_1 we have the images Q, $h_1(Q)$, $h_1 h_2(Q)$, $h_1 h_2 h_3(Q)$ and $h_1 h_2 h_3 h_4(Q)$. Each interior angle of Q appears as the interior angle of one of the five copies of Q at q_1.

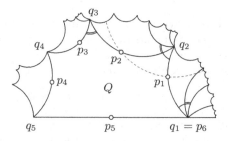

Fig. 18. The angle sum of the outer pentagon is π.

Now let's proceed with the statements (i)–(iii). If $\mathrm{tr}(p_1 \cdots p_5) = 0$, then $h_1 \cdots h_5$ is a half-turn. As $h_1 \cdots h_5$ fixes $q_1 = p_6$, this half-turn is h_6. Therefore, $h_1 h_2 h_3 h_4 = \tau$, where $\tau = h_6 h_5$ is the translation with axis through p_6, p_5 sending q_5 to q_1. Hence, the angles of the images of Q at q_1 add up to π.

If Q has angle sum π, then both Q and $h_1 h_2 h_3 h_4(Q)$ have a side on the axis of τ. It follows that $h_1 h_2 h_3 h_4(Q) = \tau(Q)$, and this can only be if $h_1 h_2 h_3 h_4 = \tau$. Hence $p_1 \cdots p_5 = \pm q_1$. Moreover, by looking at how the images of side p_1, p_2 under h_1, h_2, etc., cross the corresponding images of Q, and repeating the same exercise with p_3, p_4 we see that lines $p_1 \wedge p_2$, $p_3 \wedge p_4$,

$p_5 \wedge q_1$ have the "infinite triangle" constellation and so by Theorem 6.7, the inner pentagon is good. Trivially, (iii) implies (i). $\square$

Let us fix an outer pentagon Q with vertices $q_1, \ldots, q_5$, midpoints of sides $p_1, \ldots, p_5$ and angle sum equal to π as above and let σ be the half-turn around midpoint p_5. The union

$$\mathcal{O} = Q \cup \sigma(Q)$$

is then called a *symmetric admissible octagon*. The products (in $SL(2,\mathbb{R})$)

$$\beta_i = -p_i p_5, \quad i = 1, 2, 3, 4, \tag{89}$$

operate as hyperbolic isometries

$$x \mapsto g_i(x) = \beta_i x \beta_i^{-1} \tag{90}$$

with axes $B_i := -p_i \wedge p_5$. The choices of the signs in these definitions have been made such that we can write

$$\beta_i = b_i \mathbf{1} + B_i, \quad \text{with} \quad b_i = -\mathrm{tr}(p_i p_5) > 1.$$

Since $(p_1 \cdots p_5)^2 = p_6^2 = -\mathbf{1}$, the β_i satisfy the following relation in $SL(2,\mathbb{R})$,

$$\beta_1 \beta_2^{-1} \beta_3 \beta_4^{-1} \beta_1^{-1} \beta_2 \beta_3^{-1} \beta_4 = \mathbf{1}. \tag{91}$$

We may now interpret $\mathcal{O}$ as a domain in $\mathbb{H}$, and g_1, g_2, g_3, g_4 as members of $\mathrm{Isom}(\mathbb{H})$. Then the g_i are side pairing isometries as in Section 3.6, and the combinatorial pattern of the pairing is such that the vertices form just one cycle. Since the sum of the interior angles of $\mathcal{O}$ equals 2π, Poincaré's polygon theorem (Theorem 3.36) yields the following.

Theorem 6.18. *Let $\mathcal{O} \subset \mathbb{H}$ be an admissible symmetric octagon with opposite-sides identifying isometries g_1, g_2, g_3, g_4 as above. Then*

$$G := \langle g_1, g_2, g_3, g_4 \rangle_{\mathrm{Isom}(\mathbb{H})}$$

is a Fuchsian group with fundamental domain $\mathcal{O}$, The generators satisfy the relation

$$g_1 g_2^{-1} g_3 g_4^{-1} g_1^{-1} g_2 g_3^{-1} g_4 = \mathrm{id},$$

and all elements of $G \setminus \{\mathrm{id}\}$ are hyperbolic. $\square$

Fig. 19 shows part of the tiling of the hyperbolic plane with images of $\mathcal{O}$. The figure has been obtained by an implementation of Algorithm 3.37.

Exercise 6.19. In this exercise, $\mathcal{O}$, Q, the points p_i with half-turns h_i and $\sigma = h_5$, etc., have the same meaning as above.

(a) The group $\tilde{G} = \langle h_1, h_2, h_3, h_4, h_5 \rangle_{\mathrm{Isom}(\mathbb{H})}$ is a Fuchsian group with fundamental domain Q.

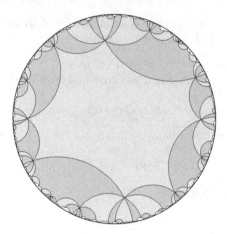

Fig. 19. A partial tiling of the unit disc obtained during the second round of Algorithm 3.37, for an octagon with opposite-side pairing.

(b) All elements of $\tilde{G}$ are elliptic or hyperbolic. The elliptic elements of $\tilde{G}$ are the half-turns around the points $g(p_i)$, $g \in G$, $i = 1, \ldots, 5$.
(c) $\{\sigma g \sigma \mid g \in G\} = G$.

The next exercise concerns the quotient surface $\mathcal{S} = \mathbb{H}/G$. We recall that the points of $\mathcal{S}$ are the orbits

$$\boldsymbol{x} = G(x) = \{g(x) \mid g \in G\}.$$

Exercise 6.20.

(a) $\mathcal{S} = \mathbb{H}/G$ is a compact hyperbolic surface of genus 2.
(b) For any $\boldsymbol{x} \in \mathbb{H}/G$ the point

$$\phi(\boldsymbol{x}) \stackrel{\text{def}}{=} G(\sigma(x)) \tag{92}$$

remains the same if we replace x by any other point $x' \in G(x)$, and therefore (92) defines a mapping $\phi : \mathcal{S} \to \mathcal{S}$.
(c) $\phi^2 = \text{id}$.
(d) The fixed points of ϕ on $\mathcal{S}$ are $\boldsymbol{p_1}, \ldots, \boldsymbol{p_6}$.
(e) $\phi : \mathcal{S} \to \mathcal{S}$ is an isometry.

(Hint: for (b) use Exercise 6.19(c). For (e) use the characterization of the distance on $\mathcal{S}$ given in Exercise 3.24.)

6.5 Fuchsian Groups of Genus 2

We will now show that the surfaces obtained in the preceding section cover *all* compact hyperbolic surfaces of genus 2. For this we first state without proof the classification of the compact hyperbolic surfaces. (For a proof we refer to [10].)

Any compact orientable hyperbolic surface can be obtained as the quotient $\mathcal{S} = \mathbb{H}/\Gamma$, where

$$\Gamma = \langle \gamma_1, \delta_1, \ldots, \gamma_g, \delta_g \rangle_{\text{Isom}(\mathbb{H})},$$

where the γ_i, δ_i are the side pairing isometries of a so-called *canonical fundamental domain*, satisfying the relation

$$\gamma_1 \delta_1 \gamma_1^{-1} \delta_1^{-1} \cdots \gamma_g \delta_g \gamma_g^{-1} \delta_g^{-1} = \text{id}.$$

The number g is uniquely determined by the surface and is called its *genus* (c.f. Exercise 3.38).

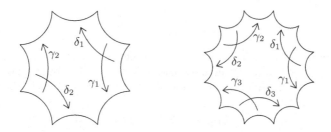

Fig. 20. Side pairings for the so-called canonical fundamental domains.

Fig. 20 shows the pasting pattern for $g = 2$ and $g = 3$. Furthermore, each couple γ_i, δ_i has crossing axes, say C_i, D_i, and each *commutator* $\gamma_i \delta_i \gamma_i^{-1} \delta_i^{-1}$ has an axis A_i which does not intersect C_i and D_i. In the case $g = 2$ — the only case to concern us here — one has $A_1 = A_2$, and the couples C_1, D_1 and C_2 and D_2 lie on either side of $A = A_1 = A_2$. In the case $g \geq 3$, the pattern is this: for each i the couple C_i, D_i lies on one side of A_i, and all axes C_j, D_j, A_j lie on the other side of A_i.

For a concise notation we will say that Γ is a *Fuchsian group of genus g* if it has generators with these properties.

Exercise 6.21. Let $G = \langle g_1, g_2, g_3, g_4 \rangle_{\text{Isom}(\mathbb{H})}$ be as in Theorem 6.18. Find generators $\gamma_1, \delta_1, \gamma_2, \delta_2$ of G (written in terms of g_1, g_2, g_3, g_4) in order to give a direct verification that G is a Fuchsian group of genus 2 in the sense of the preceding definition.
(Hint: a solution is given in formula (95).)

Theorem 6.22. *For any Fuchsian group Γ of genus 2 there exists an admissible symmetric octagon $\mathcal{O}$ with opposite-side pairing isometries g_1, g_2, g_3, g_4 satisfying*

$$g_1 g_2^{-1} g_3 g_4^{-1} g_1^{-1} g_2 g_3^{-1} g_4 = \text{id}$$

such that $\Gamma = \langle g_1, g_2, g_3, g_4 \rangle_{\text{Isom}(\mathbb{H})}$, and $\mathcal{O}$ is a fundamental domain for the action of Γ on $\mathbb{H}$.

Proof. It is again useful to work with half-turns. Let p_3 be the intersection point of C_1, D_1, and p_6 the intersection point of C_2, D_2. There exist points $p_1 \in C_1$, $p_2 \in D_1$, and $p_4 \in C_2$, $p_5 \in D_2$, such that

$$\gamma_1 = h_1 h_3, \qquad \gamma_2 = h_4 h_6,$$
$$\delta_1 = h_3 h_2, \qquad \delta_2 = h_6 h_5,$$

where h_i denotes the half-turn around p_i, $i = 1, \ldots, 6$. Then

$$\gamma_1 \delta_1 \gamma_1^{-1} \delta_1^{-1} = h_1 h_3 h_3 h_2 h_3 h_1 h_2 h_3 = (h_1 h_2 h_3)^2,$$
$$\gamma_2 \delta_2 \gamma_2^{-1} \delta_2^{-1} = h_4 h_6 h_6 h_5 h_6 h_4 h_5 h_6 = (h_4 h_5 h_6)^2.$$

Since $\gamma_1 \delta_1 \gamma_1^{-1} \delta_1^{-1} \gamma_2 \delta_2 \gamma_2^{-1} \delta_2^{-1} = \text{id}$, we get

$$(h_1 h_2 h_3)^2 = (h_6 h_5 h_4)^2.$$

As both sides belong to Γ they are hyperbolic. It follows that $h_1 h_2 h_3$ and $h_6 h_5 h_4$ are hyperbolic as well, have the same axis, the same displacement length and the same direction. This means that $h_1 h_2 h_3 = h_6 h_5 h_4$, and therefore

$$h_1 h_2 h_3 h_4 h_5 h_6 = \text{id},$$

and hexagon $p_1, \ldots, p_6$ is closing. We claim that

$$\Gamma = \langle g_1, g_2, g_3, g_4 \rangle_{\text{Isom}(\mathbb{H})}, \quad \text{with} \quad g_i = h_i h_5, \ i = 1, 2, 3, 4. \tag{93}$$

Using that $h_1 \cdots h_6 = \text{id}$ we first compute $h_3 h_4 = h_2 h_1 h_1 h_2 h_3 h_4 h_5 h_6 h_6 h_5 = h_2 h_1 h_6 h_5 = h_2 h_3 h_3 h_1 h_6 h_5 = \delta_1^{-1} \gamma_1^{-1} \delta_2$, and then get successively

$$\begin{aligned}
g_4 &= h_4 h_6 h_6 h_5 = \gamma_2 \delta_2 \\
g_3 &= h_3 h_4 h_4 h_5 = \delta_1^{-1} \gamma_1^{-1} \delta_2 \gamma_2 \delta_2 \\
g_2 &= h_2 h_3 h_3 h_5 = \delta_1^{-2} \gamma_1^{-1} \delta_2 \gamma_2 \delta_2 \\
g_1 &= h_1 h_3 h_3 h_5 = \gamma_1 \delta_1^{-1} \gamma_1^{-1} \delta_2 \gamma_2 \delta_2.
\end{aligned} \tag{94}$$

Conversely, with this list we immediately verify that

$$\begin{aligned}
\gamma_1 &= g_1 g_3^{-1} \\
\delta_1 &= g_3 g_2^{-1} \\
\gamma_2 &= g_4^2 g_3^{-1} g_2 g_1^{-1} \\
\delta_2 &= g_1 g_2^{-1} g_3 g_4^{-1}.
\end{aligned} \tag{95}$$

This proves (93).

To finish the proof of the theorem we have to provide the claimed fundamental domain. We first show that $p_1, \ldots, p_5$ is a good pentagon. For this we momentarily work with the matrix model and interpret the p_i as members of

P_M. We then already know that $p_1 \cdots p_6 = \varepsilon\mathbf{1}$ with $\varepsilon = \pm 1$, and we have to show that $\varepsilon = +1$.

Conjugating Γ with an orientation reversing isometry, if necessary, we may assume that p_1, p_2, p_3 has the orientation for which $\mathrm{tr}(p_1 p_2 p_3) > 0$. We thus write $p_1 p_2 p_3 = a\mathbf{1} + A$ with $a > 1$ and $A \in \mathrm{T0}(2, \mathbb{R})$. Then A is the axis of $h_1 h_2 h_3$ and at the same time the common axis of the commutators $\gamma_i \delta_i \gamma_i^{-1} \delta_i^{-1}$, $i = 1, 2$. As mentioned above, points p_1, p_2, p_3 lie on one side of this axis and p_4, p_5, p_6 on the other. By Example 5.3 and using (71) we have therefore

$$\mathrm{tr}(p_3 A)\mathrm{tr}(p_4 A)\mathrm{tr}(p_3 p_4) > 0.$$

Since $\mathrm{tr}(p_3 p_4) < 0$ and $\mathrm{tr}(p_3 A) = \mathrm{tr}(p_3 p_1 p_2 p_3) = -\mathrm{tr}(p_1 p_2) > 0$, we get $\mathrm{tr}(p_4 A) < 0$. Now $\mathrm{tr}(p_4 A) = \mathrm{tr}(p_1 p_2 p_3 p_4) = \varepsilon\mathrm{tr}(p_6 p_5)$, where $\mathrm{tr}(p_6 p_5) < 0$, and therefore $\varepsilon = 1$.

Since $p_1, \ldots, p_5$ is good, the outer pentagon Q with vertices $q_1 = p_6$, $q_2 = h_1(q_1)$, $q_3 = h_2(q_2)$, $q_4 = h_3(q_3)$, $q_5 = h_4(q_4)$ is convex and has angle sum π (Lemma 6.16 and Theorem 6.17). Building $\mathcal{O} = Q \cup h_5(Q)$ we now obtain an admissible symmetric octagon with opposite-side pairing isometries g_1, g_2, g_3, g_4. By (93) and Theorem 6.18, $\mathcal{O}$ is a fundamental domain of Γ, and the proof of Theorem 6.22 is complete. □

The preceding theorem together with Exercise 6.20 yields the following consequence.

Theorem 6.23. *Every compact hyperbolic surface $\mathcal{S}$ of genus 2 admits an orientation preserving isometry $\phi : \mathcal{S} \to \mathcal{S}$ satisfying $\phi^2 = \mathrm{id}$ and having exactly six fixed points.* □

For a hyperbolic surface of genus $g \geq 3$, the isometry group is trivial, in general (see for example in [10]). Hence, Theorem 6.23 expresses a peculiarity of genus 2. Also, the isometry $\phi : \mathcal{S} \to \mathcal{S}$ as given by the theorem is uniquely determined. It is called the *hyperelliptic involution*. The fixed points of ϕ are the so-called *Weierstrass points*.

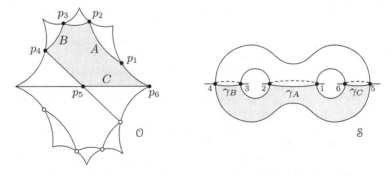

Fig. 21. A symmetric admissible octagon and the corresponding quotient surface. The points lettered 1 through 6 are the Weierstrass points.

Fig. 21 shows an admissible symmetric octagon together with an intuitive view of the corresponding quotient surface $\mathcal{S} = \mathbb{H}/\Gamma = \mathbb{H}/G$ (notation as in the proof of Theorem 6.22). The points marked $1, \ldots, 6$ are the Weierstrass points. They are the images of $p_1, \ldots, p_6$ under the natural projection $\mathcal{O} \to \mathcal{S}$. The lines marked A, B, C through p_1, p_2, respectively p_3, p_4 and p_5, p_6 are projected to the closed geodesics $\gamma_A, \gamma_B, \gamma_C$ on $\mathcal{S}$. The shaded area on $\mathcal{S}$ corresponds to the interior of the hexagon $C_6 = (p_1, \ldots, p_6)$.

We close the lecture remarking that Example 6.5 allows us to parametrize the entire space of Fuchsian groups (respectively, hyperbolic surfaces) of genus 2: by Theorem 6.22, any such group arises from an admissible symmetric octagon $\mathcal{O}$, where by Theorem 6.17, $\mathcal{O}$ is based on a good hexagon as constructed in Example 6.5. Conversely, Theorem 6.18 tells us that this construction always yields a Fuchsian group of genus 2.

The parametrization is as follows: with $a, b, c > 1$ and $\tau_\alpha, \tau_\beta, \tau_\gamma \in \mathbb{R}$, one defines lines $A, B, C \in L_M$ using (73), then points $p_1, \ldots, p_6 \in P_M$ on these lines via (83), and finally matrices $\beta_1, \beta_2, \beta_3, \beta_4 \in \mathrm{SL}(2, \mathbb{R})$ via (89). This defines the group as a group of matrices, and the Fuchsian group itself is $G = \langle g_1, g_2, g_3, g_4 \rangle_{\mathrm{Isom}(\mathbb{H})}$, where g_i is the isometry represented by β_i, $i = 1, 2, 3, 4$. The parameters $a, b, c, \tau_\alpha, \tau_\beta, \tau_\gamma$ are a version of the so-called *Fenchel-Nielsen* parameters (see, for example, [12]).

References

1. A. Aigon-Dupuy, P. Buser, M. Cibils, A. Künzle, F. Steiner: Hyperbolic octagons and Teichmüller space in genus 2. J. Math. Phys. **46**(3), 033513 (2005).
2. A. Aigon, R. Silhol: Hyperbolic hexagons and algebraic curves in genus 3. J. London Math. Soc. (2), **66**, 671–690 (2002).
3. R. Aurich, M. Sieber, F. Steiner: Quantum chaos of the Hadamard-Gutzwiller model. Phys. Rev. Lett., **61**, 483–487 (1988).
4. R. Aurich, E. B. Bogomolny, F. Steiner: Periodic orbits on the regular hyperbolic octagon. Physica D, **48**, 91–101 (1991).
5. A. F. Beardon: *The geometry of discrete groups*, volume 91 of Graduate Texts in Mathematics (Springer-Verlag, New York, 1995).
6. P. Buser: *Geometry and spectra of compact Riemann surfaces*, volume 106 of Progress in Mathematics (Birkhäuser, Boston, 1992).
7. P. Buser, R. Silhol: Some remarks on the uniformizing function in genus 2. Geom. Dedicata **115**, 121–133 (2005).
8. M. P. Do Carmo: *Riemannian geometry*, Mathematics: Theory and Applications (Birkhäuser, Boston, 1992).
9. D. B. A. Epstein, C. Petronio: An exposition of Poincaré's polyhedron theorem. Enseign. Math. (2) **40**, 113–170 (1994).
10. H. M. Farkas, I, Kra: *Riemann surfaces*, volume 71 of Graduate Texts in Mathematics, second edition (Springer-Verlag, New York, 1992).
11. W. Fenchel: *Elementary geometry in hyperbolic space*, vol. 11 of de Gruyter Studies in Mathematics (Walter de Gruyter & Co., Berlin, 1989).

12. W. Fenchel, J. Nielsen: *Discontinuous groups of isometries in the hyperbolic plane*, vol. 29 of de Gruyter Studies in Mathematics (Walter de Gruyter & Co., Berlin, 2003).

13. H. Helling: Diskrete Untergruppen von $SL(2, \mathbb{R})$. Invent. Math. **17**, 217–229 (1972).

14. H. Helling: Über den Raum der kompakten Riemannschen Flächen vom Geschlecht 2. J. Reine Angew. Math. **268/269**, 286–293 (1974).

15. J. M. Lee: *Riemannian manifolds. An introduction to curvature*, volume 176 of Graduate Texts in Mathematics (Springer-Verlag, New York, 1997).

16. W. S. Massey: *A basic course in algebraic topology*, volume 127 of Graduate Texts in Mathematics (Springer-Verlag, New York, 1991).

17. H. Poincaré: Sur les fonctions Fuchsiennes. C. R. Acad. Sci. Paris **92**, 333-335 (1881). Reprinted in: *Oeuvres de Henri Poincaré. Tome II*, Éditions Jacques Gabay, Sceaux, 1995).

18. K.-D. Semmler: A fundamental domain for the Teichmüller space of compact Riemann surfaces of genus 2. PhD Thesis, École Polytechnique Fédérale de Lausanne, 1988.

19. M. Seppälä, T. Sorvali: *Geometry of Riemann surfaces and Teichmüller spaces*, volume 169 of North-Holland Mathematics Studies (North-Holland Publishing Co., Amsterdam, 1992).

II. Selberg's Trace Formula: An Introduction

Jens Marklof

School of Mathematics, University of Bristol, Bristol BS8 1TW, UK,
j.marklof@bristol.ac.uk

The aim of this short lecture course is to develop Selberg's trace formula for a compact hyperbolic surface $\mathcal{M}$, and discuss some of its applications. The main motivation for our studies is *quantum chaos*: the Laplace-Beltrami operator $-\Delta$ on the surface $\mathcal{M}$ represents the quantum Hamiltonian of a particle, whose classical dynamics is governed by the (strongly chaotic) geodesic flow on the unit tangent bundle of $\mathcal{M}$. The trace formula is currently the only available tool to analyze the fine structure of the spectrum of $-\Delta$; no individual formulas for its eigenvalues are known. In the case of more general quantum systems, the role of Selberg's formula is taken over by the semiclassical *Gutzwiller trace formula* [11], [7].

We begin by reviewing the trace formulas for the simplest compact manifolds, the circle $\mathbb{S}^1$ (Section 1) and the sphere $\mathbb{S}^2$ (Section 2). In both cases, the corresponding geodesic flow is integrable, and the trace formula is a consequence of the *Poisson summation formula*. In the remaining sections we shall discuss the following topics: *the Laplacian on the hyperbolic plane and isometries* (Section 3); *Green's functions* (Section 4); *Selberg's point pair invariants* (Section 5); *The ghost of the sphere* (Section 6); *Linear operators on hyperbolic surfaces* (Section 7); *A trace formula for hyperbolic cylinders and poles of the scattering matrix* (Section 8); *Back to general hyperbolic surfaces* (Section 9); *The spectrum of a compact surface, Selberg's pre-trace and trace formulas* (Section 10); *Heat kernel and Weyl's law* (Section 11); *Density of closed geodesics* (Section 12); *Trace of the resolvent* (Section 13); *Selberg's zeta function* (Section 14); *Suggestions for exercises and further reading* (Section 15).

Our main references are Hejhal's classic lecture notes [13, Chapters ONE and TWO], Balazs and Voros' excellent introduction [1], and Cartier and Voros' *nouvelle interprétation* [6]. Section 15 comprises a list of references for further reading.

1 Poisson summation

The Poisson summation formula reads

$$\sum_{m \in \mathbb{Z}} h(m) = \sum_{n \in \mathbb{Z}} \int_{\mathbb{R}} h(\rho) \, e^{2\pi i n \rho} \, d\rho \qquad (1)$$

for any sufficiently nice test function $h : \mathbb{R} \to \mathbb{C}$. One may for instance take $h \in C^2(\mathbb{R})$ with $|h(\rho)| \ll (1 + |\rho|)^{-1-\delta}$ for some $\delta > 0$. (The notation $x \ll y$ means here *there exists a constant $C > 0$ such that $x \leq Cy$.*) Then both sums in (1) converge absolutely. (1) is proved by expanding the periodic function

$$f(\rho) = \sum_{m \in \mathbb{Z}} h(\rho + m) \qquad (2)$$

in its Fourier series, and then setting $\rho = 0$.

The Poisson summation formula is our first example of a *trace formula*: The eigenvalues of the positive Laplacian $-\Delta = -\frac{d^2}{dx^2}$ on the circle $\mathbb{S}^1$ of length 2π are m^2 where $m = 0, \pm 1, \pm 2, \ldots$, with corresponding eigenfunctions $\varphi_m(x) = (2\pi)^{-1/2} e^{imx}$. Consider the linear operator L acting on 2π-periodic functions by

$$[Lf](x) := \int_0^{2\pi} k(x, y) f(y) \, dy \qquad (3)$$

with kernel

$$k(x, y) = \sum_{m \in \mathbb{Z}} h(m) \, \varphi_m(x) \, \overline{\varphi}_m(y). \qquad (4)$$

Then

$$L\varphi_m = h(m)\varphi_m \qquad (5)$$

and hence the Poisson summation formula says that

$$\operatorname{Tr} L = \sum_{m \in \mathbb{Z}} h(m) = \sum_{n \in \mathbb{Z}} \int_{\mathbb{R}} h(\rho) \, e^{2\pi i n \rho} \, d\rho. \qquad (6)$$

The right hand side in turn has a geometric interpretation as a sum over the periodic orbits of the geodesic flow on $\mathbb{S}^1$, whose lengths are $2\pi|n|$, $n \in \mathbb{Z}$.

An important example for a linear operator of the above type is the resolvent of the Laplacian, $(\Delta + \rho^2)^{-1}$, with $\operatorname{Im} \rho < 0$. The corresponding test function is $h(\rho') = (\rho^2 - \rho'^2)^{-1}$. Poisson summation yields in this case

$$\sum_{m \in \mathbb{Z}} (\rho^2 - m^2)^{-1} = \sum_{n \in \mathbb{Z}} \int_{\mathbb{R}} \frac{e^{-2\pi i |n| \rho'}}{\rho^2 - \rho'^2} \, d\rho' \qquad (7)$$

and by shifting the contour to $-i\infty$ and collecting the residue at $\rho' = \rho$,

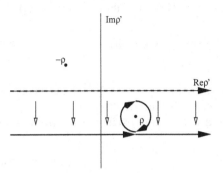

we find

$$\sum_{m\in\mathbb{Z}}(\rho^2 - m^2)^{-1} = \frac{\pi i}{\rho}\sum_{n\in\mathbb{Z}}e^{-2\pi i|n|\rho}.$$ (8)

The right hand side resembles the geometric series expansion of $\cot z$ for $\operatorname{Im} z < 0$,

$$\cot z = \frac{2 i e^{-iz}\cos z}{1 - e^{-2iz}} = i(1 + e^{-2iz})\sum_{n=0}^{\infty}e^{-2inz} = i\sum_{n\in\mathbb{Z}}e^{-2i|n|z}.$$ (9)

Hence

$$\sum_{m\in\mathbb{Z}}(\rho^2 - m^2)^{-1} = \frac{\pi}{\rho}\cot(\pi\rho),$$ (10)

which can also be written in the form

$$\frac{1}{2}\sum_{m\in\mathbb{Z}}\left[\frac{1}{\rho - m} + \frac{1}{\rho + m}\right] = \pi\cot(\pi\rho),$$ (11)

The above h is an example of a test function with particularly useful analytical properties. More generally, suppose

(i) h is analytic for $|\operatorname{Im}\rho| \le \sigma$ for some $\sigma > 0$;
(ii) $|h(\rho)| \ll (1 + |\operatorname{Re}\rho|)^{-1-\delta}$ for some $\delta > 0$, uniformly for all ρ in the strip $|\operatorname{Im}\rho| \le \sigma$.

Theorem 1.1. *If h satisfies* (i), (ii), *then*

$$\sum_{m\in\mathbb{Z}}h(m) = \frac{1}{2i}\int_{\mathcal{C}_=}h(\rho)\cot(\pi\rho)\,d\rho$$ (12)

where the path of integration $\mathcal{C}_=$ is

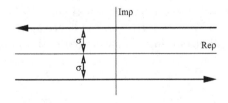

Proof. The Poisson summation formula (1) may be written in the form

$$\sum_{m\in\mathbb{Z}} h(m) = \frac{1}{2}\sum_{n\in\mathbb{Z}}\int_{\mathbb{R}} [h(\rho)+h(-\rho)]\,e^{-2\pi\mathrm{i}|n|\rho}\,d\rho. \tag{13}$$

We shift the contour of the integral to $\int_{-\infty-\mathrm{i}\sigma}^{\infty-\mathrm{i}\sigma}$. The geometric series expansion of $\cot z$ in (9) converges absolutely, uniformly for all z with fixed negative imaginary part. We may therefore exchange summation and integration,

$$\sum_{m\in\mathbb{Z}} h(m) = \frac{1}{2\mathrm{i}}\int_{-\infty-\mathrm{i}\sigma}^{\infty-\mathrm{i}\sigma} [h(\rho)+h(-\rho)]\,\cot(\pi\rho)\,d\rho. \tag{14}$$

We conclude with the observation that

$$\frac{1}{2\mathrm{i}}\int_{-\infty-\mathrm{i}\sigma}^{\infty-\mathrm{i}\sigma} h(-\rho)\,\cot(\pi\rho)\,d\rho = \frac{1}{2\mathrm{i}}\int_{\infty+\mathrm{i}\sigma}^{-\infty+\mathrm{i}\sigma} h(\rho)\,\cot(\pi\rho)\,d\rho \tag{15}$$

since $\cot z$ is odd.   □

Remark 1.2. This theorem can of course also be proved by shifting the lower contour in (12) across the poles to the upper contour, and evaluating the corresponding residues.

2 A trace formula for the sphere

The Laplacian on the sphere $\mathbb{S}^2$ is given by

$$\Delta = \frac{1}{\sin\theta}\frac{\partial}{\partial\theta}\left(\sin\theta\frac{\partial}{\partial\theta}\right) + \frac{1}{\sin^2\theta}\frac{\partial^2}{\partial\phi^2} \tag{16}$$

where $\theta\in[0,\pi)$, $\phi\in[0,2\pi)$ are the standard spherical coordinates. The eigenvalue problem

$$(\Delta+\lambda)f = 0 \tag{17}$$

is solved by the spherical harmonics $f = Y_l^m$ for integers $l = 0,1,2,\ldots$, $m = 0,\pm1,\pm2,\ldots,\pm l$, where

$$Y_l^m(\theta,\phi) = (-1)^m\left[\frac{(2l+1)}{4\pi}\frac{(l-m)!}{(l+m)!}\right]^{1/2} P_l^m(\cos\theta)\,e^{\mathrm{i}m\phi} \tag{18}$$

and P_l^m denotes the associated Legendre function of the first kind. The eigenvalue corresponding to Y_l^m is $\lambda = l(l+1)$, and hence appears with multiplicity $2l+1$. Let us label all eigenvalues (counting multiplicity) by

$$0 = \lambda_0 < \lambda_1 \le \lambda_2 \le \ldots \to \infty, \tag{19}$$

and set $\rho_j = \sqrt{\lambda_j + \frac{1}{4}} > 0$. For any even test function $h \in C^2(\mathbb{R})$ with the bound $|h(\rho)| \ll (1+|\operatorname{Re}\rho|)^{-2-\delta}$ for some $\delta > 0$ (assume this bound also holds for the first and second derivative) we have

$$\sum_{j=0}^{\infty} h(\rho_j) = \sum_{l=0}^{\infty}(2l+1)\, h(l+\tfrac{1}{2}) \tag{20}$$

$$= \sum_{l=-\infty}^{\infty} |l+\tfrac{1}{2}|\, h(l+\tfrac{1}{2}) \tag{21}$$

$$= \sum_{n=-\infty}^{\infty} \int_{\mathbb{R}} |l+\tfrac{1}{2}|\, h(l+\tfrac{1}{2}) e^{2\pi i l n}\, dl \tag{22}$$

$$= 2\sum_{n\in\mathbb{Z}}(-1)^n \int_0^{\infty} \rho\, h(\rho)\,\cos(2\pi n\rho)\, d\rho, \tag{23}$$

in view of the Poisson summation formula. We used the test function $|\rho| h(\rho)$ which is not continuously differentiable at $\rho = 0$. This is not a problem, since we check that (using integration by parts twice)

$$\int_0^{\infty} \rho\, h(\rho)\,\cos(2\pi n\rho)\, d\rho = O(n^{-2}) \tag{24}$$

hence all sums are absolutely convergent. With $\operatorname{Area}(\mathbb{S}^2) = 4\pi$, it is suggestive to write the trace formula (23) for the sphere in the form

$$\sum_{j=0}^{\infty} h(\rho_j) = \frac{\operatorname{Area}(\mathbb{S}^2)}{4\pi} \int_{\mathbb{R}} |\rho|\, h(\rho)\, d\rho + \sum_{n\neq 0}(-1)^n \int_{\mathbb{R}} |\rho|\, h(\rho)\, e^{2\pi i n\rho}\, d\rho. \tag{25}$$

As in the trace formula for the circle, the sum on the right hand side may again be viewed as a sum over the closed geodesics of the sphere which, of course, all have lengths $2\pi|n|$. The factor $(-1)^n$ accounts for the number of conjugate points traversed by the corresponding orbit.

The sum in (23) resembles the geometric series expansion for $\tan z$ for $\operatorname{Im} z < 0$,

$$\tan z = -\cot(z+\pi/2) = -i\sum_{n\in\mathbb{Z}}(-1)^n e^{-2i|n|z}. \tag{26}$$

As remarked earlier, the sum converges uniformly for all z with fixed $\operatorname{Im} z < 0$. We have in fact the uniform bound

$$\sum_{n\in\mathbb{Z}}\left|(-1)^n e^{-2i|n|z}\right| \leq 1 + 2\sum_{n=1}^{\infty} e^{2n\operatorname{Im} z} \leq 1 + 2\int_0^{\infty} e^{2x\operatorname{Im} z}\, dx = 1 - \frac{1}{\operatorname{Im} z} \tag{27}$$

which holds for all z with $\operatorname{Im} z < 0$.

Let us use this identity to rewrite the trace formula. Assume h satisfies the following hypotheses.

(i) h is analytic for $|\operatorname{Im}\rho| \leq \sigma$ for some $\sigma > 0$;

(ii) h is even, i.e., $h(-\rho) = h(\rho)$;

(iii) $|h(\rho)| \ll (1 + |\operatorname{Re}\rho|)^{-2-\delta}$ for some $\delta > 0$, uniformly for all ρ in the strip $|\operatorname{Im}\rho| \leq \sigma$.

Theorem 2.1. *If h satisfies (i), (ii), (iii), then*

$$\sum_{j=0}^{\infty} h(\rho_j) = -\frac{1}{2i} \int_{\mathcal{C}_\times} h(\rho)\, \rho \tan(\pi\rho)\, d\rho, \qquad (28)$$

with the path of integration

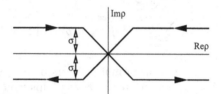

Proof. We express (23) in the form

$$\sum_{n\in\mathbb{Z}} (-1)^n \int_0^{\infty} \rho\, h(\rho)\, e^{2\pi i|n|\rho}\, d\rho + \sum_{n\in\mathbb{Z}} (-1)^n \int_0^{\infty} \rho\, h(\rho)\, e^{-2\pi i|n|\rho}\, d\rho. \qquad (29)$$

which equals

$$-\sum_{n\in\mathbb{Z}} (-1)^n \int_{-\infty}^{0} \rho\, h(\rho)\, e^{-2\pi i|n|\rho}\, d\rho + \sum_{n\in\mathbb{Z}} (-1)^n \int_0^{\infty} \rho\, h(\rho)\, e^{-2\pi i|n|\rho}\, d\rho. \qquad (30)$$

Let us first consider the second integral. We change the path of integration to $\mathcal{C}_1$:

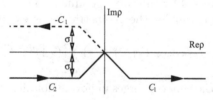

Due to the uniform bound (27) and

$$\int_{\mathcal{C}_1} \left| \rho\, h(\rho)[1 - (2\pi \operatorname{Im}\rho)^{-1}]\, d\rho \right| < \infty \qquad (31)$$

we may exchange integration and summation, and hence the second integral in (30) evaluates to

$$\sum_{n \in \mathbb{Z}} (-1)^n \int_0^\infty \rho\, h(\rho)\, e^{-2\pi i |n|\rho}\, d\rho = i \int_{\mathcal{C}_1} h(\rho)\, \rho \tan(\pi \rho)\, d\rho \qquad (32)$$

The first integral in (30) is analogous, we have

$$-\sum_{n \in \mathbb{Z}} (-1)^n \int_{-\infty}^0 \rho\, h(\rho)\, e^{-2\pi i |n|\rho}\, d\rho = -i \int_{\mathcal{C}_2} h(\rho)\, \rho \tan(\pi \rho)\, d\rho \qquad (33)$$

$$= i \int_{\mathcal{C}_2^{-1}} h(\rho)\, \rho \tan(\pi \rho)\, d\rho. \qquad (34)$$

The final result is obtained by reflecting these paths at the origin, using the fact that h is even. $\quad\square$

Remark 2.2. The poles of $\tan z$ and corresponding residues can be easily worked out from (11),

$$\pi \tan(\pi \rho) = -\pi \cot\left[\pi(\rho + \tfrac{1}{2})\right] \qquad (35)$$

$$= -\frac{1}{2} \sum_{l=-\infty}^\infty \left[\frac{1}{\rho - (l - \frac{1}{2})} + \frac{1}{\rho + (l + \frac{1}{2})}\right] \qquad (36)$$

$$= -\frac{1}{2} \sum_{l=-\infty}^\infty \left[\frac{1}{\rho + (l + \frac{1}{2})} + \frac{1}{\rho - (l + \frac{1}{2})}\right] \qquad (37)$$

(the sum has not been reordered, we have simply shifted the bracket)

$$= -\sum_{l=0}^\infty \left[\frac{1}{\rho + (l + \frac{1}{2})} + \frac{1}{\rho - (l + \frac{1}{2})}\right]. \qquad (38)$$

Note that the extra factor ρ in the integral (28), as compared to Theorem 1.1, yields the multiplicity of the eigenvalues of the sphere.

3 The hyperbolic plane

In this section we briefly summarize some basic features of hyperbolic geometry; for a detailed discussion see the lecture notes [5].

   The hyperbolic plane $\mathbb{H}^2$ may be abstractly defined as the simply connected two-dimensional Riemannian manifold with Gaussian curvature -1. Let us introduce three convenient coordinate systems for $\mathbb{H}^2$: the Poincaré disk $\mathfrak{D} = \{z : |z| < 1\}$, the upper half plane $\mathfrak{H} = \{z : \operatorname{Im} z > 0\}$ and polar coordinates $(\tau, \phi) \in \mathbb{R}_{\geq 0} \times [0, 2\pi)$. In these parametrizations, the line element ds, volume element $d\mu$ and the Riemannian distance $d(z, z')$ between two points $z, z' \in \mathbb{H}^2$ are as follows.

$\mathbb{H}^2$	ds^2	$d\mu$	$\cosh d(z, z')$
$\mathfrak{D}$	$\dfrac{4(dx^2 + dy^2)}{(1 - x^2 - y^2)^2}$	$\dfrac{4\,dx\,dy}{(1 - x^2 - y^2)^2}$	$1 + \dfrac{2\|z - z'\|^2}{(1 - \|z\|)^2(1 - \|z'\|)^2}$
$\mathfrak{H}$	$\dfrac{dx^2 + dy^2}{y^2}$	$\dfrac{dx\,dy}{y^2}$	$1 + \dfrac{\|z - z'\|^2}{2\operatorname{Im} z \operatorname{Im} z'}$
polar	$d\tau^2 + \sinh^2 \tau\, d\phi^2$	$\sinh \tau\, d\tau\, d\phi$	$\cosh \tau$ [for $z = (\tau, \phi)$, $z' = (0,0)$]

The *group of isometries* of $\mathbb{H}^2$, denoted by $\operatorname{Isom}(\mathbb{H}^2)$, is the group of smooth coordinate transformations which leave the Riemannian metric invariant. The group of *orientation preserving* isometries is called $\operatorname{Isom}^+(\mathbb{H}^2)$. We define the *length* of an isometry $g \in \operatorname{Isom}(\mathbb{H}^2)$ by

$$\ell_g = \ell(g) = \inf_{z \in \mathbb{H}^2} d(gz, z). \tag{39}$$

Those $g \in \operatorname{Isom}^+(\mathbb{H}^2)$ for which $\ell > 0$ are called *hyperbolic*. In the half plane model, $\operatorname{Isom}^+(\mathbb{H}^2)$ acts by fractional linear transformations,

$$g : \mathfrak{H} \to \mathfrak{H}, \qquad z \mapsto gz := \frac{az + b}{cz + d}, \qquad \begin{pmatrix} a & b \\ c & d \end{pmatrix} \in \operatorname{SL}(2, \mathbb{R}) \tag{40}$$

(we only consider orientation-preserving isometries here). We may therefore identify g with a matrix in $\operatorname{SL}(2, \mathbb{R})$, where the matrices g and $-g$ obviously correspond to the same fractional linear transformation. $\operatorname{Isom}^+(\mathbb{H}^2)$ may thus be identified with the group $\operatorname{PSL}(2, \mathbb{R}) = \operatorname{SL}(2, \mathbb{R})/\{\pm 1\}$. In this representation,

$$2\cosh\left(\ell_g/2\right) = \max\{|\operatorname{tr} g|, 2\}, \tag{41}$$

since every matrix $g \in \operatorname{SL}(2, \mathbb{R})$ is conjugate to one of the following three,

$$\begin{pmatrix} 1 & b \\ 0 & 1 \end{pmatrix}, \qquad \begin{pmatrix} a & 0 \\ 0 & a^{-1} \end{pmatrix}, \qquad \begin{pmatrix} \cos(\theta/2) & \sin(\theta/2) \\ -\sin(\theta/2) & \cos(\theta/2) \end{pmatrix}, \tag{42}$$

with $b \in \mathbb{R}$, $a \in \mathbb{R}_{>0}$, $\theta \in [0, 2\pi)$.

The *Laplace-Beltrami operator* (or *Laplacian* for short) Δ of a smooth Riemannian manifold with metric

$$ds^2 = \sum_{ij} g_{jk} dx^j dx^k \tag{43}$$

is given by the formula

$$\Delta = \sum_{ij} \frac{1}{\sqrt{g}} \frac{\partial}{\partial x^j} \left(\sqrt{g}\, g^{jk} \frac{\partial}{\partial x^k} \right) \tag{44}$$

where g^{jk} are the matrix coefficients of the inverse of the matrix (g_{jk}), and $g = |\det(g_{jk})|$. In the above coordinate systems for $\mathbb{H}^2$ the Laplacian takes the following form.

$\mathbb{H}^2$	Δ
$\mathfrak{D}$	$\dfrac{(1-x^2-y^2)^2}{4}\left(\dfrac{\partial^2}{\partial x^2}+\dfrac{\partial^2}{\partial y^2}\right)$
$\mathfrak{H}$	$y^2\left(\dfrac{\partial^2}{\partial x^2}+\dfrac{\partial^2}{\partial y^2}\right)$
polar	$\dfrac{1}{\sinh\tau}\dfrac{\partial}{\partial\tau}\left(\sinh\tau\dfrac{\partial}{\partial\tau}\right)+\dfrac{1}{\sinh^2\tau}\dfrac{\partial^2}{\partial\phi^2}$

One of the important properties of the Laplacian is that it commutes with every isometry $g \in \mathrm{Isom}(\mathbb{H}^2)$. That is,

$$\Delta T_g = T_g \Delta \qquad \forall g \in \mathrm{Isom}(\mathbb{H}^2). \tag{45}$$

where the left translation operator T_g acting on functions f on $\mathbb{H}^2$ is defined by

$$[T_g f](z) = f(g^{-1}z) \tag{46}$$

with $g \in \mathrm{Isom}(\mathbb{H}^2)$. Even though (45) is an intrinsic property of the Laplacian and is directly related to the invariance of the Riemannian metric under isometries, it is a useful exercise to verify (45) explicitly. To this end note that every isometry may be represented as a product of fractional linear transformations of the form $z \mapsto az$ $(a > 0)$, $z \mapsto z + b$ $(b \in \mathbb{R})$, $z \mapsto -1/z$, $z \mapsto -\bar{z}$. It is therefore sufficient to check (45) only for these four transformations.

4 Green's functions

The Green's function $G(z, w; \lambda)$ corresponding to the differential equation $(\Delta + \lambda)f(z) = 0$ is formally defined as the integral kernel of the resolvent $(\Delta + \lambda)^{-1}$, i.e., by the equation

$$(\Delta + \lambda)^{-1}f(z) = \int G(z, w; \lambda)\, f(w)\, d\mu(w) \tag{47}$$

for a suitable class of test functions f. A more precise characterization is as follows:

(G1) $G(\,\cdot\,, w; \lambda) \in \mathrm{C}^\infty(\mathbb{H}^2 - \{w\})$ for every fixed w;
(G2) $(\Delta + \lambda)G(z, w; \lambda) = \delta(z, w)$ for every fixed w;
(G3) as a function of (z, w), $G(z, w; \lambda)$ depends on the distance $d(z, w)$ only;
(G4) $G(z, w; \lambda) \to 0$ as $d(z, w) \to \infty$.

Here $\delta(z, w)$ is the Dirac distribution at w with respect to the measure $d\mu$. It is defined by the properties that

(D1) $\delta(z, w)d\mu(z)$ is a probability measure on $\mathbb{H}^2$;
(D2) $\int_{\mathbb{H}^2} f(z)\,\delta(z, w)\, d\mu(z) = f(w)$ for all $f \in \mathrm{C}(\mathbb{H}^2)$.

E.g., in the disk coordinates $z = x + iy$, $w = u + iv \in \mathfrak{D}$ we then have

$$\delta(z, w) = \frac{(1 - x^2 - y^2)^2}{4} \delta(x - u) \delta(y - v) \tag{48}$$

where $\delta(x)$ is the usual Dirac distribution with respect to Lebesgue measure on $\mathbb{R}$. In polar coordinates, where w is taken as the origin, $\tau = d(z, w)$, and

$$\delta(z, w) = \frac{\delta(\tau)}{2\pi \sinh \tau}. \tag{49}$$

Property (G2) therefore says that $(\Delta + \lambda)G(z, w; \lambda) = 0$ for $z \neq w$, and

$$\int_{d(z,w)<\epsilon} (\Delta + \lambda)G(z, w; \lambda)d\mu(z) = 1 \qquad \forall \epsilon > 0. \tag{50}$$

In view of (G3), there is a function $f \in C^\infty(\mathbb{R}_{>0})$ such that $f(\tau) = G(z, w; \lambda)$. Then

$$1 = \int_{d(z,w)<\epsilon} (\Delta + \lambda)G(z, w; \lambda)d\mu(z) \tag{51}$$

$$= 2\pi \int_0^\epsilon \left(\frac{d}{d\tau} \left(\sinh \tau \frac{d}{d\tau} \right) + \lambda \sinh \tau \right) f(\tau) \, d\tau \tag{52}$$

$$= 2\pi \sinh \epsilon f'(\epsilon) + 2\pi\lambda \int_0^\epsilon \sinh \tau \, f(\tau) \, d\tau. \tag{53}$$

Taylor expansion around $\epsilon = 0$ yields $f'(\epsilon) = 1/(2\pi\epsilon) + O(1)$ and thus $f(\epsilon) = (1/2\pi) \log \epsilon + O(1)$ as $\epsilon \to 0$. Equation (G2) is therefore equivalent to

$$\begin{cases} (\Delta + \lambda)G(z, w; \lambda) = 0, & d(z, w) > 0, \\ G(z, w; \lambda) = (1/2\pi) \log d(z, w) + O(1), & d(z, w) \to 0. \end{cases} \tag{54}$$

Proposition 4.1. *If $\rho \in \mathbb{C}$ with $\operatorname{Im} \rho < 1/2$, and $\lambda = \rho^2 + \frac{1}{4}$, then*

$$G(z, w; \lambda) = -\frac{1}{2\pi} Q_{-\frac{1}{2}+i\rho} \big(\cosh d(z, w) \big) \tag{55}$$

satisfies (G1)-(G4), where Q_ν is the Legendre function of the second kind.

Proof. With $f(\tau) = G(z, w; \lambda)$, (54) becomes

$$\left[\frac{1}{\sinh \tau} \frac{d}{d\tau} \left(\sinh \tau \frac{d}{d\tau} \right) + \lambda \right] f(\tau) = 0. \tag{56}$$

Set $r = \cosh \tau$, $\tilde{f}(\cosh \tau) = f(\tau)$, and $\lambda = -\nu(\nu + 1)$, to obtain Legendre's differential equation

$$\left[(1 - r^2) \frac{d^2}{dr^2} - 2r \frac{d}{dr} + \nu(\nu + 1) \right] \tilde{f}(r) = 0, \tag{57}$$

whose solutions are the associated Legendre functions $P_\nu(r)$ and $Q_\nu(r)$. Q_ν has the integral representation

$$Q_{-\frac{1}{2}+i\rho}(\cosh\tau) = \frac{1}{\sqrt{2}} \int_\tau^\infty \frac{e^{-i\rho t}}{\sqrt{\cosh t - \cosh\tau}}\, dt \tag{58}$$

which converges absolutely for $\operatorname{Im}\rho < 1/2$ and $\tau > 0$. From this it is evident that $Q_{-\frac{1}{2}+i\rho}(\cosh\tau) \to 0$ as $\tau \to \infty$ (see also Lemma 4.2 below), so (G4) holds. For $\tau \to 0$ (ρ fixed) it has the asymptotics required in (54), cf. the well known relation

$$Q_{-\frac{1}{2}+i\rho}(\cosh\tau) \sim -\big(\log(\tau/2) + \gamma + \psi(\tfrac{1}{2}+i\rho)\big) \tag{59}$$

where γ is Euler's constant and ψ the logarithmic derivative of Euler's Γ function. □

Lemma 4.2. *Given any $\epsilon > 0$ there is a constant $C_\epsilon > 0$ such that*

$$|Q_{-\frac{1}{2}+i\rho}(\cosh\tau)| \leq C_\epsilon(1 + |\log\tau|)\, e^{\tau(\operatorname{Im}\rho - \frac{1}{2}+\epsilon)} \tag{60}$$

uniformly for all $\tau > 0$ and $\rho \in \mathbb{C}$ with $\operatorname{Im}\rho < \frac{1}{2} - \epsilon$.

Proof. From the integral representation (58) we infer

$$|Q_{-\frac{1}{2}+i\rho}(\cosh\tau)| \leq \frac{1}{\sqrt{2}} \int_\tau^\infty \frac{e^{t\operatorname{Im}\rho}}{\sqrt{\cosh t - \cosh\tau}}\, dt \tag{61}$$

$$\leq \frac{1}{\sqrt{2}}\, e^{\tau(\operatorname{Im}\rho - \frac{1}{2}+\epsilon)} \int_\tau^\infty \frac{e^{t(\frac{1}{2}-\epsilon)}}{\sqrt{\cosh t - \cosh\tau}}\, dt \tag{62}$$

since

$$e^{t(\operatorname{Im}\rho - \frac{1}{2}+\epsilon)} \leq e^{\tau(\operatorname{Im}\rho - \frac{1}{2}+\epsilon)} \tag{63}$$

for all $t \geq \tau$, if $\epsilon > 0$ is chosen small enough. The remaining integral

$$\int_\tau^\infty \frac{e^{t(\frac{1}{2}-\epsilon)}}{\sqrt{\cosh t - \cosh\tau}}\, dt \tag{64}$$

has a $\log\tau$ singularity at $\tau = 0$ and is otherwise uniformly bounded for all $\tau \to \infty$. □

Remark 4.3. This is only a crude upper bound, but sufficient for our purposes.

To highlight the ρ dependence, we shall use

$$G_\rho(z, w) = -\frac{1}{2\pi} Q_{-\frac{1}{2}+i\rho}\big(\cosh d(z, w)\big) \tag{65}$$

instead of $G(z, w; \lambda)$.

Lemma 4.4. *Suppose* $f : \mathbb{H}^2 \to \mathbb{C}$ *with* $|f(z)| \leq A e^{\alpha d(z,o)}$, *with constants* $A, \alpha > 0$. *Then the integral*

$$\int_{\mathbb{H}^2} G_\rho(z,w) f(w) \, d\mu(w) \tag{66}$$

converges absolutely, uniformly in $\operatorname{Re} \rho$, *provided* $\operatorname{Im} \rho < -(\alpha + \frac{1}{2})$. *The convergence is also uniform in* z *in compact sets in* $\mathbb{H}^2$.

Proof. Note that $|f(w)| \leq A e^{\alpha d(o,w)} \leq A e^{\alpha d(o,z)} e^{\alpha d(z,w)}$. In polar coordinates (take w as the origin) $\tau = d(z,w)$, and $d\mu = \sinh \tau \, d\tau \, d\phi$. In view of Lemma 4.2, the integral (66) is bounded by

$$\ll_\epsilon \int_0^\infty |\log \tau| \, e^{-(\frac{1}{2} - \operatorname{Im} \rho - \epsilon - \alpha)\tau} \sinh \tau \, d\tau \tag{67}$$

which converges under the hypothesis on $\operatorname{Im} \rho$. $\square$

5 Selberg's point-pair invariants

Let H be a subgroup of $\operatorname{Isom}(\mathbb{H}^2)$. An H-point-pair invariant $k : \mathbb{H}^2 \times \mathbb{H}^2 \to \mathbb{C}$ is defined by the relations

(K1) $k(gz, gw) = k(z,w)$ for all $g \in H$, $z,w \in \mathbb{H}^2$;
(K2) $k(w,z) = k(z,w)$ for all $z,w \in \mathbb{H}^2$.

Here we will only consider point-pair invariants which are functions of the distance between z, w, such as the Green's function $G_\rho(z,w)$ studied in the previous section. Hence $H = \operatorname{Isom}(\mathbb{H}^2)$ in this case. We sometimes use the notation $k(\tau) = k(z,w)$ with $\tau = d(z,w)$. Let us consider

$$k(z,w) = \frac{1}{\pi i} \int_{-\infty}^\infty G_\rho(z,w) \rho \, h(\rho) \, d\rho \tag{68}$$

where the test function $h : \mathbb{C} \to \mathbb{C}$ satisfies the following conditions.

(H1) h is analytic for $|\operatorname{Im} \rho| \leq \sigma$ for some $\sigma > 1/2$;
(H2) h is even, i.e., $h(-\rho) = h(\rho)$;
(H3) $|h(\rho)| \ll (1 + |\operatorname{Re} \rho|)^{-2-\delta}$ for some fixed $\delta > 0$, uniformly for all ρ in the strip $|\operatorname{Im} \rho| \leq \sigma$.

For technical reasons we will sometimes use the stronger hypothesis

(H3*) $|h(\rho)| \ll_N (1 + |\operatorname{Re} \rho|)^{-N}$ for any fixed $N > 1$, uniformly for all ρ in the strip $|\operatorname{Im} \rho| \leq \sigma$.

The Fourier transform of h is

$$g(t) = \frac{1}{2\pi} \int_{\mathbb{R}} h(\rho)\, e^{-i\rho t} d\rho. \tag{69}$$

With the integral representation (58) one immediately finds

$$k(z,w) = -\frac{1}{\pi\sqrt{2}} \int_{\tau}^{\infty} \frac{g'(t)}{\sqrt{\cosh t - \cosh \tau}}\, dt, \quad \tau = d(z,w). \tag{70}$$

The analyticity of h and (H3) imply that g and its first derivative (*all derivatives* provided (H3*) holds) are exponentially decaying for $|t| \to \infty$. To see this, consider

$$g^{(\nu)}(t) = \frac{1}{2\pi} \int_{\mathbb{R}} (-i\rho)^{\nu} h(\rho)\, e^{-i\rho t} d\rho \tag{71}$$

$$= \frac{1}{2\pi} \int_{\mathbb{R}-i\sigma} (-i\rho)^{\nu} h(\rho)\, e^{-i\rho t} d\rho \tag{72}$$

$$= \frac{1}{2\pi} e^{-\sigma t} \int_{\mathbb{R}} [-i(\rho-i\sigma)]^{\nu} h(\rho-i\sigma)\, e^{-i\rho t} d\rho. \tag{73}$$

Since, due to (H3*),

$$\int_{\mathbb{R}} |(\rho - i\sigma)^{\nu} h(\rho - i\sigma)|\, d\rho < \infty \tag{74}$$

we find

$$|g^{(\nu)}(t)| \ll_{\nu} e^{-\sigma|t|}. \tag{75}$$

The point-pair invariant $k(z,w)$ gives rise to the linear operator L defined by

$$[Lf](z) := \int_{\mathbb{H}^2} k(z,w) f(w) d\mu(w). \tag{76}$$

Proposition 5.1. *Suppose* $f \in C^2(\mathbb{H}^2)$ *is a solution of* $(\Delta + \rho^2 + \frac{1}{4})f = 0$ *with* $|\operatorname{Im}\rho| \le \sigma$ *and* $|f(z)| \le Ae^{\alpha d(z,o)}$, *with constants* $A > 0, 0 \le \alpha < \sigma - \frac{1}{2}$. *Then, for* h *satisfying* (H1), (H2), (H3),

$$Lf = h(\rho)f. \tag{77}$$

Proof. We have

$$[Lf](z) = \int_{\mathbb{H}^2} k(z,w) f(w) d\mu(w) \tag{78}$$

$$= \frac{1}{\pi i} \int_{\mathbb{H}^2} \left\{ \int_{-\infty}^{\infty} G_{\rho'}(z,w)\rho'\, h(\rho')\, d\rho' \right\} f(w) d\mu(w) \tag{79}$$

$$= \frac{1}{\pi i} \int_{\mathbb{H}^2} \left\{ \int_{-\infty-i\sigma}^{\infty-i\sigma} G_{\rho'}(z,w)\rho'\, h(\rho')\, d\rho' \right\} f(w) d\mu(w) \tag{80}$$

where we have shifted the contour of integration by σ. Then

$$= \frac{1}{\pi i} \int_{-\infty-i\sigma}^{\infty-i\sigma} \left\{ \int_{\mathbb{H}^2} G_{\rho'}(z,w)f(w)d\mu(w) \right\} \rho' \, h(\rho') \, d\rho' \qquad (81)$$

since the inner integral converges absolutely, uniformly in $\mathrm{Re}\,\rho$, see Lemma 4.4. We have

$$\int_{\mathbb{H}^2} G_{\rho'}(z,w)f(w)d\mu(w) = (\Delta + \rho'^2 + \tfrac{1}{4})^{-1}f(z) = (\rho'^2 - \rho^2)^{-1}f(z) \qquad (82)$$

and thus

$$[Lf](z) = \frac{1}{\pi i}\, f(z) \int_{-\infty-i\sigma}^{\infty-i\sigma} \frac{\rho' h(\rho')}{\rho'^2 - \rho^2}\, d\rho'. \qquad (83)$$

This integral converges absolutely, cf. (H3), and is easily calculated. We shift the contour from $-i\sigma$ to $+i\sigma$ and collect residues, so that

$$\frac{1}{2\pi i} \int_{-\infty-i\sigma}^{\infty-i\sigma} \frac{2\rho' h(\rho')}{\rho'^2 - \rho^2}\, d\rho' = h(\rho) + h(-\rho) + \frac{1}{2\pi i} \int_{-\infty+i\sigma}^{\infty+i\sigma} \frac{2\rho' h(\rho')}{\rho'^2 - \rho^2}\, d\rho'. \qquad (84)$$

Since h is even, the integral on the right hand side equals the negative of the left hand side, and thus

$$\frac{1}{2\pi i} \int_{-\infty-i\sigma}^{\infty-i\sigma} \frac{2\rho' h(\rho')}{\rho'^2 - \rho^2} = h(\rho), \qquad (85)$$

which concludes the proof. □

It is useful to define the auxiliary functions $\Phi, Q : \mathbb{R}_{\geq 0} \to \mathbb{C}$ by the relations

$$\Phi\big(2(\cosh\tau - 1)\big) = k(\tau) = k(z,w), \qquad \tau = d(z,w), \qquad (86)$$

and

$$Q\big(2(\cosh t - 1)\big) = g(t). \qquad (87)$$

Lemma 5.2. *The following statements are equivalent.*

(i) h *satisfies* (H1), (H2), (H3*).
(ii) $Q \in C^\infty(\mathbb{R}_{\geq 0})$ *with*

$$\left| Q^{(\nu)}(\eta) \right| \ll_\nu \left(1 + \eta\right)^{-\sigma-\nu} \qquad \forall \eta \geq 0. \qquad (88)$$

Proof. Clearly $g \in C^\infty(\mathbb{R})$ if and only if $Q \in C^\infty(\mathbb{R}_{\geq 0})$ (this is obvious for $t \neq 0$; the problem at $t = 0$ can be resolved by expanding in Taylor series). In view of (75), the bound (88) is evident for $\nu = 0$. The νth derivative of g is of the form

$$g^{(\nu)}(t) = \sum_{j=0}^{\nu} a_{j\nu}\, e^{j|t|}\, Q^{(j)}(2(\cosh t - 1))\, (1 + O(e^{-|t|})) \qquad (89)$$

with suitable coefficients $a_{j\nu}$. Hence, by induction on ν,

$$e^{\nu|t|} \left| Q^{(\nu)}(2(\cosh t - 1)) \right| \ll_\nu \left| g^{(\nu)}(t) \right| + \left| \sum_{j=0}^{\nu-1} a_{j\nu} \, e^{j|t|} \, Q^{(j)}(2(\cosh t - 1)) \right| \tag{90}$$

$$\ll_\nu \left| g^{(\nu)}(t) \right| + \sum_{j=0}^{\nu-1} e^{j|t|} e^{(-\sigma-j)|t|} \tag{91}$$

$$\ll_\nu e^{-\sigma|t|}. \tag{92}$$

This proves (i) $\Rightarrow$ (ii). Conversely, (88) implies via (89) the exponential decay of g, which proves (H1). (H3*) follows from $g \in C^\infty(\mathbb{R})$. $\square$

The integral transform (70) reads in terms of the functions Q, Φ,

$$\Phi(\xi) = -\frac{1}{\pi} \int_\xi^\infty \frac{Q'(\eta)}{\sqrt{\eta - \xi}} \, d\eta. \tag{93}$$

In order to specify Q uniquely for a given Φ, we always assume in the following that $Q(\eta) \to 0$ for $\eta \to \infty$.

Lemma 5.3. *Consider the following conditions.*

(i) $Q \in C^\infty(\mathbb{R}_{\geq 0})$ *and* $\left| Q^{(\nu)}(\eta) \right| \ll_\nu (1 + \eta)^{-\sigma-\nu}$.

(ii) $\Phi \in C^\infty(\mathbb{R}_{\geq 0})$ *and* $\left| \Phi^{(\nu)}(\xi) \right| \ll_\nu (1 + \xi)^{-\sigma-\nu-1/2+\epsilon}$.

Then (i) *implies* (ii) *for any fixed* $\epsilon > 0$, *and* (ii) *implies* (i) *for any fixed* $\epsilon < 0$.

Proof. The νth derivative of Φ is

$$\Phi^{(\nu)}(\xi) = -\frac{1}{\pi} \frac{d^\nu}{d\xi^\nu} \int_0^\infty \frac{Q'(\eta + \xi)}{\sqrt{\eta}} \, d\eta \tag{94}$$

$$= -\frac{1}{\pi} \int_0^\infty \frac{Q^{(\nu+1)}(\eta + \xi)}{\sqrt{\eta}} \, d\eta \tag{95}$$

$$= -\frac{1}{\pi} \int_\xi^\infty \frac{Q^{(\nu+1)}(\eta)}{\sqrt{\eta - \xi}} \, d\eta. \tag{96}$$

Therefore (i) implies $\Phi \in C^\infty(\mathbb{R}_{\geq 0})$. Furthermore, from (95),

$$\left| \Phi^{(\nu)}(\xi) \right| \ll_\nu \int_0^\infty \frac{(1 + \eta + \xi)^{-\sigma-\nu-1}}{\sqrt{\eta}} \, d\eta \tag{97}$$

$$\ll_\nu (1 + \xi)^{-\sigma-\nu-1/2+\epsilon} \int_0^\infty \frac{(1 + \eta)^{-1/2-\epsilon}}{\sqrt{\eta}} \, d\eta, \tag{98}$$

for $\epsilon > 0$ small enough. The last integral converges for any $\epsilon > 0$.

The implication (ii) $\Rightarrow$ (i) follows analogously from the inversion formula

$$Q(\eta) = \int_\eta^\infty \frac{\Phi(\xi)}{\sqrt{\xi - \eta}}\, d\xi. \tag{99}$$

To show that (99) is indeed consistent with (93), write (99) in the form

$$Q(\eta) = \int_{-\infty}^\infty \Phi(\eta + \xi^2) d\xi \tag{100}$$

and thus

$$Q'(\eta) = \int_{-\infty}^\infty \Phi'(\eta + \xi^2) d\xi. \tag{101}$$

The right hand side of (93) is

$$-\frac{1}{\pi} \int_\xi^\infty \frac{Q'(\eta)}{\sqrt{\eta - \xi}}\, d\eta = -\frac{1}{\pi} \int_{\mathbb{R}} Q'(\xi + \eta^2) d\eta \tag{102}$$

$$= -\frac{1}{\pi} \int_{\mathbb{R}} \int_{\mathbb{R}} \Phi'(\xi + \eta_1^2 + \eta_2^2) d\eta_1 d\eta_2, \tag{103}$$

where we have used (101) in the last step. This equals of course

$$= -2 \int_0^\infty \Phi'(\xi + r^2) r dr = -\int_0^\infty \Phi'(\xi + r) dr = \Phi(\xi) \tag{104}$$

which yields the left hand side of (93). □

Proposition 5.4. *The following statements are equivalent.*

(i) h *satisfies* (H1), (H2), (H3*).
(ii) $k(z, w)$ *is in* $\mathrm{C}^\infty(\mathbb{H}^2 \times \mathbb{H}^2)$ *with the bound on the* ν*th derivative,*

$$\left|k^{(\nu)}(\tau)\right| \ll_\nu \mathrm{e}^{-(\sigma + 1/2 - \epsilon)\tau} \qquad \forall \tau \geq 0, \tag{105}$$

for any fixed $\epsilon > 0$.

Proof. In view of Lemmas 5.2 and 5.3, the statement (i) is equivalent to the condition for Φ, statement (ii) in Lemma 5.3. Since $k(z, w) = k(\tau) = \Phi\big(2(\cosh \tau - 1)\big)$, the proof is exactly the same as that of Lemma 5.2 with $g(t)$ replaced by $k(\tau)$, and $Q\big(2(\cosh t - 1)\big)$ by $\Phi\big(2(\cosh \tau - 1)\big)$. □

6 The ghost of the sphere

Note that for $z = w$, the kernel $k(z, w)$ has a finite value, unlike the logarithmic divergence of the Green's function $G_\rho(z, w)$. In fact,

$$k(z,z) = -\frac{1}{\pi\sqrt{2}} \int_0^\infty \frac{g'(t)}{\sqrt{\cosh t - 1}}\, dt \tag{106}$$

$$= -\frac{1}{2\pi} \int_0^\infty \frac{g'(t)}{\sinh(t/2)}\, dt \tag{107}$$

$$= \frac{1}{4\pi^2} \int_0^\infty \left\{ \int_{-\infty}^\infty \frac{\sin(\rho t)}{\sinh(t/2)} h(\rho)\, \rho\, d\rho \right\} dt \tag{108}$$

$$= \frac{1}{4\pi^2} \int_{-\infty}^\infty \left\{ \int_0^\infty \frac{\sin(\rho t)}{\sinh(t/2)}\, dt \right\} h(\rho)\, \rho\, d\rho \tag{109}$$

where changing the order of integration is justified, since

$$\int_0^\infty \left| \frac{\sin(\rho t)}{\sinh(t/2)} \right| dt \le |\rho| \int_0^\infty \frac{t}{\sinh(t/2)}\, dt \ll |\rho| \tag{110}$$

and $|h(\rho)| \ll (1+|\rho|)^{-4}$, assuming (H3*). We use the geometric series expansion

$$\frac{1}{\sinh(t/2)} = \frac{2e^{-t/2}}{1 - e^{-t}} = 2 \sum_{l=0}^\infty \exp\left[-\left(l + \tfrac{1}{2}\right) t \right], \tag{111}$$

so for $|\operatorname{Im}\rho| < 1/2$

$$\int_0^\infty \frac{\sin(\rho t)}{\sinh(t/2)}\, dt = i \sum_{l=0}^\infty \left[\frac{1}{i\rho - (l + \tfrac{1}{2})} + \frac{1}{i\rho + (l + \tfrac{1}{2})} \right] \tag{112}$$

$$= -\pi i \tan(\pi i \rho) = \pi \tanh(\pi\rho), \tag{113}$$

compare (38). We conclude

$$k(z,z) = \frac{1}{4\pi} \int_{-\infty}^\infty h(\rho) \tanh(\pi\rho)\, \rho\, d\rho. \tag{114}$$

Let us conclude this section by noting that the logarithmic divergence of the Green's function is independent of ρ, see (54). It may therefore be removed by using instead

$$G_\rho(z,w) - G_{\rho_*}(z,w) \tag{115}$$

where $\rho_* \ne \rho$ is a fixed constant in $\mathbb{C}$ with $|\operatorname{Im}\rho_*| < 1/2$. We then have from (58)

$$\lim_{w\to z} [G_\rho(z,w) - G_{\rho_*}(z,w)] = \lim_{\tau\to 0} -\frac{1}{2\pi\sqrt{2}} \int_\tau^\infty \frac{e^{-i\rho t} - e^{-i\rho_* t}}{\sqrt{\cosh t - \cosh\tau}}\, dt \tag{116}$$

$$= -\frac{1}{4\pi} \int_0^\infty \frac{e^{-i\rho t} - e^{-i\rho_* t}}{\sinh(t/2)}\, dt \tag{117}$$

$$= -\frac{1}{2\pi i} \sum_{l=0}^\infty \left[\frac{1}{\rho - i(l + \tfrac{1}{2})} - \frac{1}{\rho_* - i(l + \tfrac{1}{2})} \right], \tag{118}$$

where we have used the geometric series expansion (111) as above. The last sum clearly converges since

$$\frac{1}{\rho - \mathrm{i}(l + \frac{1}{2})} - \frac{1}{\rho_* - \mathrm{i}(l + \frac{1}{2})} = O(l^{-2}). \tag{119}$$

Remark 6.1. In analogy with the trace formula for the sphere, we may view the geometric series expansion,

$$\tanh(\pi\rho) = \sum_{n \in \mathbb{Z}} (-1)^n e^{-2\pi|n|\rho}, \tag{120}$$

cf. (26), as a sum over closed orbits on the sphere, but now with imaginary action. These orbits have an interpretation as tunneling (or ghost) orbits.

7 Hyperbolic surfaces

Let $\mathcal{M}$ be a smooth Riemannian surface (finite or infinite) of constant negative curvature which can be represented as the quotient $\Gamma \backslash \mathbb{H}^2$, where Γ is a strictly hyperbolic Fuchsian group (i.e., all elements $\gamma \in \Gamma - \{1\}$ have $\ell_\gamma > 0$). The space of square integrable functions on $\mathcal{M}$ may therefore be identified with the space of measurable functions $f : \mathbb{H}^2 \to \mathbb{C}$ satisfying the properties

$$T_\gamma f = f \quad \forall \gamma \in \Gamma \tag{121}$$

and

$$\|f\|^2 := \int_{\mathcal{F}_\Gamma} |f|^2 d\mu < \infty \tag{122}$$

where T_γ is the translation operator defined in (46) and $\mathcal{F}_\Gamma$ is any fundamental domain of Γ in $\mathbb{H}^2$. We denote this space by $L^2(\Gamma \backslash \mathbb{H}^2)$. The inner product

$$\langle f_1, f_2 \rangle = \int_{\mathcal{F}_\Gamma} f_1 \overline{f}_2 d\mu \tag{123}$$

makes $L^2(\Gamma \backslash \mathbb{H}^2)$ a Hilbert space. Similarly, we may identify $C^\infty(\Gamma \backslash \mathbb{H}^2)$ with the space of functions $f \in C^\infty(\mathbb{H}^2)$ satisfying (121) (note that more care has to be taken here when Γ contains elliptic elements). Since Δ commutes with T_γ, it maps $C^\infty(\Gamma \backslash \mathbb{H}^2) \to C^\infty(\Gamma \backslash \mathbb{H}^2)$.

To study the spectrum of the Laplacian on $\Gamma \backslash \mathbb{H}^2$, let us consider the linear operator L of functions on $\Gamma \backslash \mathbb{H}^2$,

$$[Lf](z) := \int_{\Gamma \backslash \mathbb{H}^2} k_\Gamma(z, w) f(w) d\mu(w) \tag{124}$$

with kernel

$$k_\Gamma(z, w) = \sum_{\gamma \in \Gamma} k(\gamma z, w) \tag{125}$$

with the point-pair invariant k as defined in (68). The convergence of the sum is guaranteed by the following lemma, cf. Proposition 7.2 below.

Lemma 7.1. *For every $\delta > 0$, there is a $C_\delta > 0$ such that*

$$\sum_{\gamma \in \Gamma} e^{-(1+\delta)d(\gamma z, w)} \leq C_\delta \tag{126}$$

for all $(z, w) \in \mathbb{H}^2 \times \mathbb{H}^2$.

Proof. Place a disk $\mathcal{D}_\gamma(r) = \{z' \in \mathbb{H} : d(\gamma z, z') \leq r\}$ around every point $z_\gamma = \gamma z$, and denote the area of $\mathcal{D}_\gamma(r)$ by Area(r). Then

$$e^{-(1+\delta)d(z_\gamma, w)} \leq \frac{e^{r(1+\delta)}}{\text{Area}(r)} \int_{\mathcal{D}_\gamma(r)} e^{-(1+\delta)d(z', w)} d\mu(z'). \tag{127}$$

If

$$r < \frac{1}{2} \min_{\gamma \in \Gamma - \{1\}} \ell_\gamma \tag{128}$$

the disks $\mathcal{D}_\gamma(r)$ do not overlap. (Note that $\min_{\gamma \in \Gamma - \{1\}} \ell_\gamma > 0$ since Γ is strictly hyperbolic and acts properly discontinuously on $\mathbb{H}^2$.) Therefore

$$\sum_{\gamma \in \Gamma} e^{-(1+\delta)d(\gamma z, w)} \leq \frac{e^{r(1+\delta)}}{\text{Area}(r)} \int_{\mathbb{H}^2} e^{-(1+\delta)d(z', w)} d\mu(z') \tag{129}$$

$$= \frac{2\pi e^{r(1+\delta)}}{\text{Area}(r)} \int_0^\infty e^{-(1+\delta)\tau} \sinh \tau \, d\tau. \tag{130}$$

This integral converges for any $\delta > 0$. □

Proposition 7.2. *If h satisfies (H1), (H2), (H3*), then the kernel $k_\Gamma(z, w)$ is in $C^\infty(\Gamma \backslash \mathbb{H}^2 \times \Gamma \backslash \mathbb{H}^2)$, with $k_\Gamma(z, w) = k_\Gamma(w, z)$.*

Proof. Proposition 5.4 and Lemma 7.1 show that the sum over $k(\gamma z, w)$ converges absolutely and uniformly (take $\delta = \sigma - 1/2 - \epsilon > 0$). The same holds for sums over any derivative of $k(\gamma z, w)$. Hence $k_\Gamma(z, w)$ is in $C^\infty(\mathbb{H}^2 \times \mathbb{H}^2)$. To prove invariance under Γ, note that

$$k_\Gamma(\gamma z, w) = \sum_{\gamma' \in \Gamma} k(\gamma' \gamma z, w) = \sum_{\gamma' \in \Gamma} k(\gamma' z, w) = k_\Gamma(z, w). \tag{131}$$

Thus $k_\Gamma(z, w)$ is a function on $\Gamma \backslash \mathbb{H}^2$ with respect to the first argument. Secondly

$$k_\Gamma(z, w) = \sum_{\gamma' \in \Gamma} k(\gamma' z, w) = \sum_{\gamma' \in \Gamma} k(w, \gamma' z)$$

$$= \sum_{\gamma' \in \Gamma} k(\gamma'^{-1} w, \gamma'^{-1} \gamma' z) = \sum_{\gamma' \in \Gamma} k(\gamma'^{-1} w, z) = k_\Gamma(w, z), \tag{132}$$

which proves symmetry. Both relations imply immediately $k_\Gamma(z, \gamma w) = k_\Gamma(z, w)$. □

Proposition 7.3. *Suppose* $f \in C^2(\Gamma \backslash \mathbb{H}^2)$ *is a solution of* $(\Delta + \rho^2 + \frac{1}{4})f = 0$ *with* $|\operatorname{Im} \rho| \le \sigma$ *and* $|f(z)| \le A e^{\alpha d(z,o)}$, *with constants* $A > 0, 0 \le \alpha < \sigma - \frac{1}{2}$. *Then, for* h *satisfying* (H1), (H2), (H3),

$$Lf = h(\rho)f. \tag{133}$$

Proof. Note that

$$\int_{\mathcal{F}_\Gamma} k_\Gamma(z, w)f(w)d\mu(w) = \int_{\mathbb{H}^2} k(z, w)f(w)d\mu(w) \tag{134}$$

and recall Proposition 5.1. □

8 A trace formula for hyperbolic cylinders

The simplest non-trivial example of a hyperbolic surface is a hyperbolic cylinder. To construct one, fix some $\gamma \in \operatorname{Isom}^+(\mathbb{H}^2)$ of length $\ell = \ell(\gamma) > 0$, and set $\Gamma = \mathfrak{Z}$, where $\mathfrak{Z}$ is the discrete subgroup generated by γ, i.e.,

$$\mathfrak{Z} = \{\gamma^n : n \in \mathbb{Z}\}. \tag{135}$$

We may represent $\mathfrak{Z} \backslash \mathbb{H}^2$ in halfplane coordinates, which are chosen in such a way that

$$\gamma = \begin{pmatrix} e^{\ell/2} & 0 \\ 0 & e^{-\ell/2} \end{pmatrix}. \tag{136}$$

A fundamental domain for the action of γ on $\mathfrak{H}$, $z \mapsto e^\ell z$, is given by

$$\{z \in \mathfrak{H} : 1 \le y < e^\ell\}. \tag{137}$$

It is therefore evident that $\mathfrak{Z} \backslash \mathbb{H}^2$ has infinite volume. A more convenient set of parameters for the cylinder are the coordinates $(s, u) \in \mathbb{R}^2$, with

$$x = ue^s, \qquad y = e^s, \tag{138}$$

where the volume element reads now

$$d\mu = ds\, du. \tag{139}$$

In these coordinates, the action of γ is $(s, u) \mapsto (s + \ell, u)$, and hence a fundamental domain is

$$\mathcal{F}_\mathfrak{Z} = \{(s, u) \in \mathbb{R}^2 : 0 \le s < \ell\}. \tag{140}$$

Note that

$$\cosh d(\gamma^n z, z) = 1 + \frac{|e^{n\ell}z - z|^2}{2e^{n\ell}y^2} = 1 + 2\sinh^2(n\ell/2)(1 + u^2) \tag{141}$$

and hence

$$k_3(z,z) = k(z,z) + \sum_{\substack{n\neq 0}}^{\infty} k(\gamma^n z, z) \tag{142}$$

$$= k(z,z) + 2\sum_{n=1}^{\infty} k(\gamma^n z, z) \tag{143}$$

$$= k(z,z) + 2\sum_{n=1}^{\infty} \Phi\big(4\sinh^2(n\ell/2)(1+u^2)\big). \tag{144}$$

From this we can easily work out a trace formula for the hyperbolic cylinder:

Proposition 8.1. *If h satisfies* (H1), (H2), (H3), *then*

$$\int_{3\backslash\mathbb{H}^2} \big[k_3(z,z) - k(z,z)\big]d\mu = \sum_{n=1}^{\infty} \frac{\ell\, g(n\ell)}{\sinh(n\ell/2)}. \tag{145}$$

Proof. We have

$$\int_{\mathbb{R}}\int_0^{\ell} \Phi\big(4\sinh^2(n\ell/2)(1+u^2)\big)\,ds\,du \tag{146}$$

$$= \frac{\ell}{2\sinh(n\ell/2)}\int_{\mathbb{R}} \Phi\big(4\sinh^2(n\ell/2)+\xi^2\big)\,d\xi \tag{147}$$

and with (100),

$$= \frac{\ell}{2\sinh(n\ell/2)}\, Q\big(4\sinh^2(n\ell/2)\big) \tag{148}$$

$$= \frac{\ell}{2\sinh(n\ell/2)}\, Q\big(2(\cosh(n\ell)-1)\big) \tag{149}$$

which yields the right hand side of (145), cf. (87). □

Proposition 8.2. *If h satisfies* (H1), (H2), (H3), *then*

$$\int_{3\backslash\mathbb{H}^2} \big[k_3(z,z) - k(z,z)\big]d\mu = \int_{\mathbb{R}} h(\rho)\, n_3(\rho)\, d\rho \tag{150}$$

where

$$n_3(\rho) = \frac{\ell}{\pi}\sum_{m=0}^{\infty} \big\{\exp\big[(m+\tfrac{1}{2}+\mathrm{i}\rho)\,\ell\big]-1\big\}^{-1} \tag{151}$$

is a meromorphic function in $\mathbb{C}$ with simple poles at the points

$$\rho_{\nu m} = \frac{2\pi}{\ell}\nu + \mathrm{i}\big(m+\tfrac{1}{2}\big), \qquad \nu\in\mathbb{Z},\quad m=0,1,2,\ldots, \tag{152}$$

and residues $\mathrm{res}_{\rho_{\nu m}} n_3 = 1/(\pi\mathrm{i})$.

Proof. The geometric series expansion of $1/\sinh$ (111) yields

$$\sum_{n=1}^{\infty} \frac{\ell\, g(n\ell)}{\sinh(n\ell/2)} = \frac{\ell}{\pi} \int_{\mathbb{R}} \sum_{m=0}^{\infty} \sum_{n=1}^{\infty} \exp\left[-\left(m + \tfrac{1}{2} + i\rho\right) n\ell\right] h(\rho)\, d\rho \qquad (153)$$

and using again the geometric series, this time for the sum over n,

$$\sum_{n=1}^{\infty} \exp\left[-\left(m + \tfrac{1}{2} + i\rho\right) n\ell\right] = \left\{1 - \exp\left[-\left(m + \tfrac{1}{2} + i\rho\right)\ell\right]\right\}^{-1} - 1 \qquad (154)$$

$$= \left\{\exp\left[\left(m + \tfrac{1}{2} + i\rho\right)\ell\right] - 1\right\}^{-1}. \qquad (155)$$

This proves the formula for $n_3(\rho)$. Near each pole $\rho_{\nu m}$ we have

$$n_3(\rho) \sim \frac{\ell}{\pi} \left\{\exp\left[\left(m + \tfrac{1}{2} + i\rho\right)\ell\right] - 1\right\}^{-1} \qquad (156)$$

$$\sim \frac{1}{\pi i} \frac{1}{\rho - \left[(2\pi/\ell)\nu + i\left(m + \tfrac{1}{2}\right)\right]} \qquad (157)$$

and so $\mathrm{res}_{\rho_{\nu m}} n_3 = 1/(\pi i)$. $\square$

The poles of $n_3(\rho)$ are called the *scattering poles* of the hyperbolic cylinder. A useful formula for n_3 is

$$n_3(\rho) = \frac{1}{2\pi} \sum_{n=1}^{\infty} \frac{\ell\, e^{-i\rho n\ell}}{\sinh(n\ell/2)}, \qquad \mathrm{Im}\,\rho < 1/2, \qquad (158)$$

which follows immediately from the above proof. Furthermore, by shifting the path of integration to $-i\infty$, we have the identity

$$\int_{\mathbb{R}} \frac{n_3(\rho')}{\rho^2 - \rho'^2}\, d\rho' = \frac{\pi i}{\rho} n_3(\rho), \qquad \mathrm{Im}\,\rho < 0. \qquad (159)$$

9 Back to general hyperbolic surfaces

Let us now show that the kernel $k_\Gamma(z, w)$ of a general hyperbolic surface $\Gamma\backslash\mathbb{H}^2$ (with Γ strictly hyperbolic) can be written as a superposition of kernels corresponding to hyperbolic cylinders.

Define the *conjugacy class* of any element $\gamma \in \Gamma$ as

$$\{\gamma\} := \{\tilde{\gamma} \in \Gamma : \tilde{\gamma} = g\gamma g^{-1} \text{ for some } g \in \Gamma\}. \qquad (160)$$

Clearly the length ℓ_γ is the same for all elements in one conjugacy class. The *centralizer* of γ is

$$\mathfrak{z}_\gamma := \{g \in \Gamma : g\gamma = \gamma g\}. \qquad (161)$$

Lemma 9.1. *If $\gamma \in \Gamma$ is hyperbolic, then the centralizer is the infinite cyclic subgroup*

$$\mathfrak{Z}_\gamma = \{\gamma_*^n : n \in \mathbb{Z}\},\tag{162}$$

where $\gamma_* \in \Gamma$ is uniquely determined by γ via the relation $\gamma_*^m = \gamma$ for $m \in \mathbb{N}$ as large as possible.

Proof. $\gamma \in \mathrm{PSL}(2,\mathbb{R}) \simeq \mathrm{Isom}^+(\mathbb{H}^2)$ is conjugate to a diagonal matrix

$$\begin{pmatrix} e^{\ell_\gamma/2} & 0 \\ 0 & e^{-\ell_\gamma/2} \end{pmatrix}\tag{163}$$

with $\ell_\gamma > 0$. The equation

$$\begin{pmatrix} e^{\ell_\gamma/2} & 0 \\ 0 & e^{-\ell_\gamma/2} \end{pmatrix}\begin{pmatrix} a & b \\ c & d \end{pmatrix} = \begin{pmatrix} a & b \\ c & d \end{pmatrix}\begin{pmatrix} e^{\ell_\gamma/2} & 0 \\ 0 & e^{-\ell_\gamma/2} \end{pmatrix}\tag{164}$$

has the only solution $b = c = 0$, $a = d^{-1}$. Hence the centralizer is a diagonal subgroup of (a conjugate of) Γ. Since Γ is discrete, the centralizer must be discrete, which forces it to be cyclic. $\square$

Remark 9.2. If γ is hyperbolic and Γ strictly hyperbolic, then $\{\gamma\} \neq \{\gamma^n\}$ for all $n \neq 1$. Furthermore, the centralizers of γ and γ^n coincide.

The sum in (125) can now be expressed as

$$\sum_{\gamma \in \Gamma} k(\gamma z, w) = k(z,w) + \sum_{\gamma \in H}\sum_{g \in \mathfrak{Z}_\gamma \backslash \Gamma} k(g^{-1}\gamma g z, w)\tag{165}$$

$$= k(z,w) + \sum_{\gamma \in H}\sum_{g \in \mathfrak{Z}_\gamma \backslash \Gamma} k(\gamma g z, g w)\tag{166}$$

where the respective first sums run over a set H of hyperbolic elements, which contains one representative for each conjugacy class $\{\gamma\}$. We may replace this sum by a sum over *primitive* elements. If we denote by $H_* \subset H$ the subset of primitive elements, (166) equals

$$= k(z,w) + \sum_{\gamma \in H}\sum_{g \in \mathfrak{Z}_\gamma \backslash \Gamma} k(\gamma g z, g w)\tag{167}$$

$$= k(z,w) + \sum_{\gamma \in H_*}\sum_{g \in \mathfrak{Z}_\gamma \backslash \Gamma}\sum_{n=1}^{\infty} k(\gamma^n g z, g w)\tag{168}$$

and hence, finally,

$$k_\Gamma(z,w) - k(z,w) = \frac{1}{2}\sum_{\gamma \in H_*}\sum_{g \in \mathfrak{Z}_\gamma \backslash \Gamma}\left\{ k_{\mathfrak{Z}_\gamma}(gz, gw) - k(gz, gw) \right\};\tag{169}$$

recall that

$$k_{\mathfrak{Z}_\gamma}(z,w) = \sum_{n \in \mathbb{Z}} k(\gamma^n z, w).\tag{170}$$

10 The spectrum of a compact surface

It is well known that for any compact Riemannian manifold, $-\Delta$ has positive discrete spectrum, i.e.,

$$0 = \lambda_0 < \lambda_1 \leq \lambda_2 \leq \ldots \to \infty, \tag{171}$$

with corresponding eigenfunctions $\varphi_0 = \text{const}, \varphi_1, \varphi_2, \ldots \in C^\infty(\Gamma\backslash\mathbb{H}^2)$, which satisfy

$$(\Delta + \lambda_j)\varphi_j = 0 \tag{172}$$

and form an orthonormal basis of $L^2(\Gamma\backslash\mathbb{H}^2)$. Furthermore, since Δ is real-symmetric, the φ_j can be chosen to be real-valued. We furthermore define

$$\rho_j = \sqrt{\lambda_j - \tfrac{1}{4}}, \qquad -\pi/2 \leq \arg\rho_j < \pi/2. \tag{173}$$

If $f \in C^2(\Gamma\backslash\mathbb{H}^2)$, the expansion

$$f(z) = \sum_j c_j\varphi_j(z), \qquad c_j = \langle f, \varphi_j \rangle, \tag{174}$$

converges absolutely, uniformly for all $z \in \mathbb{H}^2$. This follows from general spectral theoretic arguments, compare [13, p. 3 and Chapter THREE].

Proposition 10.1. *If h satisfies* (H1), (H2), (H3), *then*

$$L\varphi_j = h(\rho_j)\varphi_j. \tag{175}$$

Proof. Apply Proposition 7.3. Each eigenfunction φ_j is bounded so $\alpha = 0$. Furthermore, by the positivity of $-\Delta$, we have $|\operatorname{Im}\rho| \leq 1/2 < \sigma$. $\square$

Proposition 10.2. *If h satisfies* (H1), (H2), (H3*), *then*

$$k_\Gamma(z,w) = \sum_{j=0}^\infty h(\rho_j)\,\varphi_j(z)\,\overline{\varphi}_j(w), \tag{176}$$

which converges absolutely, uniformly in $z, w \in \mathbb{H}^2$.

Proof. The spectral expansion (174) of $k_\Gamma(z,w)$ as a function of z yields

$$k_\Gamma(z,w) = \sum_{j=0}^\infty c_j\varphi_j(z), \tag{177}$$

with

$$c_j = \int_{\Gamma\backslash\mathbb{H}^2} k_\Gamma(z,w)\overline{\varphi}_j(z)\,d\mu(z) = \overline{[L\varphi_j]}(w) = \overline{h}(\rho_j)\,\overline{\varphi}_j(w) = h(\rho_j)\,\overline{\varphi}_j(w). \tag{178}$$

The proof of uniform convergence follows from standard spectral theoretic arguments [13, Prop. 3.4, p.12]. $\square$

In the case $z = w$, Proposition 10.2 implies immediately the following theorem.

Theorem 10.3 (Selberg's pre-trace formula). *If h satisfies* (H1), (H2), (H3*), *then*

$$\sum_{j=0}^{\infty} h(\rho_j)\,|\varphi_j(z)|^2 = \frac{1}{4\pi} \int_{-\infty}^{\infty} h(\rho) \tanh(\pi\rho)\,\rho\,d\rho + \sum_{\gamma \in \Gamma - \{1\}} k(\gamma z, z). \quad (179)$$

which converges absolutely, uniformly in $z \in \mathbb{H}^2$.

Proof. Use (114) for the $\gamma = 1$ term.   □

Using (169), the pre-trace formula (179) becomes

$$\sum_{j=0}^{\infty} h(\rho_j)\,|\varphi_j(z)|^2 = \frac{1}{4\pi} \int_{-\infty}^{\infty} h(\rho) \tanh(\pi\rho)\,\rho\,d\rho \quad (180)$$

$$+ \frac{1}{2} \sum_{\gamma \in H_*} \sum_{g \in 3_\gamma \backslash \Gamma} \left\{ k_{3_\gamma}(gz, gz) - k(gz, gz) \right\}. \quad (181)$$

Theorem 10.4 (Selberg's trace formula). *If h satisfies* (H1), (H2), (H3*), *then*

$$\sum_{j=0}^{\infty} h(\rho_j) = \frac{\text{Area}(\mathcal{M})}{4\pi} \int_{-\infty}^{\infty} h(\rho) \tanh(\pi\rho)\,\rho\,d\rho + \sum_{\gamma \in H_*} \sum_{n=1}^{\infty} \frac{\ell_\gamma\, g(n\ell_\gamma)}{2 \sinh(n\ell_\gamma/2)}, \quad (182)$$

which converges absolutely.

(We will see in the next section (Corollary 11.2) that the condition (H3*) may in fact be replaced by (H3).)

Proof. We integrate both sides of the pre-trace formula (179) over $\Gamma \backslash \mathbb{H}^2$. By the L^2 normalization of the eigenfunctions φ_j, the left hand side of (179) yields the left hand side of (182). The first term on the right hand side is trivial, and the second term follows from the observation that

$$\sum_{g \in 3_\gamma \backslash \Gamma} \int_{\Gamma \backslash \mathbb{H}^2} f(gz)\,d\mu = \int_{3 \backslash \mathbb{H}^2} f(z)\,d\mu, \quad (183)$$

which allows us to apply Proposition 8.1 to the inner sum in (181).   □

Remark 10.5. The absolute convergence of the sum on the right hand side of (182) only requires (H1), or

$$|g(t)| \ll e^{-\sigma|t|}, \qquad \forall t > 0. \quad (184)$$

108   Jens Marklof

One way of seeing this is is that, since g is only evaluated on the discrete subset (the *length spectrum*)

$$\{\ell_\gamma : \ \gamma \in \Gamma - \{1\}\} \subset \mathbb{R}_{>0}, \tag{185}$$

we may replace g by an even $C^\infty(\mathbb{R})$ function $\tilde{g}$ (for which absolute convergence is granted) so that $g(\ell_\gamma) = \tilde{g}(\ell_\gamma)$ for all γ, and

$$|\tilde{g}(t)| \ll e^{-\sigma|t|}, \qquad \forall t > 0. \tag{186}$$

Remark 10.6. We may interpret the sum over conjugacy classes in the spirit of Propositions 8.1 and 8.2: provided h satisfies (H1), (H2), (H3*), we have

$$\sum_{j=0}^\infty h(\rho_j) = \frac{\text{Area}(\mathcal{M})}{4\pi} \int_{-\infty}^\infty h(\rho) \tanh(\pi\rho)\, \rho\, d\rho + \frac{1}{2} \sum_{\gamma \in H_*} \int_{\mathbb{R}} h(\rho)\, n_{3_\gamma}(\rho)\, d\rho. \tag{187}$$

Alternatively, replace the first term on the right hand side in (182) by (108), then

$$\sum_{j=0}^\infty h(\rho_j) = -\frac{\text{Area}(\mathcal{M})}{2\pi} \int_0^\infty \frac{g'(t)}{\sinh(t/2)}\, dt + \sum_{\gamma \in H_*} \sum_{n=1}^\infty \frac{\ell_\gamma\, g(n\ell_\gamma)}{2\sinh(n\ell_\gamma/2)}. \tag{188}$$

11 The heat kernel and Weyl's law

As a first application of the trace formula, we now prove Weyl's law for the asymptotic number of eigenvalues λ_j below a given λ,

$$N(\lambda) = \#\{j : \lambda_j \le \lambda\}, \tag{189}$$

as $\lambda \to \infty$.

Proposition 11.1 (Weyl's law).

$$N(\lambda) \sim \frac{\text{Area}(\mathcal{M})}{4\pi} \lambda, \qquad \lambda \to \infty. \tag{190}$$

Proof. For any $\beta > 0$, the test function

$$h(\rho) = e^{-\beta\rho^2} \tag{191}$$

is admissible in the trace formula. The Fourier transform is

$$g(t) = \frac{e^{-t^2/(2\beta)}}{\sqrt{4\pi\beta}}, \tag{192}$$

and so (182) reads in this special case (with $\lambda_j = \rho_j^2 + \frac{1}{4}$)

$$\sum_{j=0}^{\infty} e^{-\beta\lambda_j} = \frac{\text{Area}(\mathcal{M})}{4\pi} \int_{-\infty}^{\infty} e^{-\beta(\rho^2+\frac{1}{4})} \tanh(\pi\rho) \, \rho \, d\rho$$

$$+ \frac{e^{-\beta/4}}{\sqrt{4\pi\beta}} \sum_{\gamma\in H_*} \sum_{n=1}^{\infty} \frac{\ell_\gamma \, e^{-(n\ell_\gamma)^2/(2\beta)}}{2\sinh(n\ell_\gamma/2)}. \quad (193)$$

The sum on the right hand side clearly tends to zero in the limit $\beta \to 0$. Since $\tanh(\pi\rho) = 1 + O(e^{-2\pi|\rho|})$ for all $\rho \in \mathbb{R}$, we obtain

$$\sum_{j=0}^{\infty} e^{-\beta\lambda_j} = \frac{\text{Area}(\mathcal{M})}{4\pi\beta} + O(1), \qquad \beta \to 0. \quad (194)$$

The Proposition now follows from a classical Tauberian theorem [18]. $\square$

The sum $\sum_{j=0}^{\infty} e^{-\beta\lambda_j}$ represents of course the trace of the *heat kernel* $e^{\beta\Delta}$.

Corollary 11.2. *The condition (H3*) in Theorem 10.4 and Remark 10.6 can be replaced by (H3).*

Proof. Weyl's law implies that, for any $\delta > 0$

$$\sum_{j=0}^{\infty} (1+\lambda_j)^{-1-\delta/2} < \infty, \qquad \text{i.e.,} \qquad \sum_{j=0}^{\infty} (1+\text{Re}\,\rho_j)^{-2-\delta} < \infty. \quad (195)$$

To prove this claim, note that

$$\sum_{j=0}^{\infty} (1+\lambda_j)^{-1-\delta/2} = \int_0^{\infty} (1+x)^{-1-\delta/2} dN(x) \quad (196)$$

$$= (1+x)^{-1-\delta/2} N(x) \Big|_{x=0}^{\infty} \quad (197)$$

$$+ \left(1 + \frac{\delta}{2}\right) \int_0^{\infty} (1+x)^{-2-\delta/2} N(x) \, dx \quad (198)$$

(use integration by parts) which is finite since $N(x)$ grows linearly with x.

If h satisfies (H1), (H2), (H3), then the function $h_\epsilon(\rho) = h(\rho)e^{-\epsilon\rho^2}$ clearly satisfies (H1), (H2), (H3*) for any $\epsilon > 0$, with the additional uniform bound

$$|h_\epsilon(\rho)| \ll (1+|\text{Re}\,\rho|)^{-2-\delta}. \quad (199)$$

where the implied constant is independent of ϵ. By repeating the calculation that leads to (75), we obtain the following estimate for the Fourier transform of h_ϵ,

$$|g_\epsilon(t)| \ll e^{-\sigma|t|}, \quad (200)$$

where the implied constant is again independent of ϵ. Theorem 10.4 yields

$$\sum_{j=0}^{\infty} h_\epsilon(\rho_j) = \frac{\text{Area}(\mathcal{M})}{4\pi} \int_{-\infty}^{\infty} h_\epsilon(\rho) \tanh(\pi\rho)\, \rho\, d\rho + \sum_{\gamma \in H_*} \sum_{n=1}^{\infty} \frac{\ell_\gamma\, g_\epsilon(n\ell_\gamma)}{2\sinh(n\ell_\gamma/2)}.$$
(201)

Due to the above ϵ-uniform bounds, both sides of the trace formula converge absolutely, uniformly for all $\epsilon > 0$. We may therefore take the limit $\epsilon \to 0$ inside the sums and integral.

12 The density of closed geodesics

In the previous section we have used the trace formula to obtain Weyl's law on the distribution of eigenvalues λ_j. By using the appropriate test function, one can similarly work out the asymptotic number of primitive closed geodesic with lengths $\ell_\gamma \leq L$,

$$\Pi(L) = \#\{\gamma \in H_* : \ell_\gamma \leq L\}. \tag{202}$$

In view of Remark 10.5 we know that, for any $\delta > 0$,

$$\sum_{\gamma \in H_*} \ell_\gamma e^{-\ell_\gamma(1+\delta)} < \infty \tag{203}$$

which implies that, for any $\epsilon > 0$,

$$\Pi(L) \ll_\epsilon e^{L(1+\epsilon)}. \tag{204}$$

There is in fact an a priori geometric argument (cf. [13]) which yields this bounds with $\epsilon = 0$, but the rough estimate (204) is sufficient for the following argument.

Let us consider the density of closed geodesics in the interval $[a+L, b+L]$ where a and b are fixed and $L \to \infty$. To avoid technicalities, we will here only use smoothed counting functions

$$\sum_{\gamma \in H_*} \psi_L(\ell_\gamma) = \int_0^\infty \psi_L(t)\, d\Pi(t) \tag{205}$$

where $\psi_L(t) = \psi(t - L)$ and $\psi \in C_0^\infty(\mathbb{R})$. One may think of ψ as a smoothed characteristic function of $[a,b]$. Stronger results for true counting functions require a detailed analysis of Selberg's zeta function, which will be introduced in Section 14.

Let $\rho_0, \ldots, \rho_M$ be those ρ_j with $\text{Im}\,\rho_j < 0$. The corresponding eigenvalues $\lambda_0, \ldots, \lambda_M$ are referred to as the *small* eigenvalues.

Proposition 12.1. Let $\psi \in C_0^\infty(\mathbb{R})$. Then, for $L > 1$,

$$\int_0^\infty \psi_L(t)\, d\Pi(t) = \int_0^\infty \psi_L(t)\, d\tilde{\Pi}(t) + O\left(\frac{e^{L/2}}{L}\right), \tag{206}$$

where

$$d\widetilde{\Pi}(t) = \sum_{j=0}^{M} \frac{e^{(\frac{1}{2}+i\rho_j)t}}{t} \, dt. \tag{207}$$

Proof. The plan is to apply the trace formula with

$$g(t) = \frac{2\sinh(t/2)}{t} \left[\psi(t-L) + \psi(-t-L)\right] \tag{208}$$

which is even and, for L large enough, in $C_0^\infty(\mathbb{R})$. Hence its Fourier transform,

$$h(\rho) = \int_{\mathbb{R}} \frac{1}{t} \left(e^{(\frac{1}{2}+i\rho)t} - e^{(-\frac{1}{2}+i\rho)t}\right) \left[\psi(t-L) + \psi(-t-L)\right] dt \tag{209}$$

$$= \int_{\mathbb{R}} \frac{1}{t} \left(e^{(\frac{1}{2}+i\rho)t} - e^{(-\frac{1}{2}+i\rho)t} + e^{(-\frac{1}{2}-i\rho)t} - e^{(\frac{1}{2}-i\rho)t}\right) \psi(t-L)\, dt, \tag{210}$$

satisfies (H1), (H2), (H3). Let us begin with the integral

$$\int_{\mathbb{R}} \frac{1}{t} e^{(\frac{1}{2}+i\rho)t} \psi(t-L)\, dt = e^{(\frac{1}{2}+i\rho)L} \int_{\mathbb{R}} \frac{1}{t+L} e^{(\frac{1}{2}+i\rho)t} \psi(t)\, dt. \tag{211}$$

Repeated integration by parts yields the upper bound

$$\left| \int_{\mathbb{R}} \frac{1}{t+L} e^{(\frac{1}{2}+i\rho)t} \psi(t)\, dt \right| \ll_N \frac{1}{(1+|\rho|)^N} \int_{\mathrm{supp}\,\psi} \frac{1}{t+L} e^t\, dt \tag{212}$$

$$\ll_N \frac{1}{L(1+|\rho|)^N}. \tag{213}$$

So

$$\int_{\mathbb{R}} \frac{1}{t} e^{(\frac{1}{2}+i\rho)t} \psi(t-L)\, dt \ll_N \frac{e^{(\frac{1}{2}-\mathrm{Im}\,\rho)L}}{L(1+|\rho|)^N}. \tag{214}$$

This bound is useful for $\mathrm{Im}\,\rho = 0$. The other corresponding integrals can be estimated in a similar way, to obtain the bounds (assume $-1/2 \le \mathrm{Im}\,\rho \le 0$)

$$\int_{\mathbb{R}} \frac{1}{t} e^{(-\frac{1}{2}+i\rho)t} \psi(t-L)\, dt \ll_N \frac{1}{L(1+|\rho|)^N}, \tag{215}$$

$$\int_{\mathbb{R}} \frac{1}{t} e^{(-\frac{1}{2}-i\rho)t} \psi(t-L)\, dt \ll_N \frac{1}{L(1+|\rho|)^N}, \tag{216}$$

and

$$\int_{\mathbb{R}} \frac{1}{t} e^{(\frac{1}{2}-i\rho)t} \psi(t-L)\, dt \ll_N \frac{e^{L/2}}{L(1+|\rho|)^N}. \tag{217}$$

Therefore, using the above bound with $N = 3$, say, yields

$$\sum_{j=0}^{M} h(\rho_j) = \int_0^\infty \psi_L(t)\, d\widetilde{\Pi}(t) + O\left(\frac{e^{L/2}}{L}\right) \tag{218}$$

and

$$\sum_{j=M+1}^{\infty} h(\rho_j) - \frac{\text{Area}(\mathcal{M})}{4\pi} \int_{-\infty}^{\infty} h(\rho)\tanh(\pi\rho)\,\rho\,d\rho = O_N\left(\frac{e^{L/2}}{L}\right). \quad (219)$$

The sum of the above terms equals, by the trace formula, the expression

$$\sum_{\gamma \in H_*} \sum_{n=1}^{\infty} \frac{1}{n}\psi_L(n\ell_\gamma). \quad (220)$$

The a priori bound (204) tells us that terms with $n \geq 2$ (corresponding to repetitions of primitive closed geodesics) are of lower order. To be precise,

$$\sum_{\gamma \in H_*} \sum_{n=2}^{\infty} \frac{1}{n}\psi_L(n\ell_\gamma) \ll_\epsilon \sum_{2 \leq n \leq (L+b)/\ell_{\min}} \frac{1}{n}e^{(L+b)(1+\epsilon)/n} \ll_\epsilon e^{L(1+\epsilon)/2}, \quad (221)$$

where we assume that ψ_L is supported in $[a+L, b+L]$, and $\ell_{\min}$ is the length of the shortest primitive closed geodesic. Therefore

$$\sum_{\gamma \in H_*} \psi_L(\ell_\gamma) = \int_0^{\infty} \psi_L(t)\,d\widetilde{\Pi}(t) + O\left(e^{L(1+\epsilon)/2}\right). \quad (222)$$

The leading order term as $L \to \infty$ is

$$\sum_{\gamma \in H_*} \psi_L(\ell_\gamma) \sim \int_0^{\infty} \psi_L(t)\frac{e^t}{t}\,dt \ll \frac{e^{L+b}}{L+b} \quad (223)$$

which leads to the improved upper bound for the sum involving repetitions,

$$\sum_{\gamma \in H_*} \sum_{n=2}^{\infty} \frac{1}{n}\psi_L(n\ell_\gamma) \ll \sum_{2 \leq n \leq (L+b)/\ell_{\min}} \frac{e^{(L+b)/n}}{L+b} \ll \frac{e^{L/2}}{L}, \quad (224)$$

and hence leads to the desired improved error estimate in (222). $\square$

13 Trace of the resolvent

The trace of the resolvent $R(\lambda) = (\Delta + \lambda)^{-1}$ is formally

$$\text{Tr}\, R(\lambda) = \sum_{j=0}^{\infty} (\lambda - \lambda_j)^{-1} = \sum_{j=0}^{\infty} h(\rho_j), \qquad h(\rho') = (\rho^2 - \rho'^2)^{-1}, \quad (225)$$

where $\rho = \sqrt{\lambda - \frac{1}{4}}$ as usual. The test function h does not, however, respect condition (H3). To overcome this difficulty, we define the regularized resolvent

$$\widetilde{R}(\lambda) = (\Delta + \lambda)^{-1} - (\Delta + \lambda_*)^{-1} \tag{226}$$

for some fixed λ_*. The corresponding test function is

$$h(\rho') = (\rho^2 - \rho'^2)^{-1} - (\rho_*^2 - \rho'^2)^{-1}, \tag{227}$$

which clearly satisfies (H3), since

$$h(\rho') = \frac{\rho_*^2 - \rho^2}{(\rho^2 - \rho'^2)(\rho_*^2 - \rho'^2)} = O(\rho'^{-4}). \tag{228}$$

We have already encountered the kernel of the regularized resolvent,

$$k(z, w) = G_\rho(z, w) - G_{\rho_*}(z, w), \tag{229}$$

in Section 6. The trace of the regularized resolvent is thus

$$\operatorname{Tr} \widetilde{R}(\rho) = \sum_{j=0}^{\infty} \left[(\rho^2 - \rho_j{}^2)^{-1} - (\rho_*^2 - \rho_j{}^2)^{-1} \right] \tag{230}$$

which, for any fixed $\rho^* \notin \{\pm\rho_j\}$, is a meromorphic function in $\mathbb{C}$ with simple poles at $\rho = \pm\rho_j$. h is analytic in the strip $|\operatorname{Im}\rho'| \le \sigma$ provided $\sigma < |\operatorname{Im}\rho| < |\operatorname{Im}\rho^*|$ where $\sigma > 1/2$.

The trace formula (187) implies therefore (use formula (118) for the first term on the right hand side, and (159) for the second)

$$\operatorname{Tr} \widetilde{R}(\rho) = -\frac{\operatorname{Area}(\mathcal{M})}{2\pi\mathrm{i}} \sum_{l=0}^{\infty} \left[\frac{1}{\rho - \mathrm{i}(l + \frac{1}{2})} - \frac{1}{\rho_* - \mathrm{i}(l + \frac{1}{2})} \right]$$
$$+ \frac{\pi\mathrm{i}}{2\rho} \sum_{\gamma \in H_*} n_{3\gamma}(\rho) + C(\rho_*). \tag{231}$$

where

$$C(\rho_*) = -\frac{\pi\mathrm{i}}{2\rho_*} \sum_{\gamma \in H_*} n_{3\gamma}(\rho_*) \tag{232}$$

converges absolutely, cf. Remark 10.5.

Let us rewrite this formula as

$$\frac{1}{2\rho} \sum_{\gamma \in H_*} n_{3\gamma}(\rho) = \frac{1}{\pi\mathrm{i}} \sum_{j=0}^{\infty} \left[\frac{1}{\rho^2 - \rho_j{}^2} - \frac{1}{\rho_*^2 - \rho_j{}^2} \right]$$
$$- \frac{\operatorname{Area}(\mathcal{M})}{2\pi^2} \sum_{l=0}^{\infty} \left[\frac{1}{\rho - \mathrm{i}(l + \frac{1}{2})} - \frac{1}{\rho_* - \mathrm{i}(l + \frac{1}{2})} \right] - \frac{1}{\pi\mathrm{i}} C(\rho_*). \tag{233}$$

All quantities on the right hand side are meromorphic for all $\rho \in \mathbb{C}$, for every fixed $\rho_* \in \mathbb{C}$ away from the singularities (this is guaranteed for $|\operatorname{Im}\rho_*| > 1/2$). Therefore (233) provides a meromorphic continuation of

$$n_\Gamma(\rho) := \sum_{\gamma \in H_*} n_{3_\gamma}(\rho) \tag{234}$$

to the whole complex plane.

Proposition 13.1. *The function $n_\Gamma(\rho)$ has a meromorphic continuation to the whole complex plane, with*

(i) *simple poles at $\rho = \rho_0, \pm\rho_1, \pm\rho_2, \ldots$ with residue $\operatorname{res}_{\pm\rho_0} n_\Gamma = 1/(\pi i)$, and*

$$\operatorname{res}_{\pm\rho_j} n_\Gamma = \begin{cases} 2\mu_j/(\pi i) & \text{if } \rho_j = 0, \\ \pm\mu_j/(\pi i) & \text{if } \rho_j \neq 0, \end{cases} \tag{235}$$

where μ_j is the multiplicity of ρ_j.

(ii) *simple poles at $\rho = i(l + \frac{1}{2})$ with residue*

$$\operatorname{res}_{i(l+\frac{1}{2})} n_\Gamma = \begin{cases} \frac{\operatorname{Area}(\mathcal{M})}{2\pi^2 i} (2l+1) + \frac{1}{\pi i} & \text{if } l = 0, \\ \frac{\operatorname{Area}(\mathcal{M})}{2\pi^2 i} (2l+1) & \text{if } l > 0, \end{cases} \tag{236}$$

(iii) *the functional relation*

$$n_\Gamma(\rho) + n_\Gamma(-\rho) = -\frac{\operatorname{Area}(\mathcal{M})}{\pi} \rho \tanh(\pi\rho). \tag{237}$$

Proof. (i) and (ii) are clear. (iii) follows from the identity (112). □

14 Selberg's zeta function

Selberg's zeta function is defined by

$$Z(s) = \prod_{\gamma \in H_*} \prod_{m=0}^{\infty} \left(1 - e^{-\ell_\gamma(s+m)} \right), \tag{238}$$

which converges absolutely for $\operatorname{Re} s > 1$; this will become clear below, cf. (241) and (242). Each factor

$$\prod_{m=0}^{\infty} \left(1 - e^{-\ell_\gamma(s+m)} \right) \tag{239}$$

converges for all $s \in \mathbb{C}$, with zeros at

$$s = s_{\nu m} = -m + i(2\pi/\ell_\gamma)\nu, \qquad \nu \in \mathbb{Z}, \quad m = 0, 1, 2, \ldots. \tag{240}$$

Note that $s_{\nu m} = \frac{1}{2} + i\rho_{\nu m}$ where $\rho_{\nu m}$ are the scattering poles for the hyperbolic cylinder $3_\gamma \backslash \mathbb{H}^2$. What is more,

$$\frac{d}{ds} \log \prod_{m=0}^{\infty} \left(1 - e^{-\ell_\gamma(s+m)} \right) = -\ell_\gamma \sum_{m=0}^{\infty} \left(1 - e^{\ell_\gamma(s+m)} \right)^{-1} = \pi n_{3_\gamma}(\rho), \tag{241}$$

with $s = \frac{1}{2} + i\rho$, and thus

$$\frac{Z'}{Z}(s) = \pi n_\Gamma(\rho), \qquad \text{Re}\, s > 1 \quad \text{(i.e., Im}\, \rho < -1/2). \tag{242}$$

Recall that the genus is related to the area of $\mathcal{M}$ by $\text{Area}(\mathcal{M}) = 4\pi(g-1)$.

Theorem 14.1. *The Selberg zeta function can be analytically continued to an entire function $Z(s)$ whose zeros are characterized as follows. (We divide the set of zeros into two classes,* trivial *and* non-trivial*).*

(i) *The non-trivial zeros of $Z(s)$ are located at $s = 1$ and $s = \frac{1}{2} \pm i\rho_j$ ($j = 1, 2, 3, \ldots$) with multiplicity*

$$\begin{cases} 2\mu_j & \text{if } \rho_j = 0 \\ \mu_j & \text{if } \rho_j \neq 0. \end{cases} \tag{243}$$

The zero at $s = 1$ (corresponding to $j = 0$) has multiplicity 1.
(ii) *The trivial zeros are located at at $s = -l$, $l = 0, 1, 2, \ldots$ and have multiplicity $2g - 1$ for $l = 0$ and $2(g-1)(2l+1)$ for $l > 0$.*

Furthermore $Z(s)$ satisfies the functional equation

$$Z(s) = Z(1-s) \exp\left[4\pi(g-1) \int_0^{s-\frac{1}{2}} v \tan(\pi v)\, dv \right]. \tag{244}$$

Proof. Equation (233) yields

$$\frac{1}{2s-1} \frac{Z'}{Z}(s) = \sum_{j=0}^{\infty} \left[\frac{1}{(s-\frac{1}{2})^2 + \rho_j{}^2} + \frac{1}{\rho_*^2 - \rho_j{}^2} \right]$$

$$- 2(g-1) \sum_{l=0}^{\infty} \left[\frac{1}{s+l} - \frac{1}{l + \frac{1}{2} + i\rho_*} \right] + C(\rho_*). \tag{245}$$

Note that

$$\frac{2s-1}{(s-\frac{1}{2})^2 + \rho_j{}^2} = \frac{1}{s - (\frac{1}{2} + i\rho_j)} + \frac{1}{s - (\frac{1}{2} - i\rho_j)}, \tag{246}$$

hence the corresponding residue is 1. Furthermore

$$-2(g-1) \frac{2s-1}{s+l} \tag{247}$$

has residue $2(g-1)(2l+1)$ at $s = -l$. Statements (i) and (ii) are now evident. The functional relation follows from (237), which can be written as

$$\frac{Z'}{Z}(s) + \frac{Z'}{Z}(1-s) = 4\pi(g-1)(s-\tfrac{1}{2}) \tan[\pi(s-\tfrac{1}{2})]. \tag{248}$$

Integrating this yields

$$\log Z(s) - \log Z(1-s) = 4\pi(g-1) \int_0^{s-1/2} v \tan(\pi v)\, dv + c, \qquad (249)$$

that is

$$Z(s)/Z(1-s) = \exp\left[4\pi(g-1) \int_0^{s-1/2} v \tan(\pi v)\, dv + c \right], \qquad (250)$$

The constant of integration c is determined by setting $s = 1/2$. Notice that the exponential is independent of the path of integration. □

One important application of the zeta function is a precise asymptotics for the number of primitive closed geodesics of length less than L, $L \to \infty$; we have [13]

$$\Pi(L) = \int_1^L d\widetilde{\Pi}(t) + O\left(\frac{e^{\frac{3}{4}L}}{\sqrt{L}} \right). \qquad (251)$$

The error estimate is worse than in Proposition 12.1, since we have replaced the smooth test functions by a characteristic function. The asymptotic relation (251) is often referred to as *Prime Geodesic Theorem*, due to its similarity with the *Prime Number Theorem*. The proof of (251) in fact follows the same strategy as in the Prime Number Theorem, where the Selberg zeta function plays the role of Riemann's zeta function.

15 Suggestions for exercises and further reading

1. **Poisson summation.**

   a) The Poisson summation formula (1) reads in higher dimension d

   $$\sum_{m \in \mathbb{Z}^d} f(m) = \sum_{n \in \mathbb{Z}^d} \widehat{f}(n) \tag{252}$$

   with the Fourier transform

   $$\widehat{f}(\tau) = \int_{\mathbb{R}^d} f(\rho)\, e^{2\pi i \tau \cdot \rho} d\rho. \tag{253}$$

   Prove (252) for a suitable class of test functions f.

   b) Show that (252) can be written in the form

   $$\sum_{m \in L^*} f(m) = \mathrm{Vol}(L \backslash \mathbb{R}^d) \sum_{n \in L} \widehat{f}(n) \tag{254}$$

   where L is any lattice in $\mathbb{R}^d$ and L^* its dual lattice.

   c) Any flat torus can be represented as the quotient $L \backslash \mathbb{R}^d$, where the Riemannian metric is the usual euclidean metric. Show that the normalized eigenfunctions of the Laplacian are

   $$\varphi_m(x) = \mathrm{Vol}(L \backslash \mathbb{R}^d)^{-1/2} e^{2\pi i m \cdot x} \tag{255}$$

   for every $m \in L^*$ and work out the corresponding eigenvalues λ_j.

   d) Use (254) to derive a trace formula for

   $$\sum_{j=0}^{\infty} h(\rho_j)$$

   where $\rho_j = \sqrt{\lambda_j}$ (this formula is the famous Hardy-Voronoi formula, cf. [12]).

2. **Semiclassics.**

   a) Show that for $\rho \to \infty$

   $$G_\rho(z, w) = -\frac{1}{2\pi} \sqrt{\frac{\pi}{2\rho \sinh \tau}}\, e^{-i\rho\tau - i\pi/4} + O(\rho^{-1}), \tag{256}$$

   for all fixed $\tau = d(z, w) > 0$. Hint: divide the integral (58) into the ranges $[\tau, 2\tau)$ and $[2\tau, \infty)$. The second range is easily controlled. For the first range, use the Taylor expansion for $\cosh t$ at $t = \tau$ to expand the denominator of the integrand. Relation (256) can also be obtained from the connection of the Legendre function with the hypergeometric series $F(a, b; c; z)$,

   $$Q_\nu(\cosh \tau) = \sqrt{\pi}\frac{\Gamma(\nu+1)}{\Gamma(\nu+\frac{3}{2})}\frac{e^{-(\nu+1)\tau}}{(1-e^{-2\tau})^{1/2}} F\left(\frac{1}{2}, \frac{1}{2}; \nu+\frac{3}{2}; \frac{1}{1-e^{2\tau}}\right). \tag{257}$$

b) Show that (256) is consistent with [11, eq. (41)]. Hint: use the (s, u)-coordinates defined in (138).

c) Compare the Gutzwiller trace formula [11], [7], with the Selberg trace formula. (Analogues of the *ghost of the sphere* (Section 6) for more general systems are discussed in [2].)

3. **The Riemann-Weil explicit formula.**

a) Compare the Selberg trace formula with the Riemann-Weil explicit formula [12, eq. (6.7)], by identifying Riemann zeros with the square-root $\sqrt{\lambda_j - \frac{1}{4}}$ of eigenvalues λ_j of the Laplacian, and logs of prime numbers with lengths of closed geodesics. See [3] for more on this.

b) What is the analogue of the ghost of the sphere?

4. **Further reading.** In this course we have discussed Selberg's trace formula in the simplest possible set-up, for the spectrum of the Laplacian on a compact surface. The full theory, which is only outlined in Selberg's original paper [16], is developed in great detail in Hejhal's lecture notes [13], [14], where the following generalizations are discussed.

a) The discrete subgroup Γ may contain elliptic elements, which leads to conical singularities on the surface, and reflections (i.e., orientation reversing isometries). Technically more challenging is the treatment of groups Γ which contain parabolic elements. In this case $\Gamma \backslash \mathbb{H}^2$ is no longer compact, and the spectrum has a continuous part, cf. [14].

b) Suppose the Laplacian acts on vector valued functions $f : \mathbb{H}^2 \to \mathbb{C}^N$ which are not invariant under the action of T_γ, but satisfy

$$T_\gamma f = \chi(\gamma)f \quad \forall \gamma \in \Gamma \tag{258}$$

for some fixed unitary representation $\chi : \Gamma \to \mathrm{U}(N)$. The physical interpretation of this set-up, in the case $N = 1$, is that Aharonov-Bohm flux lines thread the holes of the surface.

c) The Laplacian may act on automorphic forms of weight α, which corresponds, in physical terms, to the Hamiltonian for a constant magnetic field B perpendicular to the surface, where the strength of B is proportional to α. See Comtet's article [8] for details.

I also recommend Balazs and Voros' Physics Reports article [1] and the books by Buser [4], Iwaniec [15] and Terras [17], which give a beautiful introduction to the theory. Readers interested in hyperbolic three-space will enjoy the book by Elstrodt, Grunewald and Mennicke [9]. Gelfand, Graev and Pyatetskii-Shapiro [10] take a representation-theoretic view on Selberg's trace formula.

Acknowledgement. The author has been supported by an EPSRC Advanced Research Fellowship and the EC Research Training Network (Mathematical Aspects of Quantum Chaos) HPRN-CT-2000-00103.

References

1. N.L. Balazs and A. Voros, Chaos on the pseudosphere, *Phys. Rep.* **143** (1986) 109-240.
2. M.V. Berry and C.J. Howls, High orders of the Weyl expansion for quantum billiards: resurgence of periodic orbits, and the Stokes phenomenon, *Proc. Roy. Soc. London Ser.* A **447** (1994) 527-555.
3. M.V. Berry and J.P. Keating, The Riemann zeros and eigenvalue asymptotics, *SIAM Rev.* **41** (1999) 236-266.
4. P. Buser, *Geometry and spectra of compact Riemann surfaces*, *Progr. Math.* **106**, Birkhäuser Boston, Inc., Boston, MA, 1992.
5. A. Aigon-Dupuy, P. Buser and K.-D. Semmler, Hyperbolic geometry, *this volume*.
6. P. Cartier and A. Voros, Une nouvelle interprétation de la formule des traces de Selberg, *The Grothendieck Festschrift*, Vol. II, 1-67, *Progr. Math.* **87**, Birkhäuser Boston, Boston, MA, 1990.
7. M. Combescure, J. Ralston and D. Robert, A proof of the Gutzwiller semiclassical trace formula using coherent states decomposition, *Comm. Math. Phys.* **202** (1999) 463-480.
8. A. Comtet, On the Landau levels on the hyperbolic plane, *Ann. Physics* **173** (1987) 185-209.
9. J. Elstrodt, F. Grunewald and J. Mennicke, *Groups acting on hyperbolic space*, Springer-Verlag, Berlin, 1998.
10. I.M. Gelfand, M.I. Graev and I.I. Pyatetskii-Shapiro, *Representation theory and automorphic functions*, Academic Press, Inc., Boston, MA, 1990 (Reprint of the 1969 edition).
11. M.C. Gutzwiller, The semi-classical quantization of chaotic Hamiltonian systems. *Chaos et physique quantique* (Les Houches, 1989), 201-250, North-Holland, Amsterdam, 1991.
12. D.A. Hejhal, The Selberg trace formula and the Riemann zeta function, *Duke Math. J.* **43** (1976) 441-482.
13. D.A. Hejhal, *The Selberg trace formula for* PSL(2, ℝ), Vol. 1. Lecture Notes in Mathematics **548**, Springer-Verlag, Berlin-New York, 1976.
14. D.A. Hejhal, *The Selberg trace formula for* PSL(2, ℝ), Vol. 2. Lecture Notes in Mathematics **1001**, Springer-Verlag, Berlin-New York, 1983.
15. H. Iwaniec, *Spectral methods of automorphic forms*, 2nd ed., *Graduate Studies in Mathematics* **53**, AMS, Providence, RI; Rev. Mat. Iberoamericana, Madrid, 2002.
16. A. Selberg, Harmonic analysis and discontinuous groups in weakly symmetric Riemannian spaces with applications to Dirichlet series, *J. Indian Math. Soc.* **20** (1956) 47-87.
17. A. Terras, *Harmonic analysis on symmetric spaces and applications* I, Springer-Verlag, New York, 1985.
18. D.V. Widder, *The Laplace Transform*, Princeton Math. Series **6**, Princeton University Press, Princeton, 1941.

III. Semiclassical Approach to Spectral Correlation Functions

Martin Sieber

School of Mathematics, University of Bristol, Bristol BS8 1TW, UK

1 Introduction

One motivation for deriving semiclassical methods for chaotic systems like the Gutzwiller trace formula was to find a substitute for the EBK quantisation rules which provide semiclassical approximations for energy levels and wave functions in integrable systems [1]. Although the Gutzwiller trace formula allows to determine energy levels semiclassically from a knowledge of classical periodic orbits, in practise only a relatively small number of the lowest energies can be determined in this way due to the exponential proliferation of the number of long periodic orbits in chaotic systems.

However, the trace formula is very powerful in applications in which one is interested not in individual levels, but in statistical properties of the spectrum in the semiclassical regime. These spectral statistics are in the centre of interest in quantum chaos, because they are found to be universal and to provide a clear signature for chaos in the underlying classical system. All numerical evidence points to an agreement of spectral statistics in generic chaotic systems with those of random matrix theory (RMT) [2]. The aim of the semiclassical method is to find a theoretical explanation and, possibly, a proof for this connection between quantum chaos and random matrix theory. Similarly as on the quantum side, what is needed for this task is not a knowledge of individual periodic orbits, but of statistical properties of long periodic orbits.

The first steps in this direction were made in the seminal works of Hannay and Ozorio de Almeida, and Berry. Using the mean distribution of long periodic orbits [3] Berry showed that two-point statistics agree with RMT in the regime of long-range correlations in the energy spectrum [4]. To go beyond Berry's so-called diagonal approximation requires the evaluation of correlations between periodic orbits. Vice versa, the conjectured connection between quantum chaos and random matrix theory can be used to predict the existence of universal correlations between very long periodic orbits [5, 6]. So in some sense quantum mechanics reveals fundamental universal properties of periodic orbits that were not known before.

In recent years there has been a rapid development of methods to derive these periodic orbit correlations on the classical side. Proving that these correlations do indeed exist would show the validity of the random matrix hypothesis. The purpose of this short lecture series is to give an introduction into the evaluation of correlations between periodic orbits. This is done by discussing in detail the simplest example of periodic orbit correlations which involves orbits with two loops. This example reveals the basic mechanism that is also behind more complicated correlations.

In accordance with the topic of this school, the calculations are done for the motion on compact Riemann surfaces of constant negative curvature. For these systems the derivation is more transparent and can be done with more rigour than for general chaotic systems. Some of the special features that facilitate the derivation in these systems are the existence of the exact Selberg trace formula, the non-existence of conjugate points, the uniform hyperbolicity, and the tessellation properties of representations of the surfaces on the hyperbolic plane.

The course is structured as follows. In the second section the semiclassical approximation of the form factor is derived and its properties are discussed. The diagonal approximation is evaluated and the two-loop orbits which contribute to the leading off-diagonal approximation are introduced. The third section contains an introduction to phase space methods for Riemann surfaces and an evaluation of transition probabilities which are used to calculate the number of two-loop pairs. In section 4 the contribution of the two-loop orbit pairs to the form factor are evaluated. Finally, section 5 contains a discussion and an overview over further developments.

2 The spectral form factor

The spectral form factor is a two-point statistics that is convenient for semiclassical evaluations, because its argument is directly related to the time along orbits. In general chaotic systems the semiclassical approximation to the form factor is derived by applying Gutzwiller's trace formula. In the case of Riemann surfaces the existence of the exact Selberg trace formula gives the opportunity to understand the effect of the approximations involved in arriving at the semiclassical form factor.

The spectral form factor is defined as the Fourier transform of the two-point correlation function of the density of states

$$K(\tau) = \left\langle \int_{-\infty}^{\infty} \frac{d\eta}{\bar{d}(E)} \, \langle d_{\mathrm{osc}}(E + \eta/2) \, d_{\mathrm{osc}}(E - \eta/2) \rangle_E \, e^{2\pi i \eta \, \bar{d}(E) \tau} \right\rangle_\tau . \tag{1}$$

Here $\bar{d}(E)$ and $d_{\mathrm{osc}}(E)$ are the mean and oscillatory part of the density of states, $d(E)$, respectively,

$$d(E) = \sum_n \delta(E - E_n) = \bar{d}(E) + d_{\mathrm{osc}}(E) . \tag{2}$$

The mean density of states is asymptotically given by Weyl's law $\bar{d}(E) \sim A/(4\pi)$ as $E \to \infty$, where A is the area of the Riemann surface. The energy average in (1) is performed over an energy window ΔE which is classically small (i.e. small in comparison to E) and large in comparison to the mean spacing between levels $1/\bar{d}(E)$. In addition, a local average is performed over a small window $\Delta\tau$ in order to suppress strong fluctuations that would occur otherwise. The dependence of the form factor $K(\tau)$ on E (and the choice of the smoothing) is omitted in its argument.

The random matrix hypothesis predicts that the form factor of generic chaotic systems with time-reversal symmetry approaches the form factor of the Gaussian Orthogonal Ensemble (GOE) of random matrices in the semiclassical limit, $\lim_{E\to\infty} K(\tau) = K^{\mathrm{GOE}}(\tau)$, where [7]

$$K^{\mathrm{GOE}}(\tau) = \begin{cases} 2\tau - \tau \log(1 + 2\tau) & \text{if } \tau < 1 \\ 2 - \tau \log \frac{2\tau+1}{2\tau-1} & \text{if } \tau > 1 . \end{cases} \tag{3}$$

Requirements for this assumption to hold are that $\bar{d}(E)\,\Delta E \to \infty$, $\Delta E/E \to 0$, and $\Delta\tau \to 0$ in this limit.

One difficulty is to define what is meant by "generic". In the following we require that the multiplicity of lengths of periodic orbits is typically two. This means that there are typically only two periodic orbits whose lengths are identical, an orbit and its time-reverse. This condition excludes systems with symmetries, and it excludes arithmetic systems for which the mean multiplicity increases exponentially with their lengths. These arithmetic systems have spectral statistics that are described by a Poisson process, and not by one of the random matrix ensembles [8–10]. It is, in fact, not possible to require the multiplicity to be exactly two, because it is known that for Riemann surfaces with constant negative curvature the multiplicity of lengths is unbounded [11]. We assume that lengths with higher multiplicity than two are so sparse that they do not modify the calculations in the semiclassical limit.

In the following we use the Selberg trace formula to derive the semiclassical expression for the form factor. The energy spectrum for Riemann surfaces is the spectrum of minus the Laplace-Beltrami operator $-\Delta\Psi = E\Psi$. For compact Riemann surfaces with constant negative curvature the trace formula is given by (see [12])

$$\sum_n h(p_n) = \frac{A}{4\pi} \int_{-\infty}^{\infty} dp' \, h(p') \, p' \, \tanh(\pi p') + \sum_\gamma A_\gamma \, g(L_\gamma) , \tag{4}$$

where p_n are momentum eigenvalues which are related to the energies by $E_n = p_n^2 + 1/4$, and

$$A_\gamma = \frac{L_\gamma}{2R_\gamma \sinh L_\gamma/2} , \qquad g(x) = \frac{1}{\pi} \int_0^\infty dp \, h(p) \, \cos px . \tag{5}$$

The sum on the right-hand side of (4) runs over all periodic orbits γ with length L_γ and repetition number R_γ. (Repetitions of a primitive periodic orbit count as distinct orbits.) The function $h(p)$ has to satisfy certain conditions (for details see [12]). It has to be an even function in p. It has to be analytic in a strip $|\operatorname{Im} p| < 1/2 + \eta$ for some $\eta > 0$, which guarantees the convergence of the periodic orbit sum. Finally, it has to fall off sufficiently fast for large p, $|h(p)| = \mathcal{O}(|p|^{-2-\delta})$ as $|p| \to \infty$, for some $\delta > 0$. This guarantees the convergence of the sum over energies.

One example of an allowed function is

$$h_\varepsilon(p') = \frac{1}{\sqrt{2\pi\varepsilon^2}} \left(\exp\left\{ -\frac{(p-p')^2}{2\varepsilon^2} \right\} + \exp\left\{ -\frac{(p+p')^2}{2\varepsilon^2} \right\} \right) \tag{6}$$

in terms of which one can define a smoothed density of states $d_\varepsilon(E) := (2|p|)^{-1} \sum_n h_\varepsilon(p_n)$, where $E = p^2 + 1/4$.

The smoothed density of states has the property that $\lim_{\varepsilon \to 0} d_\varepsilon(E) = d(E)$. Clearly the function $h_\varepsilon(p')$ that is obtained in the limit $\varepsilon \to 0$, $h_0(p') = \delta(p'-p) + \delta(p'+p)$ is not an allowed function. Hence we cannot use the Selberg trace formula directly to evaluate (1). Let us look instead at the quantity $K_\varepsilon(\tau)$ which is obtained by replacing the spectral density in (1) by the smoothed density $d_\varepsilon(\tau)$. We then can express $K_\varepsilon(\tau)$ in terms of classical periodic orbits by using Selberg's trace formula, and evaluate the integral in (1) in leading semiclassical order. Then we take the limit $\varepsilon \to 0$. This leads to a well-defined, absolutely convergent sum over periodic orbits for the particular choice of averaging that we will use in the following. Since these steps involve a semiclassical approximation and the interchange of limit and integral we will look carefully at the final result in order to see the effect of these steps.

The semiclassical form factor that is obtained in this way is

$$K_{\mathrm{sc}}(\tau) = \frac{1}{L_H} \left\langle \sum_{\gamma,\gamma'} A_\gamma A_{\gamma'} e^{i p (L_\gamma - L_{\gamma'})} \delta\left(L - \frac{L_\gamma + L_{\gamma'}}{2} \right) \right\rangle_{E,\tau} , \tag{7}$$

where $\tau = L/L_H$, and $L_H = pA$ is the "Heisenberg length". We choose to perform the energy average by a Gaussian in the variable p with width Δp, and the τ-average by a Gaussian in the variable L with width ΔL. In more detail,

$$\langle f(x) \rangle_x = \int_{-\infty}^\infty dx' \, \frac{1}{\sqrt{2\pi}\,\Delta x} \exp\left\{ -\frac{(x-x')^2}{2\,\Delta x^2} \right\} f(x') , \tag{8}$$

where x stands for p or L. This has the additional advantage that for the particular choice $\Delta p \, \Delta L = 1/2$ the form factor can be written as absolute square of a single sum [13]

$$K_{\mathrm{sc}}(\tau) = \sqrt{\frac{2}{\pi}} \frac{\Delta p}{L_H} \left| \sum_\gamma A_\gamma \exp\left\{ i p L_\gamma - \Delta p^2 (L - L_\gamma)^2 \right\} \right|^2 . \tag{9}$$

The conditions for the semiclassical limit after (3) then translate into $p \to \infty$ with $\Delta p/p \to 0$ and $p\Delta p \to \infty$.

Let us consider expression (9) in more detail. The Selberg trace formula is exact, but we made an approximation by evaluating the integral in (1) asymptotically for large p and, after introducing a smoothing parameter ε, interchanged limit $\varepsilon \to 0$ and integral. The main question is whether (9) is a good approximation to (1) and has the same semiclassical limit.

A numerical evaluation of (9) in [14] showed, somewhat surprisingly, that $K_{\mathrm{sc}}(\tau)$ is an exponentially increasing function in τ. This exponential increase can be understood, however, by using the trace formula again in order to express (9) in terms of the spectrum. One finds that the semiclassical form factor is exactly identical to the form factor of the *momentum spectrum*. Hence the effect of the approximations involved is to replace the energy spectrum by the momentum spectrum. One main difference is that there is at least one imaginary momentum eigenvalue which is $p_0 = i/2$, corresponding to the ground state $E_0 = 0$. There might be finitely many other imaginary eigenvalues. These imaginary momentum eigenvalues are the reason for the exponential increase of $K_{\mathrm{sc}}(\tau)$. To suppress them requires a stronger condition for performing the semiclassical limit $\Delta p/\sqrt{p} \to 0$ as $p \to \infty$ instead of $\Delta p/p \to 0$ [14]. This is a necessary condition for $\lim_{p \to \infty} K_{\mathrm{sc}}(\tau) = K_{\mathrm{GOE}}(\tau)$.

2.1 The diagonal approximation

The contributions of pairs of periodic orbits in (7) and (9) are suppressed by the averaging, and also by cancellations of oscillatory contributions from different pairs of periodic orbits which are uncorrelated. The main contribution comes from a relatively small number of pairs of orbits which are correlated. Berry concluded that the most important contribution for small τ comes from pairs of orbits with identical lengths [4]. This corresponds to the diagonal approximation in which only pairs of orbits are included which are either identical or related by time inversion. The diagonal approximation for (9) has the form

$$K_{\mathrm{sc}}^{(1)}(\tau) = \sqrt{\frac{2}{\pi}} \frac{\Delta p}{L_H} 2 \sum_{\gamma} A_{\gamma}^2 \exp\{-2\,\Delta p^2\,(L - L_{\gamma})^2\}\,. \qquad (10)$$

If the semiclassical limit is performed with τ fixed and $p \to \infty$ it follows that $L \to \infty$. Hence the contributions come from very long orbits and we can replace the amplitude in (5) by

$$A_{\gamma}^2 \to \frac{L_{\gamma}^2}{e^{L_{\gamma}}}\,. \qquad (11)$$

The sum in (10) is evaluated asymptotically for large length by using the prime geodesic theorem

$$N(L) = \#\{L_{\gamma} \leq L\} \sim \frac{e^L}{L} \qquad \text{as} \qquad L \to \infty\,. \qquad (12)$$

The mean density of periodic orbits in a length interval is then obtained taking a derivative

$$\bar{\rho}(L) \sim \frac{e^L}{L} \qquad \text{as} \qquad L \to \infty, \tag{13}$$

and the sum over periodic orbits is replaced by an integral

$$\sum_{\gamma} \to \int dL' \, \bar{\rho}(L') . \tag{14}$$

Hence one finds

$$
\begin{aligned}
K_{\mathrm{sc}}^{(1)}(\tau) &\sim \sqrt{\frac{2}{\pi}} \frac{2\,\Delta p}{L_H} \int_{-\infty}^{\infty} dL' \, L' \, \exp\{-2\,\Delta p^2 \, (L - L')^2\} \\
&\sim \sqrt{\frac{2}{\pi}} \frac{2\,\Delta p\, L}{L_H} \sqrt{\frac{\pi}{2}} \frac{1}{\Delta p} \\
&\sim 2\tau
\end{aligned}
\tag{15}
$$

as $L \to \infty$. The calculations in (15) involve an application of the method of steepest descent with the condition that $L\,\Delta p \propto p\,\Delta p \to \infty$ as $L \to \infty$, in accordance with our previous conditions for performing the semiclassical limit. The result is indeed the first term in the Taylor expansion of the spectral form factor for the GOE ensemble $K_{\mathrm{GOE}}(\tau) = 2\tau - 2\tau^2 + \dots$.

2.2 Off-diagonal terms

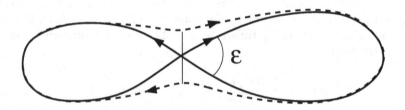

Fig. 1. A pair of two-loop orbits.

The assumption behind the evaluation of off-diagonal terms is that correlated orbits follow each other very closely almost everywhere in position space. They consist of different orbit segments, or loops, along which both orbits are almost identical. These segments might be traversed in opposite directions in systems with time-reversal symmetry. The orbits differ in the way in which the segments are connected.

The simplest example is that of a pair of orbits with two long loops as shown schematically in figure 1. The two loops are connected by two long orbit stretches which are almost parallel. (The schematic picture 1 does not really show this. Imagine ε being small.) This makes it possible that the two loops are connected in different ways for the two orbits [15, 16].

Consider now one of the two orbits. We look more closely at the encounter region where the two segments of the orbit which connect the two loops are almost parallel. (More precisely, they are anti-parallel if the sense of traversal is taken into account, but in this section we disregard the direction in which they are traversed.) We investigate how the two segments approach each other and separate again in the encounter region. For this purpose we consider the relative motion along one segment by linearising the motion around the other segment.

In the vicinity of a geodesic on a surface of constant curvature K the linearised motion is described by the Jacobi equation [1]. It has the form

$$\frac{d^2\rho}{dl^2} + K\,\rho = 0 \ . \tag{16}$$

Here l is the distance along the trajectory and ρ measures the distance perpendicular to the trajectory. The solution for $K = -1$ and initial conditions $\rho(0) = \rho_0$ and $\rho'(0) = \rho'_0$ is

$$\rho = \rho_0 \cosh l + \rho'_0 \sinh l \ . \tag{17}$$

Introducing a conjugate momentum variable $\sigma = \rho'$ leads to

$$\begin{pmatrix} \rho \\ \sigma \end{pmatrix} = M \begin{pmatrix} \rho_0 \\ \sigma_0 \end{pmatrix} \ , \quad \text{where} \quad M = \begin{pmatrix} \cosh l & \sinh l \\ \sinh l & \cosh l \end{pmatrix} \ . \tag{18}$$

Eigenvalues and eigenvectors are

$$\lambda_u = e^l, \quad \mathbf{e}_u = \begin{pmatrix} 1 \\ 1 \end{pmatrix}, \quad \lambda_s = e^{-l}, \quad \mathbf{e}_s = \begin{pmatrix} 1 \\ -1 \end{pmatrix} \ . \tag{19}$$

It is convenient to express the distance vector (in phase space) in terms of the eigenvectors

$$\begin{pmatrix} \rho \\ \sigma \end{pmatrix} = s\,\mathbf{e}_s + u\,\mathbf{e}_u \ , \quad \text{where} \quad s = s_0\,e^{-l}, \quad u = u_0\,e^l \ . \tag{20}$$

Here u and s are coordinates along the unstable and stable directions, respectively, and u_0 and s_0 are their initial values.

One sees that, in general, a neighbouring trajectory separates exponentially in positive and negative l-direction from the central trajectory. In between there is a closest encounter where the distance in phase space is minimal. If we define the distance as the square root of $\rho^2 + \sigma^2 = 2(s^2 + u^2)$ the closest encounter occurs when

$$0 = \frac{d}{dl}(s^2 + u^2) = -2s^2 + 2u^2 \ . \tag{21}$$

Solutions are either $u = s$, corresponding to $\rho = 2u$ and $\sigma = 0$, or $u = -s$, corresponding to $\rho = 0$ and $\sigma = -2u$. The orbits in figure 1 provide examples for these two cases. For the inner orbit the two segments become closest at the crossing where $\rho = 0$ and $\sigma = \varepsilon$. For the outer orbit the two segments become closest when they are parallel, i.e. $\sigma = 0$ and $\rho \neq 0$.

The encounter region in figure 1 is thus characterised by the fact that one orbit has an intersection with a small crossing angle ε, and the other orbit does not self-intersect. In the following we count the number of two-loop orbit pairs by counting the number of self-intersections with small crossing angles ε along periodic orbits. We thus assume that for every self-intersection with small ε there exists a partner periodic orbit without self-intersection. This is true in the linearised approximation [15], and is evident in symbolic dynamics [17]. The fact that orbits self-intersect either once or do not self-intersect in the encounter region is a special property of the uniformly hyperbolic dynamics. In general chaotic systems the two almost parallel segments of an orbit can, for example, cross several times before they separate again and the number of self-intersections does not agree with the number of orbit pairs [18–20].

In the remaining part of this lecture course we calculate the contribution of the two-loop orbit pairs to the form factor. This requires two main ingredients. One is the length difference of the two orbits that form a pair, and the other is the number of pairs. The latter is calculated in the next section.

The length difference can be evaluated by determining the length difference in the linearised approximation [15]. For Riemann surfaces there is a simpler way that uses hyperbolic geometry [17]. Consider a hyperbolic triangle whose three sides A, B and C are formed by geodesics. The sine-law in hyperbolic geometry has the form

$$\frac{\sin \alpha}{\sinh A} = \frac{\sin \beta}{\sinh B} = \frac{\sin \gamma}{\sinh C}. \tag{22}$$

In figure 1 one can recognise four (approximate) hyperbolic triangles that are formed by the two orbits and the thin line that connects them at the crossing. Applying the sine-law with angles $\alpha = \pi/2$, $\beta = (\pi - \varepsilon)/2$, $\gamma = 0$ results in $\cos \varepsilon/2 = \sinh B / \sinh A \approx \exp(B - A)$. This leads to the following formula for the length difference that is valid for all angles ε

$$\Delta L \approx -4 \ln \cos \frac{\varepsilon}{2}. \tag{23}$$

The error is exponentially small in the lengths of the loops.

3 The number of self-intersections

The aim of section 3 is to calculate the average number of self-intersections along arbitrary (non-periodic) trajectories of length L. The calculation can be done rigorously by using specific properties of Riemannian geometry. The intention is to transfer this result in section 4 to periodic orbits by using the uniform distribution of periodic orbits in phase space. The final result of this section can be obtained more quickly by using heuristic arguments such as in subsection 3.3 (see also e.g. [21]), but we want to avoid making any assumptions in this section.

We start by listing some properties of hyperbolic geometry that will be needed in the following. For details see the lectures of Buser [22].

We consider the upper-half plane model of the hyperbolic plane $\mathbb{H} = \{z = x + i\,y \,|\, y > 0\}$. In this model the line element and the volume element are given by

$$ds^2 = \frac{dx^2 + dy^2}{y^2} \,, \qquad d\tilde{\mu} = \frac{dx\,dy}{y^2} \,, \tag{24}$$

where a tilde is used to distinguish $d\tilde{\mu}$ from the phase space measure that is introduced later.

Of particular importance are the orientation preserving isometries of the hyperbolic plane. In the upper half plane model they take the form of fractional linear transformations.

$$z \longmapsto \frac{a\,z + b}{c\,z + d} \,, \tag{25}$$

where the associated matrices are elements of the group

$$\mathrm{SL}(2, \mathbb{R}) = \left\{ g = \begin{pmatrix} a & b \\ c & d \end{pmatrix} \,\middle|\, a, b, c, d \text{ real; } \det(g) = 1 \right\} \,. \tag{26}$$

Matrices which differ by an overall sign correspond to the same fractional linear transformation. Hence the group of orientation-preserving isometries can be identified with the group

$$\mathrm{PSL}(2, \mathbb{R}) = \mathrm{SL}(2, \mathbb{R})/\{\pm \mathbb{I}\} \,, \tag{27}$$

where $\mathbb{I}$ is the identity matrix.

Phase space (resp. a surface of constant energy in phase space) is represented by the unit tangent bundle of $\mathbb{H}$. It is parametrised by the coordinate $z = x + i\,y$ and an angle θ which describes the direction of motion at the point z.

The transformation of θ under the fractional linear transformation (25) is found by considering the transformation of an infinitesimally modified point $z + dz$. This leads to

$$\theta \longmapsto \theta - 2\arg(c\,z + d) \,. \tag{28}$$

The invariant measure on the unit tangent bundle is

$$d\mu = d\tilde{\mu}\,d\theta = \frac{dx\,dy\,d\theta}{y^2} \,. \tag{29}$$

3.1 Parametrisation of phase space

In the following a particular parametrisation of points on the unit tangent bundle is introduced which greatly simplifies the evaluation of coordinate transformations and time evolution. It exploits a one-to-one relationship that exists between elements of PSL(2, ℝ) and points on the unit tangent bundle. This leads to a very convenient method to evaluate transition probabilities in phase space which will be used to find the number of self-intersections.

Consider a matrix $g \in \mathrm{SL}(2, \mathbb{R})$ with elements g_{ij}. It can be uniquely written in the form

$$g = \begin{pmatrix} 1 & x \\ 0 & 1 \end{pmatrix} \begin{pmatrix} y^{1/2} & 0 \\ 0 & y^{-1/2} \end{pmatrix} \begin{pmatrix} \cos\theta/2 & \sin\theta/2 \\ -\sin\theta/2 & \cos\theta/2 \end{pmatrix}, \tag{30}$$

where $z = x + iy \in \mathbb{H}$, and $\theta \in [0, 4\pi)$. The relation between x, y, θ and the matrix elements of g is given by

$$x = \frac{g_{11}\,g_{21} + g_{12}\,g_{22}}{g_{21}^2 + g_{22}^2}, \quad y = \frac{1}{g_{21}^2 + g_{22}^2}, \quad \theta = -2\arg(g_{22} + i\,g_{21}). \tag{31}$$

The matrix g changes by an overall sign if θ changes by 2π. Hence by identifying θ-values which differ by 2π we obtain a one-to-one relationship between elements of PSL(2, ℝ) and points on the unit tangent bundle.

The identification of a point (z, θ) with an element in PSL(2, ℝ) has further advantages. Consider the following matrix product

$$g' := \begin{pmatrix} a & b \\ c & d \end{pmatrix} g = \begin{pmatrix} 1 & x' \\ 0 & 1 \end{pmatrix} \begin{pmatrix} y'^{1/2} & 0 \\ 0 & y'^{-1/2} \end{pmatrix} \begin{pmatrix} \cos\theta'/2 & \sin\theta'/2 \\ -\sin\theta'/2 & \cos\theta'/2 \end{pmatrix}. \tag{32}$$

A short calculation shows that

$$z' = x' + iy' = \frac{a\,z + b}{c\,z + d}, \quad \theta' = \theta - 2\arg(c\,z + d). \tag{33}$$

These transformation rules are identical to those in equations (25) and (28). This means that the action of an isometry on a point in the unit tangent bundle is simply given by a matrix multiplication in this representation. Note that the point $(z, \theta) = (i, 0)$ is represented by the identity matrix $\mathbb{I}$. Hence a general point (z, θ) is represented by the matrix g that corresponds to the isometry which maps $(i, 0)$ onto (z, θ).

As a consequence the time evolution also simplifies to a matrix multiplication. The point $(i, 0)$ evolves in time t (with unit speed) to

$$(i, 0) \longmapsto (i\,e^t, 0) \;\hat{=}\; \begin{pmatrix} e^{t/2} & 0 \\ 0 & e^{-t/2} \end{pmatrix}. \tag{34}$$

The time evolution of a general point is obtained by applying the isometry g

$$g \longmapsto g\,\Phi^t\,, \quad \text{where} \quad \Phi^t = \begin{pmatrix} e^{t/2} & 0 \\ 0 & e^{-t/2} \end{pmatrix}\,. \tag{35}$$

We will avoid using the time variable t in the following, because it is convenient to express all quantities in terms of the lengths of trajectories. We will represent time t by the length l that is covered during the time evolution.

Finally, we need delta-functions on the unit-tangent bundle. They are introduced by

$$\int \delta_0(g',g)\,f(g')\,d\mu(g') = f(g)\,, \qquad d\mu(g') = \frac{dx'\,dy'\,d\theta'}{y'^2}\,. \tag{36}$$

where $f(g)$ is a continuous function of g. The delta-function is symmetric and invariant under the application of isometries

$$\delta_0(g,g') = \delta_0(g',g) = \delta_0(h\,g',h\,g) = \delta_0(g'\,h,g\,h)\,. \tag{37}$$

It can be factorised in the form

$$\delta_0(g',g) = y^2\,\delta(x-x')\,\delta(y-y')\,\delta_p(\theta-\theta') \tag{38}$$

where δ_p denotes the 2π-periodic delta-function.

So far we have considered the motion on the full hyperbolic plane, but we are interested in the motion on compact smooth Riemann surfaces. These surfaces are represented by a compact area on the hyperbolic plane with the property that one can tessellate the full hyperbolic plane with copies of it. They are associated with a discrete subgroup Γ of $\mathrm{PSL}(2,\mathbb{R})$ whose elements map one copy onto all other copies. For a compact smooth Riemann surface this subgroup contains only hyperbolic elements (with the exception of the identity). The hyperbolic elements correspond to boosts which are analogs of translations in the Euclidean plane and are characterised by $|\operatorname{Tr} g| > 2$. The boosts in Γ cannot be arbitrarily small, in mathematical terms Γ is a strictly hyperbolic Fuchsian group [22]. A delta-function on a Riemann surface is then defined by periodising the original delta-function by summing over all copies

$$\delta(g,g') = \sum_{\gamma \in \Gamma} \delta_0(\gamma\,g,g')\,. \tag{39}$$

3.2 The average number of self-intersections

Consider an arbitrary trajectory of length L. To each self-intersection of this trajectory there is a corresponding loop which starts and ends at the self-intersection. In the following l denotes the length of the loop, ε its opening angle, and l' the distance along the trajectory up to the starting point of the loop.

The average number of self-intersections with intersection angle in an interval $[\varepsilon,\varepsilon+d\varepsilon]$ along trajectories of length L is then given by $P(\varepsilon,L)\,d\varepsilon$ where the density of self-intersections $P(\varepsilon,L)$ is

$$P(\varepsilon, L) = \frac{1}{\Sigma} \int d\mu(g) \int_0^L dl \int_0^{L-l} dl' \, \delta(g\Phi^{l'} g_\varepsilon, g\Phi^{l'+l}) \, |\sin \varepsilon| \, . \qquad (40)$$

Here the average is taken over all initial conditions on the unit tangent bundle, and $\Sigma = \int d\mu(g) = 2\pi A$. The matrix g_ε rotates the θ-coordinate by $\varepsilon - \pi$.

$$g_\varepsilon = \begin{pmatrix} \cos \frac{\varepsilon-\pi}{2} & \sin \frac{\varepsilon-\pi}{2} \\ -\sin \frac{\varepsilon-\pi}{2} & \cos \frac{\varepsilon-\pi}{2} \end{pmatrix} = \begin{pmatrix} \sin \frac{\varepsilon}{2} & -\cos \frac{\varepsilon}{2} \\ \cos \frac{\varepsilon}{2} & \sin \frac{\varepsilon}{2} \end{pmatrix} \, . \qquad (41)$$

Equation (40) can be understood in the following way: $g\Phi^{l'}$ is the starting point of a loop, and the integral gives a contribution if after an evolution of length l the trajectory arrives at a point in phase space that is obtained from the starting point by a rotation by $\varepsilon - \pi$. The $|\sin \varepsilon|$-term in (40) arises from a Jacobian that is necessary in order that the l- and l'- integrals correctly count the number of self-intersections. More accurately, with an additional integral over some ε-interval $\Delta\varepsilon$, the l- and l'-integrals yield a one for any self-intersection with intersection angle in $\Delta\varepsilon$. The origin of the $|\sin \varepsilon|$-term is most easily seen by considering the transformation from the arguments of the delta-function in x- and y- coordinates, $x(l + l') - x(l')$ and $y(l + l') - y(l')$, to the l- and l'- coordinates with the Jacobian

$$|J| = \left| \frac{dx(l+l')}{dl} \frac{dy(l')}{dl'} - \frac{dy(l+l')}{dl} \frac{dx(l')]}{dl'} \right| \, . \qquad (42)$$

This can be recognised as cross-product of two direction-vectors with length y, hence $|J| = y^2 |\sin \varepsilon|$. (The y^2 term of J is obtained by the factorisation of the delta-function.)

After interchanging integrals and changing the integration variable $g\Phi^{l'} \mapsto g$ one obtains

$$P(\varepsilon, L) = \int_0^L dl \, (L - l) \, p(\varepsilon, l) \, |\sin \varepsilon| \, , \qquad (43)$$

where $p(\varepsilon, l)$ is defined as

$$p(\varepsilon, l) = \frac{1}{\Sigma} \int d\mu(g) \, \delta(g \, g_\varepsilon, g \, \Phi^l) \, . \qquad (44)$$

It can be interpreted as the probability density to form a loop with opening angle ε and length l. It is different from zero only if there is a trajectory of length l which forms a loop with opening angle ε.

In the following it will be shown that $p(\varepsilon, l)$ and hence $P(\varepsilon, L)$ can be expressed in terms of the periodic orbits of the Riemann surface. A substitution $g \mapsto g \, g_\varepsilon^{-1}$ leads to

$$p(\varepsilon, l) = \frac{1}{\Sigma} \int d\mu(g) \, \delta(g, g \, g_\varepsilon^{-1} \, \Phi^l) \, , \qquad (45)$$

where

$$g_\varepsilon^{-1} \Phi^l = \begin{pmatrix} e^{l/2} \sin\frac{\varepsilon}{2} & e^{-l/2} \cos\frac{\varepsilon}{2} \\ -e^{l/2} \cos\frac{\varepsilon}{2} & e^{-l/2} \sin\frac{\varepsilon}{2} \end{pmatrix}. \tag{46}$$

The eigenvalues of this matrix are real with modulus $|\lambda| = e^{\pm \tilde{L}/2}$ where

$$\cosh\frac{\tilde{L}}{2} = \frac{1}{2}\left|\operatorname{Tr} g_\varepsilon^{-1} \Phi^l\right| = \cosh\frac{l}{2} \sin\left|\frac{\varepsilon}{2}\right|, \qquad -\pi < \varepsilon \leq \pi. \tag{47}$$

The matrix $g_\varepsilon^{-1} \Phi^l$ can be diagonalised by a similarity transformation $g_\varepsilon^{-1} \Phi^l = h^{-1} \Phi^{\tilde{L}} h$ and we find

$$\begin{aligned} p(\varepsilon, l) &= \frac{1}{\Sigma} \int \delta(g, g\, h^{-1} \Phi^{\tilde{L}} h)\, d\mu(g) \\ &= \frac{1}{\Sigma} \int \delta(g\, h^{-1}, g\, h^{-1} \Phi^{\tilde{L}})\, d\mu(g) \\ &= \frac{1}{\Sigma} \int \delta(g, g\, \Phi^{\tilde{L}})\, d\mu(g). \end{aligned} \tag{48}$$

The last expression is identical to $p(\pi, \tilde{L})$, the probability density to return in phase space. It is different from zero only if there exists a periodic orbit γ with length $L_\gamma = \tilde{L}$.

The above calculation shows that there is a unique relationship between a loop of length l and opening angle ε and a periodic orbit of length $\tilde{L}$ given by (47). Geometrically this signifies that any loop can be continuously deformed into a periodic orbit [16]. This relationship is not one-to-one, but there is a one-parameter family of loops with the same length l and angle ε whose initial points lie on a closed curve which are related by (47) to a periodic orbit.

The relationship between a transition probability density in phase space and periodic orbits is not restricted to loops, but holds more generally. In order to see this it is convenient to introduce a second parametrisation of $\mathrm{PSL}(2, \mathbb{R})$. A matrix g can be uniquely decomposed into

$$g = \begin{pmatrix} e^{r/2} & 0 \\ 0 & e^{-r/2} \end{pmatrix} \begin{pmatrix} 1 & s \\ 0 & 1 \end{pmatrix} \begin{pmatrix} 1 & 0 \\ u & 1 \end{pmatrix} = \begin{pmatrix} e^{r/2}(1 + u\,s) & e^{r/2}\,s \\ e^{-r/2}\,u & e^{-r/2} \end{pmatrix}. \tag{49}$$

Using (31) one finds

$$y = \frac{e^r}{u^2 + 1}, \qquad x = e^r\left(\frac{u}{u^2 + 1} + s\right), \qquad \theta = -2\arg(1 + i\,u). \tag{50}$$

Let us denote the particular form of the matrix on the right-hand side of equation (49) by g_{rsu}. The interpretation of the coordinates r, s and u can be understood by letting the time-evolution matrix act on it

$$g_{rsu}\begin{pmatrix} e^{l/2} & 0 \\ 0 & e^{-l/2} \end{pmatrix} = \begin{pmatrix} e^{(r+l)/2} & 0 \\ 0 & e^{-(r+l)/2} \end{pmatrix} \begin{pmatrix} 1 & s\,e^{-l} \\ 0 & 1 \end{pmatrix} \begin{pmatrix} 1 & 0 \\ u\,e^l & 1 \end{pmatrix}. \tag{51}$$

Hence r is a coordinate along a trajectory, and s and u are coordinates on the stable and unstable manifolds, respectively. For infinitesimal distances s and u agree with the previously introduced coordinates in section 2.2. The phase space measure is given by

$$\frac{dx\,dy\,d\theta}{y^2} = 2\,dr\,ds\,du \;. \tag{52}$$

Analogous to the calculation for $p(\varepsilon, l)$ one can express also the probability density for an arbitrary transition in phase space in terms of periodic orbits.

$$\hat{p}(r,s,u,l) = \int \delta(g\,g_{rsu}, g\Phi^l)\,d\mu(g) = \int \delta(g, g\,\Phi^{\tilde{L}})\,d\mu(g) = p(\pi, \tilde{L})\,, \tag{53}$$

where

$$g_{rsu}^{-1}\Phi^l = \begin{pmatrix} e^{-(r-l)/2} & -s\,e^{(r-l)/2} \\ -u\,e^{(-r-l)/2} & (1+u\,s)\,e^{(r-l)/2} \end{pmatrix} = h^{-1}\Phi^{\tilde{L}}\,h\,. \tag{54}$$

Now

$$\cosh\frac{\tilde{L}}{2} = \left|\cosh\frac{r-l}{2} + \frac{u\,v}{2}\exp\frac{r-l}{2}\right|\,. \tag{55}$$

This means that the transition probability density $\hat{p}(r,s,u,l)$ can be expressed in terms of the probability density to return $p(\pi, \tilde{L})$ and hence in terms of periodic orbits.

The remaining step for the evaluation of $p(\varepsilon, l)$ consists of an evaluation of the probability density to return

$$p(\pi, \tilde{L}) = \frac{1}{\Sigma}\int d\mu(g)\,\delta(g, g\Phi^{\tilde{L}}) = \frac{1}{\Sigma}\int d\mu(g)\sum_{\gamma\in\Gamma}\delta_0(\gamma\,g, g\Phi^{\tilde{L}})\,. \tag{56}$$

The steps in this calculation follow very closely those in the derivation of the trace formula [12] and we give here only the final result. The sum over the elements in γ is split into a sum over all conjugacy classes and a sum over all elements within each conjugacy class. The sum over the elements within a conjugacy class is used to replace the integral over the area of the surface by an integral over a strip in the full hyperbolic plane. The final integral over the delta- function is conveniently done in local coordinates r, s and u. The result is

$$p(\pi, \tilde{L}) = \frac{1}{\Sigma}\sum_{\gamma}\frac{L_\gamma\,\delta(\tilde{L} - L_\gamma)}{R_\gamma\,(2\sinh L_\gamma/2)^2}\,, \tag{57}$$

where the sum runs over all periodic orbits γ with length L_γ and repetition number R_γ.

Finally, by using the relation between loop length and periodic orbit length

$$\cosh\frac{l_\gamma(\varepsilon)}{2}\sin\left|\frac{\varepsilon}{2}\right| = \cosh\frac{L_\gamma}{2}\,, \tag{58}$$

we arrive at the following expression for $p(\varepsilon, l) = p(\pi, \tilde{L})$

$$p(\varepsilon, l) = \frac{1}{\Sigma} \sum_{\gamma} B_{\gamma}(\varepsilon)\, \delta(l - l_{\gamma}(\varepsilon))\,, \tag{59}$$

where

$$B_{\gamma}(\varepsilon) = \frac{L_{\gamma}}{4R_{\gamma} \sinh L_{\gamma}/2 \sqrt{\sinh^2 L_{\gamma}/2 + \cos^2 \varepsilon/2}}\,. \tag{60}$$

The integration in (43) yields the result

$$P(\varepsilon, L) = \frac{|\sin \varepsilon|}{2\pi A} \sum_{l_{\gamma}(\varepsilon) < L} (L - l_{\gamma}(\varepsilon))\, B_{\gamma}(\varepsilon) \tag{61}$$

for the average density of self-intersections with angle ε along trajectories of length L.

3.3 Asymptotic expansion of the density of self-intersections

In this section the asymptotic behaviour of $P(\varepsilon, L)$ for large L is determined. But before we do this let us look at the relation between loop length and periodic orbit length in (58) in more detail. For large L_{γ} this simplifies to

$$l_{\gamma}(\varepsilon) \sim L_{\gamma} - 2 \log \sin \frac{|\varepsilon|}{2}\,, \tag{62}$$

and for small angles

$$l_{\gamma}(\varepsilon) \sim L_{\gamma} - 2 \log \frac{|\varepsilon|}{2}\,. \tag{63}$$

The logarithmic divergence in (63) has a simple physical interpretation. Consider a loop with a very small angle ε, and consider the two legs of the loop as two trajectories with initial conditions that differ by a small ε. The separation of the two legs during time evolution is governed by the unit Lyapunov exponent, and it has to be at least of order one for the two legs to be able to meet and form a closed loop. This leads to an estimate for the minimum length that a loop with angle ε must have

$$|\varepsilon|\, \exp(l_{min}/2) = \mathcal{O}(1)\,, \tag{64}$$

and $l_{min} \sim -2 \log c\varepsilon$ for some constant c in agreement with (63). Hence the logarithmic divergence of the loop length is due to the existence of a minimal loop length for small ε.

Now we go back to $P(\varepsilon, L)$ and use (61) to determine its asymptotic behaviour for large L. We will need not only the leading order term, but the next-to-leading order term as well. For this purpose the sum over periodic orbits is split into a sum $L_\gamma < L^*$ and a sum $L_\gamma \geq L^*$. The first sum is evaluated exactly, whereas the second sum is evaluated asymptotically by using the prime geodesic theorem (12).

$$
\begin{aligned}
P(\varepsilon, L) &= \frac{\sin |\varepsilon|}{2\pi A} \sum_{l_\gamma(\varepsilon) < L} B_\gamma(\varepsilon)\,(L - l_\gamma(\varepsilon)) \\
&\sim \frac{\sin |\varepsilon|}{2\pi A} \sum_{L_\gamma < L_*} B_\gamma(\varepsilon)\,(L - l_\gamma(\varepsilon)) \\
&\quad + \frac{\sin |\varepsilon|}{2\pi A} \int_{L_*}^{L + 2\log \sin |\varepsilon|/2} dL' \left(L - L' + 2\log \sin \frac{|\varepsilon|}{2} \right) \\
&\sim \frac{\sin |\varepsilon|}{2\pi A} \sum_{L_\gamma < L_*} B_\gamma(\varepsilon)\,(L - l_\gamma(\varepsilon)) \\
&\quad + \frac{L^2}{4\pi A} \sin |\varepsilon| \left(1 - \frac{2L^*}{L} + \frac{4}{L} \log \sin \frac{|\varepsilon|}{2} \right) + \mathcal{O}(1) .
\end{aligned}
\tag{65}
$$

Note that the correction to the prime geodesic theorem is exponentially small as $L \to \infty$ and can be neglected. One can let L^* go to infinity as well as $L \to \infty$ and obtains the following two terms in the asymptotic expansion of $P(\varepsilon, L)$ for $L \to \infty$

$$
P(\varepsilon, L) \sim \frac{L^2}{4\pi A} \sin |\varepsilon| \left(1 + \frac{2C(\varepsilon)}{L} + \frac{4}{L} \log \sin \frac{|\varepsilon|}{2} \right) ,
\tag{66}
$$

where

$$
C(\varepsilon) = \lim_{L_* \to \infty} \left(\sum_{L_\gamma < L_*} B_\gamma(\varepsilon) - L_* \right) .
\tag{67}
$$

The limit in (67) converges uniformly in ε.

As will become clear later we actually need the asymptotic behaviour in the joint limit $L \to \infty$ and $\varepsilon \propto L^{-1/2} \to 0$. This can be obtained from (66), further corrections do not contribute

$$
P(\varepsilon, L) \sim \frac{L^2}{4\pi A} \sin |\varepsilon| \left(1 + \frac{4}{L} \log c|\varepsilon| \right) ,
\tag{68}
$$

where $c = \frac{1}{2} \exp(C(0)/2)$.

4 Contribution to the form factor

In the following we use the information about $P(\varepsilon, L)$ from the last section in order to evaluate the contribution of the pairs of two-loop orbits to the form factor in (9). Let us look at the leading order term as $L \propto p \to \infty$. The orbits have a length difference ΔL given by (23). As a consequence the amplitudes of the orbits differ slightly too, however this difference contributes to the next-to-leading order (see below). The number of orbit pairs are counted by counting the self-intersections of periodic orbits. As will be seen below, only self-intersections with small ε contribute. We are led to

$$K_{\text{sc}}^{(2)}(\tau) \sim \sqrt{\frac{2}{\pi}} \frac{4\,\Delta p}{L_H} \, \text{Re} \int d\varepsilon \sum_\gamma A_\gamma^2 e^{ip\Delta L(\varepsilon)} \, e^{-2\Delta p^2 (L-L_\gamma)^2} \, P_\gamma(\varepsilon) \qquad (69)$$

where $P_\gamma(\varepsilon)\,d\varepsilon$ is the number of self-intersections along the orbit γ with angle in $[\varepsilon, \varepsilon + d\varepsilon]$.

Long periodic orbits are uniformly distributed on the unit tangent bundle. As a consequence, averages over periodic orbits can be replaced by averages over the unit tangent bundle in the asymptotic limit of long orbit lengths. This property is applied in order to effectively replace the average number of self-intersections along periodic orbits by the average number of self-intersections along non-periodic trajectories within the integral in (69)

$$\sum_\gamma A_\gamma^2 P_\gamma(\varepsilon) \to \int dL' \, L' \, P(L', \varepsilon) \, . \qquad (70)$$

Doing this we arrive at

$$\begin{aligned}
K_{\text{sc}}^{(2)}(\tau) &\sim \sqrt{\frac{2}{\pi}} \frac{4\,\Delta p}{L_H} \, \text{Re} \int d\varepsilon \int dL' \, e^{ip\Delta L(\varepsilon)} \, e^{-2\Delta p^2 (L-L')^2} \, L' \, P(L', \varepsilon) \\
&\sim \frac{2L^3}{\pi p A^2} \, \text{Re} \int_0^\infty d\varepsilon \, e^{ip\varepsilon^2/2} \, \sin\varepsilon \\
&\sim \frac{2\tau^3}{\pi} p A \, \text{Re}\, i
\end{aligned}$$

where we evaluated the integral over L' by the method of steepest decent, using the leading order approximation of $P(L, \varepsilon)$ in (66), and the integral over ε in stationary phase approximation. In this step $\sin\varepsilon$ was replaced by ε. First we notice that the important contribution to the ε-interval comes from an interval of order $1/\sqrt{p}$. This means that we have to consider the joint limit $L \propto p \propto \varepsilon^{-2} \to \infty$ instead of a limit where ε is considered constant.

Second, assuming that (70) is still valid under the new limit, we see that the leading order contribution vanishes. Hence one has to consider the next-to-leading order contribution to $K^{(2)}(\tau)$ in the semiclassical limit. This arises from several corrections: the next-to-leading order term for $P(L, \varepsilon)$, the difference in the amplitudes of the orbits, the next-to-leading order correction to $\Delta L(\varepsilon)$, and the next term in the expansion of $\sin\varepsilon$ for small ε. Looking at the L-dependence of the contributions one sees that only the first contribution contributes to a τ^2 term whereas the other three contribute to a τ^3 term of the form factor.

Let us consider the first correction. We would like to use the uniform distribution of periodic orbits in order to conclude that the number of self-intersections along periodic orbits has the same asymptotic behaviour (68) as that along non-periodic trajectories. However, the argument of the uniform distribution can only be applied to obtain the leading term in a large L expansion, and not the next-to-leading order term for a joint large L small ε expansion. We assume here that periodic and non-periodic trajectories have a similar asymptotic behaviour in the considered limit, and that we can effectively make the replacement

$$\sum_\gamma A_\gamma^2 \, P_\gamma(\varepsilon) \to \int dL' \, L' \, \frac{L'^2}{4\pi A} \sin |\varepsilon| \left(1 + \frac{4}{L'} \log \tilde{c}|\varepsilon| \right) \qquad (71)$$

where the constant $\tilde{c}$ may possibly be different from the constant in the case of non-periodic trajectories. A justification is that the $\log|\varepsilon|$ term originates from the minimum loop length that exists for both, periodic and non-periodic trajectories.

Inserting (71) into (69) and following the same steps as before we arrive at

$$K_{\text{sc}}^{(2)}(\tau) \sim \frac{8L^2}{\pi p A^2} \, \text{Re} \int_0^\infty d\varepsilon \, e^{ip\varepsilon^2/2} \, \varepsilon \, \log(\tilde{c}\varepsilon) \, . \qquad (72)$$

The integral can be evaluated by substituting $\varepsilon = e^{i\pi/4} \, \varepsilon'$ and rotating the integration contour by $\pi/4$. This leads to

$$K_{\text{sc}}^{(2)}(\tau) \sim -\frac{8L^2}{\pi p A^2} \, \text{Im} \int_0^\infty d\varepsilon' \, e^{-p\varepsilon'^2/2} \, \varepsilon' \, \log(\tilde{c}\varepsilon' e^{i\pi/4}) = -2\tau^2 \, . \qquad (73)$$

This result agrees with the second term in the Taylor expansion of the GOE form factor (3) at $\tau = 0$.

Finally let us look at the other three corrections that contribute to a τ^3-term. Using the definition of the amplitude (5), the next-to-leading term of the length difference (23) for small ε, and the next-to-leading term in the Taylor expansion of $\sin \varepsilon$ we obtain

$$\frac{2L^3}{\pi p A^2} \, \text{Re} \int_0^\infty d\varepsilon \, e^{ip(\varepsilon^2/2 + \varepsilon^4/48)} \left(1 + \frac{1}{4}\varepsilon^2 \right)\left(\varepsilon - \frac{1}{6}\varepsilon^3 \right)$$

$$\approx \frac{2L^3}{\pi p A^2} \, \text{Re} \int_0^\infty d\varepsilon' \, e^{ip\varepsilon'^2/2} \, \varepsilon'(1 + \mathcal{O}(\varepsilon'^4)) \qquad (74)$$

where a substitution $\varepsilon' = \varepsilon + \varepsilon^3/48$ has been made. The different corrections of relative order ε'^2 cancel each other and hence the term (74) vanishes in the semiclassical limit [23]. So although each of the three corrections contributes to a τ^3 term their joint contribution vanishes. In summary, we find that the two-loop orbit pairs contribute a τ^2 term to the spectral form factor that is in agreement with random matrix theory.

5 Discussion

In these notes we have discussed the simplest form of correlations between different periodic orbits on compact Riemann surfaces. It was shown that these orbit pairs are responsible for a τ^2-term of the spectral form factor which is in agreement with the τ^2-term of the GOE form factor of random matrix theory. Specific properties of Riemann surfaces were used in order to arrive at this result. Many of the steps in the derivation can be done quicker by using heuristic arguments, however, a main emphasis was on using rigorous methods. One remaining assumption is that the large L small ε asymptotics of the number of self-intersections has the same form for general trajectories and periodic orbits (see equation (71)).

During recent years there have been a number of developments (some of them after this lecture series was given) which extend the present results in several directions.

— For quantum graphs the τ^2 and τ^3 terms of the form factor were derived semiclassically and shown to be in agreement with random matrix theory [24–26]. Recently it was proved by a different method that uses a supersymmetric σ-model that the full form factor of individual quantum graphs does indeed agree with random matrix theory [27, 28].

— The transport through an open chaotic cavity was investigated in [29]. Here one has to consider correlations between trajectories that go from an entry lead to an exit lead of the cavity. It was shown that the leading-order off-diagonal terms give a weak-localisation correction to the conductance which is in quantitative agreement with results from random matrix theory. This result was generalised to all higher-order terms in [30], and to the treatment of shot noise in [31].

— Generalisations of periodic-orbit correlations to non-uniformly hyperbolic systems were considered in [18–20]. One main difference to the uniformly hyperbolic case is that self-intersections are not appropriate for counting the number of orbit pairs, because there is no one-to-one correspondence between self-intersections with small ε and pairs or periodic orbits. Instead one considers encounter regions with a finite length.

— Universality classes for systems with spin were treated in [32–34], and transitions between universality classes in [18, 35, 36].

— Higher dimensions and fluctuations of matrix elements were considered in [37].

— Higher order terms in the Taylor expansion of the form factor for general chaotic systems were calculated semiclassically in the following articles: the τ^3 term was obtained in [21] and, finally, all higher orders of τ in [38, 39].

One main remaining problem is to show that other pairs of periodic orbits, which have been omitted in the evaluation of off-diagonal contributions, do indeed not contribute to the form factor in the semiclassical limit.

Another major problem is to obtain the form factor semiclassically in the regime beyond its singularity, which in the GOE-case occurs at $\tau = 1$. In this region the form factor has a different functional form, see (3). Possibly this requires to identify a further kind of periodic orbit correlations. A first step to evaluate the form factor in this regime was taken in [40].

Acknowledgement. I am indebted to Jens Marklof for introducing me to the methods of section 3. I would like to thank in particular my collaborator Klaus Richter with whom I have had many stimulating discussions. It is a pleasure to acknowledge many fruitful discussions with Petr Braun, Fritz Haake, Stefan Heusler, Sebastian Müller, Zeev Rudnik, Dominique Spehner, and Marco Turek.

References

1. M. C. Gutzwiller: *Chaos in Classical and Quantum Mechanics*, Springer, New York, (1990).
2. O. Bohigas, M. J. Giannoni and C. Schmit: *Characterization of Chaotic Quantum Spectra and Universality of Level Fluctuation Laws*, Phys. Rev. Lett. **52** (1984) 1–4.
3. J. H. Hannay and A. M. Ozorio de Almeida: *Periodic Orbits and a Correlation Function for the Semiclassical Density of States*, J. Phys. A **17** (1984) 3429–3440.
4. M. V. Berry: *Semiclassical Theory of Spectral Rigidity*, Proc. R. Soc. Lond. A **400** (1985) 229–251.
5. N. Argaman, F. Dittes, E. Doron, J. Keating, A. Kitaev, M. Sieber and U. Smilansky: *Correlations in the Actions of Periodic Orbits Derived from Quantum Chaos*, Phys. Rev. Lett. **71** (1993) 4326–4329.
6. U. Smilansky and B. Verdene: *Action Correlations and Random Matrix Theory*, J. Phys. A **36** (2003) 3525–3549.
7. F. Haake: *Quantum Signatures of Chaos*, Springer, Berlin, (2000), 2nd edn.
8. E. Bogomolny, B. Georgeot, M. J. Giannoni and C. Schmit: *Chaotic Billiards Generated by Arithmetic Groups*, Phys. Rev. Lett. **69** (1992) 1477–1480.
9. J. Bolte, G. Steil and F. Steiner: *Arithmetic Chaos and Violations of Universality in Energy Level Statistics*, Phys. Rev. Lett. **69** (1992) 2188–2191.
10. E. Bogomolny, B. Georgeot, M. J. Giannoni and C. Schmit: *Arithmetic Chaos*, Phys. Rep. **291** (1997) 219–324.
11. B. Randol: *The Length Spectrum of a Riemann Surface is Always of Unbounded Multiplicity*, Proc. of the AMS **78** (19080) 455–456.
12. J. Marklof: *Selberg's Trace Formula: An Introduction*, this volume.
13. N. Argaman, Y. Imry and U. Smilansky: *Semiclassical Analysis of Spectral Correlations in Mesoscopic Systems*, Phys. Rev. B **47** (1993) 4440–4457.
14. R. Aurich and M. Sieber: *An Exponentially Increasing Spectral Form Factor $K(\tau)$ for a Class of Strongly Chaotic Systems*, J. Phys. A **27** (1994) 1967–1979.
15. M. Sieber and K. Richter: *Correlations between Periodic Orbits and their Rôle in Spectral Statistics*, Physica Scripta **T90** (2001) 128–133.
16. M. Sieber: *Leading Off-Diagonal Approximation for the Spectral Form Factor for Uniformly Hyperbolic Systems*, J. Phys. A **35** (2002) L613–L619.

17. P. A. Braun, S. Heusler, S. Müller and F. Haake: *Statistics of Self-Crossings and Avoided Crossings of Periodic Orbits in the Hadamard-Gutzwiller Model*, Eur. Phys. J. B **30** (2002) 189–206.
18. M. Turek and K. Richter: *Leading Off-Diagonal Contribution to the Spectral Form Factor of Chaotic Quantum Systems*, J. Phys. A **36** (2003) L455–L462.
19. D. Spehner: *Spectral Form Factor of Hyperbolic Systems: Leading Off-Diagonal Approximation*, J. Phys. A **36** (2003) 7269–7290.
20. S. Müller: *Classical Basis for Quantum Spectral Fluctuations in Hyperbolic Systems*, Eur. Phys. J. B **34** (2003) 305–319.
21. S. Heusler, S. Müller, P. Braun and F. Haake: *Universal Spectral Form Factor for Chaotic Dynamics*, J. Phys. A **37** (2004) L31–L37.
22. P. Buser: *Hyperbolic Geometry, Fuchsian Groups and Riemann Surfaces*, this volume.
23. S. Heusler: *Universal Spectral Fluctuations in the Hadamard-Gutzwiller Model and beyond*, Ph.D. thesis, Universität Essen, (2003).
24. G. Berkolaiko, H. Schanz and R. S. Whitney: *Leading Off-Diagonal Correction to the Form Factor of Large Graphs*, Phys. Rev. Lett. **88** (2002) 104101-1 – 104101-4.
25. G. Berkolaiko, H. Schanz and R. S. Whitney: *Form Factor for a Family of Quantum Graphs: An Expansion to Third Order*, J. Phys. A **36** (2003) 8373–8392.
26. G. Berkolaiko: *Form Factor for Large Quantum Graphs: Evaluating Orbits with Time-Reversal*, Waves Random Media **14** (2004) S7–S27.
27. S. Gnutzmann and A. Altland: *Universal Spectral Statistics in Quantum Graphs*, Phys. Rev. Lett. **93** (2004) 194101-1 – 194101-4.
28. S. Gnutzmann and A. Altland: *Spectral Correlations of Individual Quantum Graphs*, Phys. Rev. E **72** (2005) 056215-1 – 056215-14.
29. K. Richter and M. Sieber: *Semiclassical Theory of Chaotic Quantum Transport*, Phys. Rev. Lett. **89** (2002) 206801-1 – 206801-4.
30. S. Heusler, S. Müller, P. Braun, and F. Haake: *Semiclassical Theory of Chaotic Conductors*, Phys. Rev. Lett. **96** (2006) 066804-1 – 066804-4.
31. P. Braun, S. Heusler, S. Müller, and F. Haake: *Semiclassical Prediction for Shot Noise in Chaotic Cavities*, J. Phys. A **39** (2006) L159-L165.
32. S. Heusler: *The Semiclassical Origin of the Logarithmic Singularity in the Symplectic Form Factor*, J. Phys. A **34** (2001) L483–L490.
33. J. Bolte and J. Harrison: *The Spin Contribution to the Form Factor of Quantum Graphs*, J. Phys. A **36** (2003) L433–L440.
34. J. Bolte and J. Harrison: *The Spectral Form Factor for Quantum Graphs*, in: G. Berkolaiko, R. Carlson, S. Fulling, P. Kuchment (eds.): *Quantum Graphs and Their Applications*, Contemporary Mathematics, vol. 415 (AMS, Providence, 2006).
35. T. Nagao and K. Saito: *Form Factor of a Quantum Graph in a Weak Magnetic Field*, Phys. Lett. A **311** (2003) 353–358.
36. K. Saito and T. Nagao: *Spectral Form Factor for Chaotic Dynamics in a Weak Magnetic Field*, Phys. Lett. A **352** (2006) 380-385.
37. M. Turek, D. Spehner, S. Müller, and K. Richter: *Semiclassical Form Factor for Spectral and Matrix Element Fluctuations of Multidimensional Chaotic Systems*, Phys. Rev. E **71** (2005) 016210-1 – 016210-15.

38. S. Müller, S. Heusler, P. Braun, F. Haake and A. Altland: *Semiclassical Foundation of Universality in Quantum Chaos*, Phys. Rev. Lett. **93** (2004) 014103-1 – 014103-4.

39. S. Müller, S. Heusler, P. Braun, F. Haake and A. Altland: *Periodic-Orbit Theory of Universality in Quantum Chaos*, Phys. Rev. E **72** (2005) 046207-1 – 046207-30.

40. E. B. Bogomolny and J. P. Keating: *Gutzwiller's Trace Formula and Spectral Statistics: Beyond the Diagonal Approximation*, Phys. Rev. Lett. **77** (1996) 1472-1475.

IV. Transfer Operators, the Selberg Zeta Function and the Lewis-Zagier Theory of Period Functions

Dieter H. Mayer

Institut für Theoretische Physik, TU-Clausthal, D-38678 Clausthal-Zellerfeld, Germany, dieter.mayer@tu-clausthal.de

Summary. In these lectures we discuss the transfer operator approach to the Selberg zeta function for the geodesic flow on the unit tangent bundle of a modular surface $\Gamma \setminus \mathbb{H}$. Thereby Γ denotes a subgroup of the full modular group $\mathrm{SL}(2, \mathbb{Z})$ of finite index and $\mathbb{H}$ is the hyperbolic plane. It turns out that this function can be expressed in terms of the Fredholm determinant of the classical transfer operator $\mathcal{L}_\beta$ of this flow when appropriately extended to the complex "temperature"-plane β.

The work of J. Lewis and D. Zagier, respectively our own work for the full modular group $\mathrm{SL}(2, \mathbb{Z})$ has shown that the eigenfunctions of this transfer operator at those β-values which belong to the zeroes of the Selberg function are closely related to automorphic forms for this group, both holomorphic and real analytic. For negative integer values of β they coincide with the period polynomials of Eichler, Shimura and Manin which justifies to call them quite generally period functions.

Of special interest are the eigenfunctions of the transfer operator $\mathcal{L}_\beta$ for β-values on the critical line $\Re\,\beta = \frac{1}{2}$. They are related to the Maass wave forms, that means the eigenfunctions of the hyperbolic Laplacian on the surface. In the language of quantum mechanics these functions are the eigenstates of the Schrödinger operator for a free particle moving on such a surface with constant negative curvature. It is known that the corresponding classical system, namely the geodesic flow on the surface, is highly chaotic. The theory of period functions establishes therefore an interesting relation between classical and quantum physics: the eigenfunctions of the classical transfer operator can be used to construct the eigenstates of the quantum system. This correspondence does not make any use of the trace formulas commonly used in the theory of quantum chaos.

It is well known that this quantum system has infinitely many internal symmetries described by the so called Hecke operators. Through the Lewis-Zagier correspondence for $\mathrm{SL}(2, \mathbb{Z})$ these operators act also on the eigenfunctions of the transfer operator and hence are expected to describe some new kind of symmetries for the classical system. It turns out that these operators can be derived from the transfer operators for certain modular subgroups of the full modular group.

1 Introduction

In these lectures we discuss the so called transfer operator approach to Selberg's zeta function for modular groups and its connection to the Lewis-Zagier theory of period functions. Interest in this function stems from the fact that it has both a dynamical interpretation and belongs at the same time to the family of number theoretic zeta functions like the famous Riemann zeta function. Dynamical zeta functions share many of the properties of the arithmetic zeta functions which in special cases like the Selberg or the Artin-Weil zeta function for algebraic varieties are even dynamical functions.

Obviously the classical methods of number theory cannot be applied in general in the theory of dynamical zeta functions so that zeta functions like the one by Selberg can serve as a testing ground for the applicability and efficiency of new dynamical methods. One of the few new dynamical methods developed recently for general dynamical zeta functions is the transfer operator method. This operator can be regarded as a generalized Perron-Frobenius operator well known from the ergodic theory of dynamical systems. The transfer operator has its origin in statistical mechanics [16] and in quantum field theory and is used there to calculate so called partition functions. These are sums of exponential functions, the so called Boltzmann factors, over finite configuration spaces and are of utmost importance in statistical mechanics. It was D. Ruelle who realized that the zeta functions of both discrete and continuous time dynamical systems can be expressed in the so called thermodynamic formalism [30] in terms of partition functions of 1-dim. lattice spin systems and hence can be investigated by these transfer operators. His method has been used to determine the analytic properties of some of these functions and to characterize their zeroes and poles by the spectrum of the transfer operator. Obviously such a spectral interpretation is one of the big problems also for all arithmetic zeta functions: we mention only the recent work of A. Connes [9] on the Riemann zeta function and more general L-functions of number theory.

Another characteristic property of zeta functions are the functional equations. Unfortunately, even for dynamical zeta functions which have their origin in arithmetics like the Selberg function and for which the functional equations are well known it was not possible up to now to derive these equations through the transfer operator. This is obviously a challenging problem and its solution would add to the inherent importance of the transfer operator method in the general theory of zeta functions.

The dynamical system underlying the Selberg zeta function is the geodesic flow on the unit tangent bundle of a surface of constant negative curvature. In physical terms this is just a particle moving freely on such a surface with constant kinetic energy. The quantized version of this system is described by the free Schrödinger operator which up to some multiplicative constant involving the mass and Planck's constant h is just the Laplace-Beltrami operator for the hyperbolic metric.

In Selberg's classical approach to his zeta function via the trace formula it is shown that this function has a meromorphic continuation to the entire complex plane for surfaces with finite hyperbolic volume like all the modular surfaces. The zeroes of this function fall into two classes: the trivial zeroes s_n corresponding to the zeroes $\zeta_R(2s_n) = 0$ of the Riemann zeta function ζ_R and the zeroes s_n on the line $\Re s = \frac{1}{2}$ corresponding to the eigenvalues $\lambda_n = s_n(1 - s_n)$ of the Laplace-Beltrami operator $-\triangle_{LB}$ on the corresponding surface. These eigenvalues, more precisely their statistics, played a central role in recent years in the theory of quantum chaos [33]. This theory is concerned with the relation between quantum and classical systems with chaotic behaviour.

The special interest in geodesic flows on surfaces of constant negative curvature stems from the fact that they are well understood both classically and quantum mechanically: their quantum mechanics can be directly related to well known problems of analytic number theory like the existence and explicit determination of Maass wave forms or the existence of Hecke operators. Hence deep results in number theory have found new interpretations in quantum chaos and on the other hand numerical results from quantum chaos have lead to new questions and new approaches to open problems in number theory [33].

For general dynamical systems and their zeta functions however number theoretic methods are obviously not available and one is forced to develop new approaches which obviously have to be of a more dynamical nature. One such dynamical approach is the transfer operator method [2]. It has been applied with some success to the class of uniformly hyperbolic systems and is up to now the only way to extend the domain of analyticity of their dynamical zeta functions.

In the first lecture we will discuss in some detail this transfer operator approach to the Selberg zeta function for the modular groups, that means subgroups of the full modular group SL(2, $\mathbb{Z}$) of finite index. Recently this approach has been extended by us to the Hecke triangle groups [26]. It turns out that for these groups there exists a simple symbolic dynamics which allows an explicit expression for the transfer operator and its Fredholm determinant from which many of the properties of the Selberg function can be read off. Knowing this explicit relation between the Selberg zeta function and the Fredholm determinant of the transfer operator one arrives at a new spectral interpretation of the poles and zeroes of the Selberg function.

Through the work of Lewis and Zagier [17], [18], [19] for the full modular group it became clear that certain eigenfunctions of the transfer operator are closely related to the holomorphic forms of even weight, respectively the real analytic forms of zero weight for the group SL(2, $\mathbb{Z}$). These eigenfunctions are called period functions by these authors since they generalize the period polynomials of Eichler, Shimura and Manin [12], [38] and the period functions of Zagier for holomorphic cusp respectively non-cusp forms.

In the second lecture we will discuss this theory of Lewis and Zagier and their main result establishing a 1-1 correspondence between the eigenfunctions of the transfer operator L_β with eigenvalue $\lambda = \pm 1$ on the line $\Re \beta = \frac{1}{2}$ and the Maass wave forms for $SL(2, \mathbb{Z})$. The main ingredient for this result is an extension of a converse theorem of Maass for L-functions to the period functions.

An important internal symmetry of the Laplace-Beltrami operator for the modular groups manifests itself through the existence of an infinite family of selfadjoint operators commuting among themselves and with the Laplace operator. These are the Hecke operators [1]. Through the Lewis-Zagier correspondence these operators can be transferred to the space of period functions. The resulting operators have been determined in the case of $SL(2, \mathbb{Z})$ for the period polynomials by Choie and Zagier [8] and recently for the period functions by T. Mühlenbruch [28]. By using the transfer operators for the Hecke congruence subgroups $\Gamma_0(n) \subset SL(2, \mathbb{Z})$ and certain of their eigenfunctions we could derive for any of these groups an infinite family of linear operators [15] mapping eigenspaces of the transfer operator into themselves. For the group $\Gamma_0(1) = SL(2, \mathbb{Z})$ these operators are closely related to the Hecke operators of Mühlenbruch so that we expect that also our operators for $\Gamma_0(n)$ are more or less the Hecke operators for these groups acting on the space of period functions [1].

An open problem for these and more general subgroups Γ of the full modular group is a converse theorem establishing for any eigenfunction of the transfer operator for Γ with eigenvalue $\lambda = \pm 1$ on the line $\Re \beta = \frac{1}{2}$ the existence of exactly one Maass wave form for this group [2]. The main problem here seems to be the corresponding converse theorem for L-functions which to our knowledge has been shown only within the Jacquet-Langlands adelic approach. One would rather need a classical form of this result to connect it with the theory of period functions. How one can associate to a Maass wave form a period function for the modular groups on the other hand is more or less well understood.

In the third lecture we will discuss certain eigenfunctions of the transfer operator for $\Gamma_0(n)$ and derive the corresponding Hecke-like operators for this group.

Most of the results discussed in these lectures have their origin in published work by John Lewis and Don Zagier (see References for details) and in our own work with Cheng Chang, Joachim Hilgert and Hossein Movasati (see References for details) supported by the German Research Foundation DFG through the research group "Zeta Functions and Locally Symmetric Spaces".

Highly appreciated is the help of Waltraud Hintze in the preparation of a written version of these notes.

---

[1] This has been shown recently by Fraczek et al. in [13]
[2] This has been solved recently by Deitmar and Hilgert in [11]

2 The geodesic flow on a modular surface $\Gamma \backslash \mathbb{H}$

We denote by $\mathbb{H}$ the Poincaré upper half plane

$$\mathbb{H} = \{ z \in \mathbb{C} \; : \; z = x + iy, \quad y > 0 \} \tag{1}$$

equipped with the hyperbolic metric

$$ds^2 = y^{-2}(\mathrm{d}x^2 + \mathrm{d}y^2). \tag{2}$$

The group of isometries of this space is

$$\mathrm{PSL}(2, \mathbb{R}) = \mathrm{SL}(2, \mathbb{R})/\{+1, -1\} \tag{3}$$

with $\mathrm{SL}(2, \mathbb{R}) = \left\{ g = \begin{pmatrix} a & b \\ c & d \end{pmatrix}, \quad a, b, c, d \in \mathbb{R}, \quad \det g = 1 \right\}$. This group acts on $\mathbb{H}$ through the Möbius transformations

$$gz = \frac{az + b}{cz + d}, \quad g = \begin{pmatrix} a & b \\ c & d \end{pmatrix} \in \mathrm{SL}(2, \mathbb{R}). \tag{4}$$

The discrete subgroup $\mathrm{SL}(2, \mathbb{Z}) \subset \mathrm{SL}(2, \mathbb{R})$ with

$$\mathrm{SL}(2, \mathbb{Z}) = \{ g \in \mathrm{SL}(2, \mathbb{R}), \quad a, b, c, d \in \mathbb{Z} \}$$

is called the full modular group.

A subgroup $\Gamma \subset \mathrm{SL}(2, \mathbb{Z})$ of finite index $[\mathrm{SL}(2, \mathbb{Z}) : \Gamma] < \infty$ is called a modular group.

When acting on the hyperbolic 2-space $\mathbb{H}$ we identify the groups Γ and $\mathrm{P}\Gamma = \Gamma/\{+1, -1\}$ if $-1 \in \Gamma$.

The full modular group $\mathrm{SL}(2, \mathbb{Z})$ has two generators for which we take the two elements Q and T with

$$Qz = -\frac{1}{z} \quad \text{and} \quad Tz = z + 1. \tag{5}$$

They fulfill the relations
$$Q^2 = (QT)^3 = 1.$$

The orbit space $\Gamma \backslash \mathbb{H}$ of the modular group Γ is called a modular surface where $z_1, z_2 \in \mathbb{H}$ are identified as

$$z_1 \sim z_2 \Leftrightarrow z_2 = gz_1$$

for some $g \in \Gamma$.

It is well known that these modular surfaces are not compact but have finite hyperbolic area.

A closed region $F_\Gamma \subset \mathbb{H}$ is called a fundamental domain for the group Γ if for all $z \in \mathbb{H}$ there exists $g \in \Gamma$ and $z_0 \in F_\Gamma$ with $z = gz_0$. If $z_1, z_2 \in F_\Gamma$

and $z_1 = gz_2$ for some $g \in \Gamma$, $g \neq id$, then z_1, z_2 are both on the boundary ∂F_Γ of F_Γ.

Obviously the modular surface $\Gamma \setminus \mathbb{H}$ can be identified with the space $F_\Gamma / \sim$ where points in F_Γ equivalent under Γ are identified.

Denote by $\overline{\mathbb{H}} := \mathbb{H} \cup \{\infty\} \cup \mathbb{Q}$ the compactified hyperbolic surface where $\mathbb{Q}$ denotes the rational numbers. The points ∞ and $\frac{p}{q} \in \mathbb{Q}$ are called cusps.

Extending the action of the modular group Γ to the real line $\{ z = x + iy :\ y = 0 \}$ these cusps decompose into several equivalence classes. A cusp of the modular group Γ is just one such equivalence class. Geometrically these are the points on the boundary of the modular surface at infinity.

The geodesic lines in the hyperbolic 2-space are the half circles based on the real axis. For infinite diameter these are the lines parallel to the imaginary axis $\Re z = 0$.

An oriented geodesic γ will be denoted by the two base points $\gamma_{-\infty}, \gamma_{+\infty} \in \mathbb{R} \cup \{\infty\}$ on the real line: we hence write $\gamma = (\gamma_{-\infty}, \gamma_{+\infty})$. Denote next by $\mathcal{A}$ the following set of geodesics

$$\mathcal{A} := \{\, \gamma = (\gamma_{-\infty}, \gamma_{+\infty})\ :\ 0 < |\gamma_{-\infty}| \leq 1 \leq |\gamma_{+\infty}|, \quad \gamma_{-\infty}\gamma_{+\infty} < 0 \,\}. \quad (6)$$

It is not too difficult to see that for almost every geodesic γ in $\mathbb{H}$ there exists an element $g \in \mathrm{SL}(2, \mathbb{Z})$ with $g\gamma \in \mathcal{A}$.

This shows that almost every geodesic in $\mathrm{SL}(2, \mathbb{Z}) \setminus \mathbb{H}$ has a representative in the set $\mathcal{A}$.

Denote by $\pi^\Gamma : \mathbb{H} \to \Gamma \setminus \mathbb{H}$ the covering map $\pi^\Gamma(z) = \Gamma z$.

Obviously for every geodesic γ in $\mathbb{H}$ $\pi^\Gamma \gamma$ defines a geodesic in $\Gamma \setminus \mathbb{H}$: if $\gamma_1 = g\gamma_2$ for some $g \in \Gamma$ then $\pi^\Gamma \gamma_1 = \pi^\Gamma \gamma_2$. The geodesic flow ϕ_t on $\mathbb{H}$ is the flow along the geodesics with velocity v with $|v| = 1$.

If $S\mathbb{H}$ denotes the unit tangent bundle of $\mathbb{H}$ then the geodesic flow $\phi_t :\ S\mathbb{H} \to S\mathbb{H}$ is given as

$$\phi_t(z, v) = (z_t, v_t), \quad (7)$$

where z_t lies a hyperbolic distance t away from z on the geodesic through z with unit tangent vector v and v_t is the unit tangent vector at this geodesic in the point z_t.

Obviously for the projection $\pi_*^\Gamma : S\mathbb{H} \to S(\Gamma \setminus \mathbb{H})$ the induced flow $\pi_*^\Gamma \phi_t (\pi_*^\Gamma)^{-1} : S(\Gamma \setminus \mathbb{H}) \to S(\Gamma \setminus \mathbb{H})$ is the geodesic flow on the unit tangent bundle of the modular surface $\Gamma \setminus \mathbb{H}$. We will denote this geodesic flow by ϕ_t^Γ.

It is well known that the geodesic flow on surfaces of constant negative curvature is uniformly hyperbolic or Anosov and hence allows a description in terms of symbolic dynamics. The surprising fact for the modular surfaces is that this symbolic dynamics can be described in explicit form, whereas for general Anosov systems or Axiom-A systems only existence of such a description is known.

2.1 Symbolic dynamics of the geodesic flow for modular surfaces

If $\Gamma \subset \mathrm{SL}(2,\mathbb{Z})$ has finite index $\mu = [\mathrm{SL}(2,\mathbb{Z}) : \Gamma]$ then $M_\Gamma := \Gamma \setminus \mathbb{H}$ is an in general ramified μ-fold covering of the modular surface $M := \mathrm{SL}(2,\mathbb{Z}) \setminus \mathbb{H}$. Denote then by $\pi_\Gamma : M_\Gamma \to M$ the covering map

$$\pi_\Gamma(\Gamma h) = \mathrm{SL}(2,\mathbb{Z})h. \tag{8}$$

Hence also $\pi_{\Gamma*} : SM_\Gamma \to SM$ maps the orbits of the geodesic flow on SM_Γ onto the orbits of the geodesic flow on SM. To get the symbolic dynamics of the geodesic flow on SM_Γ one only has to lift the symbolic dynamics of this flow on SM to SM_Γ. The symbolic dynamics for the flow $\phi_t : SM \to SM$ has been determined by Series [34].

To get a Poincaré section for this flow consider the set

$$S = \{\, z = i\varrho \ : \ \varrho \geq 1 \,\} \subset \mathbb{H}. \tag{9}$$

To any $z \in S$ attach the tangent vectors

$$C_z = \{\, v \in T_z\mathbb{H} \ : \ |v| = 1 \text{ such that either } \gamma(z,v) \text{ or } Q\gamma(z,v) \in \mathcal{A} \,\},$$

where $\gamma(z,v)$ denotes the geodesic determined by z and v. Then the set

$$X = \bigcup_{\substack{z \in S \\ v \in C(z)}} \pi_*(z,v) \subset SM \tag{10}$$

is a Poincaré section for the geodesic flow on SM. Indeed almost all orbits of ϕ_t hit X transversally. The map $\varrho : X \to \mathcal{A}$ defined as

$$\varrho(\pi_*(z,v)) = \gamma(z,v), \tag{11}$$

respectively

$$\varrho(\pi_*(z,v)) = Q\gamma(z,v), \tag{12}$$

is one-to-one up to the two geodesics $\gamma_\pm$ with

$$\gamma_\pm = (\pm 1, \mp 1)$$

which are mapped onto each other by Q and hence correspond to the same point in X.

An orbit $\phi_t(z,v)$ in SM is periodic iff the geodesic $\gamma(z,v)$ in $\mathbb{H}$ is fixed by an element $g \in \mathrm{SL}(2,\mathbb{Z})$, that means for $\gamma = (\gamma_{-\infty}, \gamma_{+\infty})$

$$g\gamma_{\pm\infty} = \gamma_{\pm\infty}. \tag{13}$$

If $\pi_*(z,v) \in X$ is a point on this periodic orbit then $\pi_*(z,v)$ is a periodic point of the Poincaré map $P : X \to X$ with

$$Px = \phi_{r(x)}x, \tag{14}$$

where $r : X \to \mathbb{R}_+$ denotes the recurrence time for the points in the Poincaré section X, that means

$$r(x) = \min\{\, t > 0 \ : \ \phi_t(x) \in X \quad \text{for} \quad x \in X\,\}. \tag{15}$$

To determine this Poincaré map consider for $x \in X$, $x = (z, v)$, the orbit $\gamma = (\gamma_{-\infty}, \gamma_{+\infty}) \in \mathcal{A}$ with $\varrho(\pi_+(z, v)) = \gamma$. Set $\epsilon = 1$ if $\gamma_{-\infty} < 0 < \gamma_{+\infty}$, respectively $\epsilon = -1$ if $\gamma_{+\infty} < 0 < \gamma_{-\infty}$. Then the oriented orbit $\gamma = (\gamma_{-\infty}, \gamma_{+\infty})$ can be identified with

$$\gamma = (x_1, x_2, \epsilon), \quad (x_1, x_2) \in I_2 = \{\, 0 \leq x_1 \leq 1, \ 0 \leq x_2 \leq 1\,\}, \quad \epsilon \in \{1, -1\}, \tag{16}$$

such that $\gamma_{-\infty} = -\epsilon x_2$, $\gamma_{+\infty} = \epsilon x_1^{-1}$. If $n < \gamma_{+\infty} < n + 1$, then the orbit γ' defined as

$$\gamma' = QT^{-n}\gamma$$

belongs to the set $\mathcal{A}$ since $\gamma' = (\gamma'_{-\infty}, \gamma'_{+\infty})$ with $0 < \gamma'_{-\infty} < 1$ and $\gamma'_{+\infty} < -1$. This orbit hence determines again a point $x' \in X$ and hence $x' = Px$.

In the parametrization $\gamma = (x_1, x_2, \epsilon)$ of the elements γ in $\mathcal{A}$ one finds for $\gamma' = (x'_1, x'_2, \epsilon')$

$$(x'_1, x'_2, \epsilon') = P(x_1, x_2, \epsilon) = \left(\frac{1}{x_1} - n, \ \frac{1}{x_2 + n}, \ -\epsilon\right) \quad \text{with} \quad n = \left[\frac{1}{x_1}\right]. \tag{17}$$

It is not difficult to convince oneself that exactly the same formula holds also in the case $\gamma_{+\infty} < -1$.

If $T_G : I \to I$ denotes the Gauss map

$$T_G x_1 = \frac{1}{x_1} \quad \text{mod } 1 \tag{18}$$

on the unit interval $I = \{\, x_1 \ : \ 0 < x_1 < 1\,\}$ then the Poincaré map for the geodesic flow on SM can be written as

$$P(x_1, x_2, \epsilon) = \left(T_G x_1, \ \frac{1}{x_2 + [\frac{1}{x_1}]}, \ -\epsilon\right). \tag{19}$$

Hence P can be considered a natural extension of the Gauss map of the unit interval by the group $\mathbb{Z}_2$. Obviously the closed orbits of $\phi_t : SM \to SM$ can be characterized now by the periodic points of the Poincaré map P in (19). Because of the twist in the variable ϵ there exist only periodic points of even period $2k$, hence the coordinate x_1 of a point x in the section X must be a periodic point under the Gauss map of period $2k$: But $T_G^{2k} x_1 = x_1$ iff x_1 has a periodic continued fraction expansion of period $2k$, that means

$$x_1 = \left[\overline{n_1, \cdots, n_{2k}}\right], \quad n_i \in \mathbb{N}, \quad 1 \leq i \leq 2k,$$

where $\left[\overline{n_1,\cdots,n_{2k}}\right]$ denotes the continued fraction $[n_1, n_2, \cdots]$ with $n_i = n_{i+2k}$ for all $i \in \mathbb{N}$.

A simple calculation then shows that $x = (x_1, x_2, \epsilon)$ is a periodic point of P of period $2k$ iff $x_2 = \left[\overline{n_{2k}, n_{2k-1}, \cdots, n_1}\right]$. Hence the periodic points of the Poincaré map P of the geodesic flow $\phi_t : SM \to SM$ are in 1-1 correspondence with the set

$$\left\{ (\left[\overline{n_1, \cdots, n_{2k}}\right], \epsilon), \quad n_i \in \mathbb{N}, \quad k \in \mathbb{N}, \quad \epsilon = \pm 1 \right\}.$$

To determine the periods of the corresponding closed orbits one has to determine their recurrence times with respect to the Poincaré section X. These are given by the geodesic distances between subsequent hits of the Poincaré section by the orbit: if $\gamma = (\gamma_{-\infty}, \gamma_{+\infty})$ with $n \le \gamma_{+\infty} < n + 1$, then denote by $z_1 = \xi$ the intersection point of γ with the line $\Re z = 0$ and by $z_2 = \eta$ the intersection of γ with the line $\Re z = n$. Then the geodesic distance between these two points is given by [10]

$$r(x) = \log \frac{|\gamma_{-\infty} - \eta||\gamma_\infty - \xi|}{|\gamma_{-\infty} - \xi||\gamma_\infty - \eta|} \tag{20}$$

when γ hits X in the point $x = (z_1, v_1)$ and v_1 is the tangent vector to γ in z_1.

If the base points $\gamma_{\pm\infty}$ of γ are given as

$$\gamma_{-\infty} = -[n_0, n_{-1}, n_{-2}, \cdots], \quad \gamma_{+\infty} = n_1 + [n_2, n_3, \cdots], \tag{21}$$

one finds

$$r(x) = -\frac{1}{2}\log([n_1, n_2, \cdots][n_0, n_{-1}, n_{-2}, \cdots][n_2, n_3, \cdots][n_1, n_0, n_{-1}, \cdots]). \tag{22}$$

For a periodic orbit γ with

$$\gamma_\infty^{-1} = \left[\overline{n_1, \ldots, n_{2k}}\right] \quad \text{and} \quad \gamma_{-\infty} = -\left[\overline{n_{2k}, n_{2k-1}, \ldots, n_1}\right]$$

one then derives for the period $l(\gamma)$

$$l(\gamma) = -\frac{1}{2}\sum_{r=1}^{2k}\log\Big\{ \left[\overline{n_r, n_{r+1}, \ldots, n_{2k}, n_1, \ldots, n_{r-1}}\right] \cdot$$
$$\cdot \left[\overline{n_{r-1}, n_{r-2}, \ldots, n_1, n_{2k}, \ldots, n_r}\right] \cdot$$
$$\cdot \left[\overline{n_{r+1}, n_{r+2}, \ldots, n_{2k}, n_1, \ldots, n_r}\right] \cdot$$
$$\cdot \left[\overline{n_r, n_{r-1}, \ldots, n_1, n_{2k}, \cdots n_{r+1}}\right]\Big\}.$$

Using a result by Legendre, namely

$$\prod_{r=1}^{2k} \left[\overline{n_r, n_{r+1}, \ldots n_{2k}, n_1, \ldots, n_{r-1}}\right] = \prod_{r=1}^{2k} \left[\overline{n_r, n_{r-1}, \ldots, n_1, n_{2k}, \ldots, n_{r+1}}\right],$$

we finally get for the period $l(\gamma)$ of the periodic orbit γ:

$$l(\gamma) = \log \prod_{r=1}^{2k} \left[\overline{n_r, n_{r+1}, \ldots, n_{2k}, n_1, \ldots, n_{r-1}} \right]. \tag{23}$$

2.2 The Poincaré map for the geodesic flow on M_Γ

If $\Gamma \subset \mathrm{SL}(2, \mathbb{Z})$ has finite index $\mu = [\mathrm{SL}(2, \mathbb{Z}) : \Gamma]$ consider the coset decomposition

$$\mathrm{SL}(2, \mathbb{Z}) = \bigcup_{i=1}^{\mu} \Gamma g_i \tag{24}$$

with $g_i \in \mathrm{SL}(2, \mathbb{Z})$, $1 \le i \le \mu$. Define next for Γ the set of oriented geodesics $\mathcal{A}_\Gamma$ as

$$\mathcal{A}_\Gamma := \bigcup_{i=1}^{\mu} g_i \mathcal{A}, \tag{25}$$

where $\mathcal{A}$ was defined in (6). If $\pi_\Gamma : M_\Gamma \to M$ is the natural projection of M_Γ onto M and $\pi_{\Gamma*}$ its extension to the unit tangent bundles SM_Γ and SM then a Poincaré section $X_\Gamma \subset SM_\Gamma$ for the geodesic flow on SM_Γ is obviously given by

$$X_\Gamma = (\pi_{\Gamma*})^{-1} X, \tag{26}$$

where X is the Poincaré section for the geodesic flow on SM as defined in (10).

There is again a 1–1 correspondence between X_Γ and the set $\mathcal{A}_\Gamma$ in (25) besides a finite number of geodesics in $\mathcal{A}_\Gamma$ corresponding to the same point in X_Γ if Q belongs to Γ. To determine the Poincaré map $P_\Gamma : X_\Gamma \to X_\Gamma$ take a point $x \in X_\Gamma$ corresponding to the geodesic $\gamma \in \mathcal{A}_\Gamma$. For $g_i \in \mathrm{SL}(2, \mathbb{Z})$ with $g_i^{-1}\gamma \in \mathcal{A}_\Gamma$ we know that $QT^{-n\epsilon}g_i^{-1}\gamma \in \mathcal{A}_\Gamma$ when $g_i^{-1}\gamma = (\gamma_{-\infty}, \gamma_{+\infty}, \epsilon)$ and $\gamma_{+\infty} = \epsilon x_1^{-1}$ with $n = [\frac{1}{x_1}]$.

In general however $QT^{-n\epsilon}g_i^{-1}$ does not belong to Γ and hence the orbit $QT^{-n\epsilon}g_i^{-1}\gamma$ does not define the same orbit on M_Γ. Since however

$$\mathrm{SL}(2, \mathbb{Z}) = \bigcup_{i=1}^{\mu} g_i^{-1} \Gamma \tag{27}$$

there exists a unique $g_j \in \mathrm{SL}(2, \mathbb{Z})$, $1 \le j \le \mu$ such that

$$g_j QT^{-n\epsilon}g_i^{-1} \in \Gamma. \tag{28}$$

But then $g_j QT^{-n\epsilon}g_i^{-1}\gamma \in \mathcal{A}_\Gamma$ describes the same orbit in SM_Γ. Hence it defines also a unique point $x' \in X_\Gamma$ which is just the image of $x \in X_\Gamma$ under the Poincaré map P_Γ.

Any geodesic γ in $\mathcal{A}_\Gamma$ can be characterized as

$$\gamma = (x_1, x_z, \epsilon, \Gamma g_i),$$

if $g_i^{-1}\gamma \in \mathcal{A}_\Gamma$ and $g_i^{-1}\gamma = (x_1, x_2, \epsilon)$ as defined in (16). The Poincaré map $P_\Gamma : X_\Gamma \to X_\Gamma$ then has the explicit form

$$P_\Gamma(x_1, x_2, \epsilon, \Gamma g_i) = \left(T_G x_1, \frac{1}{x_2 + [\frac{1}{x_1}]}, -\epsilon, \Gamma g_j \right) \tag{29}$$

with g_j uniquely determined by

$$\Gamma g_i T^{n\epsilon} Q = \Gamma g_j.$$

Obviously $P_\Gamma : X_\Gamma \to X_\Gamma$ can be considered an extension of the map $P : X \to X$ by the set $\Gamma \backslash \mathrm{SL}(2, \mathbb{Z})$. For $\Gamma = \mathrm{SL}(2, \mathbb{Z})$ this set consists just of one point and X_Γ can be identified with X.

3 The transfer operator for the geodesic flow on SM_Γ

In this section we will derive the transfer operator for a general subgroup $\Gamma \subset \mathrm{SL}(2, \mathbb{Z})$ with finite index $\mu = [\mathrm{SL}(2, \mathbb{Z}) : \Gamma]$. For this we recall the general definition of the Ruelle transfer operator $\mathcal{L}_T$ for a map $T : M \to M$ and an observable $A : M \to \mathbb{C}$ [2] where we do not specify at the moment the space of functions f on which the operator's action is well defined:

$$(\mathcal{L}_T f)(x) = \sum_{y \in T^{-1}x} \exp\left(A(y)\right) f(y). \tag{30}$$

For uniformly hyperbolic systems like Anosov or Axiom A systems it is known from the work of R. Bowen and D. Ruelle (see [3], [30]) that the observable

$$A(x) = -\log |DT_{ex}(x)| \tag{31}$$

is of special interest for the ergodic theory of these systems, where T_{ex} denotes the restriction of the map T to its expanding directions. In the case of the Poincaré map $P_\Gamma : X_\Gamma \to X_\Gamma$ one then finds

$$P_{\Gamma,ex}(x_1, \epsilon, \Gamma g_i) = (T_G x_1, -\epsilon, \Gamma g_j), \tag{32}$$

where g_j is given as in (28). Here we used the fact that the expanding direction of P_Γ consists of 2μ copies of the x_1 coordinate. But then one sees that

$$P_{\Gamma,ex}^{-1}(x_1, \epsilon, \Gamma g) = \left\{ \left(\frac{1}{x+n}, -\epsilon, \Gamma g Q T^{n\epsilon} \right), \quad n \in \mathbb{N} \right\}. \tag{33}$$

The generalized transfer operator $\mathcal{L}_\beta^\Gamma$ for the geodesic flow $\Phi_t^\Gamma : SM_\Gamma \to SM_\Gamma$ hence has the form

$$(\mathcal{L}_\beta^\Gamma f)(x,\epsilon,\Gamma g) = \sum_{n=1}^\infty \frac{1}{(x+n)^{2\beta}} f\left(\frac{1}{x+n}, -\epsilon, \Gamma g Q T^{n\epsilon}\right), \qquad (34)$$

since $-\log|DP_{\Gamma,ex}(x,\epsilon,\Gamma g)| = \log x^2$. We have introduced in the definition (30) of the transfer operator a complex parameter β which in statistical mechanics describes the inverse temperature. In the space of vector valued functions

$$\underline{f} : I \times \mathbb{Z}_2 \to \mathbb{C}^\mu \qquad (35)$$

with $I = [0,1]$ the unit interval, the operator $\mathcal{L}_\beta^\Gamma$ in (34) induces the operator $\mathcal{L}_\beta^{\chi_\Gamma}$ with

$$(\mathcal{L}_\beta^{\chi_\Gamma} \underline{f})(x,\epsilon) = \sum_{n=1}^\infty \frac{1}{(x+n)^{2\beta}} \chi_\Gamma(QT^{n\epsilon})\underline{f}\left(\frac{1}{x+n}, -\epsilon\right), \qquad (36)$$

where $\chi_\Gamma : \mathrm{SL}(2,\mathbb{Z}) \to \mathrm{End}\,\mathbb{C}^\mu$ denotes the representation of the group $\mathrm{SL}(2,\mathbb{Z})$ induced from the trivial 1-dimensional representation of the group Γ: if $\omega_\Gamma : \mathrm{SL}(2,\mathbb{Z}) \to \mathbb{R}$ is the characteristic function of the group $\Gamma \subset \mathrm{SL}(2,\mathbb{Z})$ with

$$\omega_\Gamma(g) = \begin{cases} 1 & \text{if } g \in \Gamma, \\ 0 & \text{if } g \notin \Gamma, \end{cases}$$

then the representation $\chi_\Gamma : \mathrm{SL}(2,\mathbb{Z}) \to \mathrm{End}\,\mathbb{C}^\mu$ is defined as

$$\chi_\Gamma(g) = (\omega_\Gamma(g_i g g_j^{-1}))_{1\le i,j\le \mu}. \qquad (37)$$

One realizes immediately that the matrix $\chi_\Gamma(g)$ is for any $g \in \mathrm{SL}(2,\mathbb{Z})$ a permutation matrix and hence there exists for any $g \in \mathrm{SL}(2,\mathbb{Z})$ a number $r > 0$ with $\chi_\Gamma(g^r) = id$. In general the representation χ_Γ is not irreducible. If

$$\chi_\Gamma = \oplus_i \chi_i$$

is its decomposition into the irreducible components χ_i, then also the transfer operator $\mathcal{L}_\beta^{\chi_\Gamma}$ in (36) can be decomposed as

$$\mathcal{L}_\beta^{\chi_\Gamma} = \bigoplus \mathcal{L}_\beta^{\Gamma,\chi_i}. \qquad (38)$$

Well understood is the case when Γ is a normal subgroup of $\mathrm{SL}(2,\mathbb{Z})$. Then the representation χ_Γ is isomorphic to the right regular representation of the factor group $\Gamma \backslash \mathrm{SL}(2,\mathbb{Z})$ and hence $\chi_\Gamma(h) = id$ for all $h \in \Gamma$. Furthermore one has

$$\chi_\Gamma = \oplus_i n_i \chi_i, \qquad (39)$$

where the sum is over all irreducible unitary representations χ_i of the group $\Gamma \backslash \mathrm{SL}(2,\mathbb{Z})$ of dimension n_i. An example of such a normal subgroup is the principal congruence subgroup

$$\Gamma(2) = \left\{ g = \begin{pmatrix} a & b \\ c & d \end{pmatrix} \in \mathrm{SL}(2,\mathbb{Z}) \; : \; \begin{pmatrix} a & b \\ c & d \end{pmatrix} \mod 2 = \begin{pmatrix} 1 & 0 \\ 0 & 1 \end{pmatrix} \right\}.$$

In this case $\Gamma(2) \setminus \mathrm{SL}(2,\mathbb{Z})$ is isomorphic to S_3, the permutation group on three elements. Hence one gets [4]

$$\chi_{\Gamma(2)} = \chi_1 \oplus \chi_{-1} \oplus \chi_2 \oplus \chi_2, \tag{40}$$

where χ_1 and χ_{-1} are the two 1-dimensional representations and χ_2 is the 2-dimensional irreducible representation of S_3. For the Hecke congruence subgroup

$$\Gamma = \Gamma_0(2) = \left\{ g = \begin{pmatrix} a & b \\ c & d \end{pmatrix} \in \mathrm{SL}(2,\mathbb{Z}) \; : \; c \mod 2 = 0 \right\}$$

on the other hand one finds [4] $\mu = [\,\mathrm{SL}(2,\mathbb{Z}) : \Gamma_0(2)\,] = 3$ and

$$\chi_{\Gamma_0(2)} = \chi_1 \oplus \chi_2. \tag{41}$$

Hence in an appropriate basis the operators $\mathcal{L}_\beta^{\Gamma(2)}$ and $\mathcal{L}_\beta^{\Gamma_0(2)}$ have certain eigenfunctions and eigenvalues in common. Indeed all the eigenfunctions and eigenvalues of the operator $\mathcal{L}_\beta^{\Gamma_0(2)}$ show up also for the operator $\mathcal{L}_\beta^{\Gamma(2)}$. This obviously is related to the fact that $\Gamma(2)$ is a subgroup of $\Gamma_0(2)$.

Up to now we did not care about the spaces on which the above transfer operator are really well defined. This we will discuss in the next section.

4 Spectral properties of the transfer operators

For $D = \{ z \; : \; |z - 1| < \frac{3}{2} \}$ denote by $B(D)$ the Banach space

$$B(D) = \{ f : D \to \mathbb{C} \; : \; f \text{ holomorphic and continuous on } \overline{D} \} \tag{42}$$

with the sup norm $\| f \| = \sup_{z \in \overline{D}} |f(z)|$, and by B the direct sum

$$B = \bigoplus_{i=1}^{2\mu} B(D). \tag{43}$$

Obviously the transfer operator $\mathcal{L}_\beta^\Gamma$ is then well defined on this Banach space for all β-values with $\Re \beta > \frac{1}{2}$ and depends in this domain holomorphically on β. Choose next any $\kappa \in \mathbb{N} \cup \{0\}$. Since there exists to the element $T \in \mathrm{SL}(2,\mathbb{Z})$ a number $r > 0$ with $\chi_\Gamma(T^r) = id$, one finds by a simple Taylor expansion

$$
\left(\mathcal{L}_\beta^{\chi_\Gamma} \underline{f}\right)(z,\epsilon) = \sum_{n=1}^{\infty}\sum_{m=1}^{r}\left(\frac{1}{z+m+r(n-1)}\right)^{2\beta}\chi_\Gamma(QT^{m\epsilon})\cdot
$$

$$
\cdot\left\{\underline{f}\left(\frac{1}{z+m+r(n-1)},-\epsilon\right)\right.
$$

$$
-\sum_{l=0}^{\kappa}\frac{\underline{f}^{(l)}(0,-\epsilon)}{l!}\frac{1}{(z+m+r(n-1))^l}\bigg\}
$$

$$
+\sum_{l=0}^{\kappa}\frac{1}{r^{2\beta+l}}\sum_{m=1}^{r}\chi_\Gamma(QT^{m\epsilon})\frac{\underline{f}^{(l)}(0,-\epsilon)}{l!}\zeta\left(2\beta+l,\frac{z+m}{r}\right),
$$

where $\zeta(s,z)$ denotes the Hurwitz zeta function

$$
\zeta(s,z) = \sum_{n=0}^{\infty}\frac{1}{(z+n)^s}, \quad z\neq 0,-1,-2,\dots. \tag{44}
$$

Since $\kappa\in\mathbb{N}\cup\{0\}$ was arbitrary and the Hurwitz zeta function is meromorphic in the entire s-plane with the only pole at $s=1$ one sees from the above expression that the transfer operator $\mathcal{L}_\beta^{\chi_\Gamma}$ has a meromorphic extension to the entire complex β-plane with possible poles only at $\beta = \beta_l = \frac{1-l}{2}$, $l = 0,1,2,\dots$. From the above expression one also sees that $\mathcal{L}_\beta^{\chi_\Gamma}$ can be written as

$$
\mathcal{L}_\beta^{\chi_\Gamma} = \mathcal{L}_\beta^{\chi_\Gamma,\kappa} + \mathcal{A}_\beta^{\chi_\Gamma,\kappa}, \tag{45}
$$

with $\mathcal{A}_\beta^{\chi_\Gamma,\kappa}$ a finite rank operator mapping onto the linear span of Hurwitz functions. For $\{\underline{e}_i\}_{1\leq i\leq 2\mu}$ an orthonormal basis of $\mathbb{C}^{2\mu}$ the elements $\underline{f}$ in B can be written as

$$
\underline{f} = \sum_{i=1}^{2\mu}\underline{e}_i\otimes f_i \tag{46}
$$

with $f_i\in B(D)$. Then one finds

$$
\left(\mathcal{L}_\beta^{\chi_\Gamma,\kappa}\underline{f}\right)(z) = \sum_{i=1}^{2\mu}\sum_{m=1}^{r}\chi_m\underline{e}_i\otimes\mathcal{L}_{\beta,m}^\kappa f_i(z) \tag{47}
$$

with

$$
\chi_m = \begin{pmatrix} 0 & \chi_\Gamma(QT^m) \\ \chi_\Gamma(QT^{-m}) & 0 \end{pmatrix} \tag{48}
$$

and

$$
(\mathcal{L}_{\beta,m}^\kappa f)(z) = \sum_{n=1}^{\infty}\frac{1}{(z+m+r(n-1))^{2\beta}}\cdot
$$

$$
\cdot\left[f\left(\frac{1}{z+m+r(n-1)}\right)\right.
$$

$$
-\sum_{l=0}^{\kappa}\frac{f^{(l)}(0)}{l!}\frac{1}{(z+m+r(n-1))^l}\bigg]. \tag{49}
$$

Hence we find

$$\mathcal{L}_\beta^{\Gamma,\kappa} = \sum_{m=1}^{r} \chi_m \otimes \mathcal{L}_{\beta,m}^{\kappa}, \qquad (50)$$

which shows that the operator $\mathcal{L}_\beta^{\Gamma,\kappa}$ is nuclear of order zero [14] and hence trace class for all $\beta \neq \beta_l$ since the operators $\mathcal{L}_{\beta,m}^{\kappa}$ have this property (see [7]). Hence also the transfer operators $\mathcal{L}_\beta^{\chi_\Gamma}$ define a meromorphic family of trace class operators in the complex β-plane.

5 The Selberg zeta function

For the geodesic flow $\Phi_t : SM \to SM$ on the unit tangent bundle of the modular surface $M = \mathrm{SL}(2,\mathbb{Z}) \setminus \mathbb{H}$ and $\chi : \mathrm{SL}(2,\mathbb{Z}) \to \mathrm{End}(V)$ an arbitrary finite dimensional representation of the full modular group $\mathrm{SL}(2,\mathbb{Z})$ the dynamical zeta function $\zeta(\beta, \chi)$ of Ruelle and Smale is defined as [32]

$$\zeta(\beta, \chi) = \prod_\gamma \det\left(1 - \chi(\sigma_\gamma)e^{-\beta l(\gamma)}\right)^{-1}, \qquad (51)$$

where the product is over the periodic orbits γ of Φ_t with prime period $l(\gamma)$ and σ_γ is a representative of the hyperbolic conjugacy class in $\mathrm{SL}(2,\mathbb{Z})$ fixing a geodesic lift $\tilde{\gamma}$ of γ in $\mathbb{H}$, that means $\sigma_\gamma\tilde{\gamma} = \tilde{\gamma}$ with $\pi\tilde{\gamma} = \gamma$. For the group $\mathrm{SL}(2,\mathbb{Z})$ we have seen that the periodic orbits γ have lifts $\tilde{\gamma}$ with $\tilde{\gamma} = (\tilde{\gamma}_{-\infty}, \tilde{\gamma}_\infty) = (x_1, x_2, \epsilon)$ with $x_1 = [\overline{n_1, \ldots, n_{2k}}]$, $x_2 = [\overline{n_{2k}, \ldots, n_1}]$ and $\epsilon = \pm 1$ for some $k \in \mathbb{N}$. Obviously the element $\sigma_\gamma = QT^{n_{2k}\epsilon}QT^{-n_{2k-1}\epsilon} \cdot \ldots \cdot QT^{-n_1\epsilon}$ fixes this geodesic $\tilde{\gamma}$ since $\sigma_\gamma\tilde{\gamma}_{\pm\infty} = \tilde{\gamma}_{\pm\infty}$. The period $l(\gamma)$ of the corresponding periodic orbit γ is given in (23) as

$$l(\gamma) = \sum_{s=0}^{2k-1} r(P^s(x)) = \log \prod_{s=1}^{2k} [\overline{n_s, n_{s+1}, \ldots, n_1, \ldots, n_{s-1}}]. \qquad (52)$$

To connect the dynamical zeta function for the geodesic flow with the transfer operator one has to consider the dynamical zeta function for the Poincaré map $P : X \to X$. Indeed, if $T : M \to M$ is any map of a manifold M with $|\mathrm{Fix}\, T^m| := |\{x \in M : T^m x = x\}| < \infty$ for all m and $\mathbb{A} : M \to \mathrm{End}(\mathbb{C}^\mu)$ a map of M into the endomorphisms of $\mathbb{C}^\mu$ the dynamical zeta function $\zeta_R(z, \mathbb{A})$ of Ruelle for the map T is defined as [31]

$$\zeta_R(z, \mathbb{A}) := \exp \sum_{n=1}^{\infty} \frac{z^n}{n} Z_n(T, \mathbb{A}) \qquad (53)$$

with $Z_n(T, \mathbb{A})$ the partition functions

$$\mathbb{Z}_n(T, \mathbb{A}) := \sum_{\xi \in \mathrm{Fix}\, T^n} \mathrm{tr} \prod_{k=0}^{n-1} \mathbb{A}(T^k(\xi)). \qquad (54)$$

In [31] Ruelle proved the following Lemma:

Lemma 5.1. *The zeta function ζ_R in (53) has in its domain of definition the following representation*

$$\zeta_R(z, \mathbb{A}) = \prod_\gamma \det \left(1 - z^{n(\gamma)} \prod_{k=0}^{n(\gamma)-1} \mathbb{A}(T^k \xi_\gamma) \right)^{-1}, \qquad (55)$$

where the product is over the primitive periodic orbits of T in M with prime period $n(\gamma)$ and ξ_γ is an arbitrary point in γ.

In the special case of the Poincaré map $P : X \to X$ of the geodesic flow Φ_t with respect to the Poincaré section X in (10) and $\mathbb{A}$ given by $\mathbb{A} = \mathbb{A}_{\beta,\chi_\Gamma}(x) = \chi_\Gamma(QT^{N(x)}) \exp(-\beta r(x))$ with $x = (x_1, x_2, \epsilon) \in X$, $N(x) = -\epsilon[\frac{1}{x}]$ and $r : X \to \mathbb{R}_+$ the recurrence time function in (22) one finds for $z = 1$

$$\zeta_R(1, \mathbb{A}) = \prod_{\gamma_P} \det \left(1 - \prod_{l=0}^{2k-1} \chi_\Gamma(QT^{N(P^l x_\gamma)}) \exp \left(-\beta r(P^l x_\gamma) \right) \right)^{-1}$$

$$= \prod_{\gamma_P} \det \left(1 - \chi_\Gamma(\sigma_\gamma) \exp \left(-\beta l(\gamma) \right) \right)^{-1}.$$

But this is just the Ruelle-Smale zeta function $\zeta(\beta, \chi_\Gamma)$ in (51) if $n(\gamma_P) = 2k$ and $l(\gamma_P) = \sum_{l=0}^{2k-1} r(P^l x_\gamma)$, since we have seen that there exists a 1–1 correspondence between the prime periodic orbits γ_P of the Poincaré map and the prime periodic orbits γ of the geodesic flow $\Phi_t : SM \to SM$ such that $l(\gamma_P)$ is exactly the period $l(\gamma)$ of the periodic orbit γ and $\sigma_\gamma \in \mathrm{SL}(2,\mathbb{Z})$ is the hyperbolic element fixing the lift $\tilde{\gamma}$ of γ.

The Selberg zeta function Z_S^χ for the geodesic flow on the unit tangent bundle SM of the modular surface M and a representation $\chi : \mathrm{SL}(2,\mathbb{Z}) \to \mathrm{End}(V)$ is defined as

$$Z_S^\chi(\beta) = \prod_\gamma \prod_{k=0}^\infty \det \left(1 - \chi(\sigma_\gamma) \mathrm{e}^{-(\beta+k)l(\gamma)} \right), \qquad (56)$$

where the product is again over the prime periodic orbits γ of $\Phi_t : SM \to SM$ and $\sigma_\gamma \in \mathrm{SL}(2,\mathbb{Z})$ fixes the lift $\tilde{\gamma}$ of γ.

The Selberg zeta function Z_S^Γ on the other hand for the geodesic flow $\Phi_t^\Gamma : SM_\Gamma \to SM_\Gamma$ for any modular subgroup $\Gamma \subset \mathrm{SL}(2,\mathbb{Z})$ of finite index is defined as

$$Z_S^\Gamma(\beta) = \prod_\gamma \prod_{k=0}^\infty \left(1 - \mathrm{e}^{-(\beta+k)l(\gamma)} \right), \qquad (57)$$

where γ now denotes a prime periodic orbit of the geodesic flow on the unit tangent bundle SM_Γ of the modular surface M_Γ with prime period $l(\gamma)$.

The transfer operators $\mathcal{L}_\beta$ are quite generally sums of so called composition operators of the general form

$$(\mathcal{L}f)(z) = \phi(z)f(\psi(z)) \tag{58}$$

defined on a Banach space B of functions $f = f(z)$ holomorphic on some disc D with ϕ holomorphic on this disc and ψ some holomorphic map $\psi : D \to D$ with $\psi(\overline{D}) \subset D$. Such operators are well known to be nuclear of order zero [14] and the trace of $\mathcal{L}$ is given by the trace formula [24]

$$\operatorname{tr}\mathcal{L} = \frac{\phi(z^*)}{1 - \psi'(z^*)}, \tag{59}$$

where z^* is the unique fixed point of the map ψ in D.

Applying this trace formula we get [7]

Proposition 5.2. *The partition functions* $Z_n(P, \mathbb{A}_{\beta,\chi_\Gamma})$, *respectively* $Z_n(P_\Gamma, \mathbb{A}_{\beta,1})$ *in (54), can be expressed as*

$$Z_n(P, \mathbb{A}_{\beta,\chi_\Gamma}) = \operatorname{tr}(\mathcal{L}_\beta^{\chi_\Gamma})^n - \operatorname{tr}(-\mathcal{L}_{\beta+1}^{\chi_\Gamma})^n,$$

respectively

$$Z_n(P_\Gamma, \mathbb{A}_{\beta,1}) = \operatorname{tr}(\mathcal{L}_\beta^\Gamma)^n - \operatorname{tr}(-\mathcal{L}_{\beta+1}^\Gamma)^n$$

with $\mathcal{L}_\beta^{\chi_\Gamma}$ *and* $\mathcal{L}_\beta^\Gamma$ *defined in (42) and (44).*

The Ruelle zeta functions $\zeta_R(z, \mathbb{A}_{\beta,\chi_\Gamma})$ and $\zeta_R(z, \mathbb{A}_{\beta,1})$ therefore can be written as

$$\zeta_R(z, \mathbb{A}_{\beta,\chi_\Gamma}) = \frac{\det(1 - z\mathcal{L}_{\beta+1}^{\chi_\Gamma})}{\det(1 - z\mathcal{L}_\beta^{\chi_\Gamma})},$$

respectively

$$\zeta_R(z, \mathbb{A}_{\beta,1}) = \frac{\det(1 - z\mathcal{L}_{\beta+1}^\Gamma)}{\det(1 - z\mathcal{L}_\beta^\Gamma)}.$$

Since the Selberg functions Z_S^Γ, respectively $Z_S^{\chi_\Gamma}$ in (56) and (57) are products of Ruelle-Smale functions with shifted arguments and these functions can be expressed by Lemma 5.1 in terms of the Ruelle zeta functions in (53) one finally gets

Theorem 5.3. *The Selberg zeta function* Z_S^Γ *for the geodesic flow* Φ_t^Γ *on the unit tangent bundle* SM_Γ *of the modular surface* M_Γ *coincides with the Selberg zeta function* $Z_S^{\chi_\Gamma}$ *of the geodesic flow on the unit tangent bundle* SM *of the modular surface* M *with representation* χ_Γ. *Both functions can be expressed as Fredholm determinants of the transfer operators* $\mathcal{L}_\beta^\Gamma$, *respectively* $\mathcal{L}_\beta^{\chi_\Gamma}$, *as*

$$Z_S^\Gamma(\beta) = \det(1 - \mathcal{L}_\beta^\Gamma) = \det(1 - \mathcal{L}_\beta^{\chi_\Gamma}) = Z_S^{\chi_\Gamma}(\beta).$$

If $\chi_\Gamma = \oplus_i \chi_i$ is the decomposition of the representation χ_Γ into its irreducible components χ_i then one finds the factorization

$$Z_S^{\chi_\Gamma}(\beta) = \prod_i Z_S^{\chi_i}(\beta)$$

which was derived also earlier by A. Venkov and P. Zograf [35] using the Selberg trace formula.

When acting on the functions

$$\underline{f} = \underline{f}(z) = \begin{pmatrix} f(z,1) \\ f(z,-1) \end{pmatrix}$$

the transfer operator $\mathcal{L}_\beta^{\chi_\Gamma}$ can be written in matrix form

$$\mathcal{L}_\beta^{\chi_\Gamma} = \begin{pmatrix} 0 & \mathcal{L}_{\beta,-} \\ \mathcal{L}_{\beta,+} & 0 \end{pmatrix}$$

with

$$\mathcal{L}_{\beta,\pm}\underline{f}(z) = \sum_{n=1}^{\infty} \frac{1}{(z+n)^{2\beta}} \chi_\Gamma(QT^{\pm n}) \underline{f}\left(\frac{1}{z+n}\right).$$

Hence

$$\det(1 - \mathcal{L}_\beta^{\chi_\Gamma}) = \det(1 - \mathcal{L}_{\beta,+}\mathcal{L}_{\beta,-}) = \det(1 - \mathcal{L}_{\beta,-}\mathcal{L}_{\beta,+}).$$

Obviously $\mathcal{L}_{\beta,+} = \mathcal{L}_{\beta,-}$ iff $\chi_\Gamma(T^2) = 1$. In this case one finds

$$\det(1 - \mathcal{L}_\beta^{\chi_\Gamma}) = \det(1 - \mathcal{L}_{\beta,+}) \det(1 + \mathcal{L}_{\beta,+}) \tag{60}$$

and hence

$$Z_S^\Gamma(\beta) = \det(1 - \mathcal{L}_{\beta,+}) \det(1 + \mathcal{L}_{\beta,+}). \tag{61}$$

Examples of modular groups for which the Selberg zeta function Z_S^Γ factorizes in this way are for instance $\mathrm{SL}(2,\mathbb{Z})$, $\Gamma(2)$, $\Gamma_0(2)$ and $\Gamma^0(2)$. It turns out that there is a slightly different transfer operator $\mathcal{L}_\beta^{MM}$ introduced by Y. Manin and M. Marcolli in [21] for which the above factorization holds for a much wider class of modular surfaces.

5.1 The Manin-Marcolli operator

Consider modular subgroups $\Gamma \subset \mathrm{SL}(2,\mathbb{Z})$ with the property

$$\Gamma \begin{pmatrix} 1 & 0 \\ 0 & -1 \end{pmatrix} = \begin{pmatrix} 1 & 0 \\ 0 & -1 \end{pmatrix} \Gamma. \tag{62}$$

For such a subgroup Γ define the extended group $\widetilde{\Gamma}$ with $\widetilde{\Gamma} = \Gamma \cup \Gamma \begin{pmatrix} -1 & 0 \\ 0 & -1 \end{pmatrix} \subset \mathrm{SL}(2,\mathbb{Z})$ and the extended subgroup $\overline{\Gamma}$ of $\mathrm{GL}(2,\mathbb{Z})$ with

$$\overline{\varGamma} = \widetilde{\varGamma} \cup \widetilde{\varGamma} \begin{pmatrix} 1 & 0 \\ 0 & -1 \end{pmatrix}. \tag{63}$$

Then one shows that

$$\overline{\varGamma} \setminus \mathrm{GL}(2,\mathbb{Z}) \cong \varGamma \setminus \mathrm{SL}(2,\mathbb{Z})$$

and hence there is a 1–1 correspondence between the residue classes of $\mathrm{GL}(2,\mathbb{Z})$ modulo $\overline{\varGamma}$ and the residue classes of $\mathrm{SL}(2,\mathbb{Z})$ modulo $\varGamma$.

Manin and Marcolli considered in [21] the following transfer operator $\mathcal{L}_\beta^{MM}$ for the modular group $\varGamma \subset \mathrm{SL}(2,\mathbb{Z})$:

$$(\mathcal{L}_\beta^{MM}\underline{f})(z) = \sum_{n-1}^{\infty} \left(\frac{1}{z+n}\right)^{2\beta} \chi_{\overline{\varGamma}} \begin{pmatrix} -n & 1 \\ 1 & 0 \end{pmatrix} \underline{f}\left(\frac{1}{z+n}\right), \tag{64}$$

where $\chi_{\overline{\varGamma}} : \mathrm{GL}(2,\mathbb{Z}) \to \mathrm{End}\,\mathbb{C}^\mu$ is the induced representation of $\mathrm{GL}(2,\mathbb{Z})$ induced from the trivial representation of $\overline{\varGamma}$. There is a simple relation between this operator $\mathcal{L}_\beta^{MM}$ and the transfer operators $\mathcal{L}_{\beta,\pm}$:

$$(\mathcal{L}_\beta^{MM})^2 = \chi_{\overline{\varGamma}} \begin{pmatrix} 0 & 1 \\ 1 & 0 \end{pmatrix} \mathcal{L}_{\beta,+} \mathcal{L}_{\beta,-} \chi_{\overline{\varGamma}} \begin{pmatrix} 0 & 1 \\ 1 & 0 \end{pmatrix}, \tag{65}$$

that means $(\mathcal{L}_\beta^{MM})^2$ and $\mathcal{L}_{\beta,+}\mathcal{L}_{\beta,-}$ are conjugate and hence one gets for the Selberg zeta functions $Z_S^\varGamma$ respectively $Z_S^{\chi_\varGamma}$:

$$Z_S^\varGamma(\beta) = Z_S^{\chi_\varGamma}(\beta) = \det(1 - (\mathcal{L}_\beta^{MM})^2) = \det(1 - \mathcal{L}_\beta^{MM})\det(1 + \mathcal{L}_\beta^{MM}). \tag{66}$$

The Hecke congruence subgroups $\varGamma_0(n)$ fulfill property (62) and hence their Selberg zeta functions have the factorization property as given in (66). From our discussion of the spectral properties of the operator $\mathcal{L}_{\beta,+}$ and $\mathcal{L}_{\beta,-}$ it follows that also the operator $\mathcal{L}_\beta^{MM}$ is meromorphic in β and nuclear of order zero away from its poles. Hence relation (66) gives a new proof of the fact that for the modular groups $\varGamma \subset \mathrm{SL}(2,\mathbb{Z})$ the Selberg zeta function is meromorphic in the entire β-plane. Its zeroes on the other hand correspond to those values of β where $\lambda = 1$ or $\lambda = -1$ belongs to the spectrum of the operator $\mathcal{L}_\beta^{MM}$. Unfortunately we did not succeed up to now to derive the location of the zeroes of Selbergs zeta function in the β-plane alone from the spectral properties of the operator $\mathcal{L}_\beta^{MM}$.

6 The period functions of Lewis and Zagier

The transfer operator approach to the Selberg zeta function relates the analytic properties of this function to spectral properties of the transfer operator. Both are purely classical objects: the Selberg function can be considered a generating function for the length spectrum of the geodesic flow whereas the transfer operator encodes further details of the dynamical properties of the flow like the Poincaré map and the recurrence time function. It is therefore not surprising that the transfer operator provides one with more informations also for the corresponding quantum system. Indeed, through the work of J. Lewis and D. Zagier and our own work for the group $SL(2,\mathbb{Z})$ it became clear that the eigenfunctions with eigenvalues $\lambda = \pm 1$ of the transfer operator $\mathcal{L}_\beta$ for β-values corresponding to nontrivial zeroes of the Selberg function are directly related to the Maass wave forms for this group. Furthermore it was found that such eigenfunctions for β-values corresponding to trivial zeroes of $Z_S(\beta)$ are related to the theory of both holomorphic and real analytic modular forms namely to the period polynomials of Eichler, Shimura and Manin for holomorphic cusp forms. To explain this connection in more detail consider any eigenfunction $\underline{f}$ of the Manin-Marcolli operator in (64) with eigenvalue λ:

$$\mathcal{L}_\beta^\Gamma \underline{f} = \lambda \underline{f},$$

where $\underline{f}$ is a holomorphic function in the domain $\mathbb{C} \setminus (-\infty, -1]$. If $\underline{\phi}$ denotes the function $\underline{\phi}(z) = \underline{f}(z-1)$ then this function fulfills the functional equation

$$\lambda \big(\underline{\phi}(z) - \chi_{\overline{F}}(T)\underline{\phi}(z+1) \big) = z^{-2\beta} \chi_{\overline{F}}\left(\begin{pmatrix} -1 & 1 \\ 0 & 1 \end{pmatrix} \right) \underline{\phi}\left(\frac{z+1}{z} \right). \qquad (67)$$

For the following discussion it is more convenient to work also with the space of functions $\psi = \psi(z, \overline{\Gamma} g)$ with

$$\psi(z, \overline{\Gamma} g) = \phi_{\overline{\Gamma} g}(z), \qquad (68)$$

if $\underline{\phi} = (\phi_{\overline{\Gamma} g})_{g=g_i,\ 1 \le i \le \mu}$. For any subgroup $\overline{\Gamma} \subset GL(2,\ \mathbb{Z})$ the group $GL(2,\mathbb{Z})$ acts as follows on the functions $\psi : \mathbb{C} \setminus \mathbb{R} \times \overline{\Gamma} \setminus GL(2,\mathbb{Z}) \to \mathbb{C}$ for $h = \begin{pmatrix} a & b \\ c & d \end{pmatrix} \in GL(2,\mathbb{Z})$:

$$(\psi_{|\beta} h)(z, \overline{\Gamma} g) = (cz+d)^{-2\beta} \psi(hz, \overline{\Gamma} g T^{-1} h^{-1} T). \qquad (69)$$

We call this action a T-twisted action of the group $GL(2,\mathbb{Z})$ contrary to the more common action without the conjugation by T on the cosets $\overline{\Gamma} g$. Using the twisted action in (69) one can write the Lewis equation (67) for the group $\Gamma \subset SL(2,\mathbb{Z})$ for $\lambda = \pm 1$ simply as

$$\psi = \psi_{|\beta} T + \lambda \psi_{|\beta} TM \qquad (70)$$

with $T = \begin{pmatrix} 1 & 1 \\ 0 & 1 \end{pmatrix}$ and $M = \begin{pmatrix} 0 & 1 \\ 1 & 0 \end{pmatrix}$. In the special case of the group $\Gamma = SL(2,\mathbb{Z})$ one finds $\overline{\Gamma} = GL(2,\ \mathbb{Z})$ and eq. (70) becomes in explicit form

$$\psi(z) = \psi(z+1) + \lambda z^{-2\beta} \psi\left(\frac{z+1}{z} \right). \qquad (71)$$

A simple calculation then shows that this equation is equivalent to the following pair of equations

$$\psi(z) = \lambda z^{-2\beta} \psi\left(\frac{1}{z}\right), \tag{72}$$

$$\psi(z) = \psi(z+1) + (z+1)^{-2\beta} \psi\left(\frac{z}{z+1}\right). \tag{73}$$

Lewis and Zagier proved in [19] the following theorem for the group $\mathrm{SL}(2,\mathbb{Z})$.

Theorem 6.1. *For $\beta \in \mathbb{C}$ with $\Re\beta = \frac{1}{2}$ there exists a 1-1 correspondence between the Maass cusp forms ϕ for $\mathrm{SL}(2,\mathbb{Z})$ with $-\Delta_{LB}\phi = \beta(1-\beta)\phi$ and the real analytic solutions of the Lewis functional equation (71) for the same β-value with asymptotic behaviour $\psi(x) = o(\frac{1}{x})$ for $x \to 0$ and $\psi(x) = o(1)$ for $x \to \infty$.*

Remark 6.2. Obviously the space of solutions of the Lewis equation (71) with the asymptotic behaviour as in Theorem 6.1 coincides with the eigenspaces to the eigenvalues $\lambda = \pm 1$ of the transfer operator $\mathcal{L}_\beta$ for the group $\mathrm{SL}(2,\mathbb{Z})$. To see this one only has to iterate the functional equation and use the asymptotic behaviour of the Hurwitz zeta function in the half plane $\Re\beta > 0$. $\quad\square$

In his Ph.D.-thesis [22] F. Martin extended the theory of period functions of Lewis and Zagier to holomorphic modular forms of weight 1 for arbitrary subgroups $\Gamma \subset \mathrm{SL}(2,\mathbb{Z})$. Thereby he used also a Lewis-type functional equation of the following form

$$\psi = \psi_{\|\beta}T + \psi_{\|\beta}MTM, \tag{74}$$

where the action $\|\beta$ of the group $\mathrm{GL}(2,\mathbb{Z})$ is similar to its action $|\beta$ in (69) but without the T-twist when acting on the residue classes. Obviously with ψ also the function $\psi_{\|\beta}M$ solves equation (74). The functional equation (74) is closely related to the functional equation (70) as shown by

Lemma 6.3. *The function ψ solves equation (70) iff ψ solves the pair of equations*

$$\psi = \psi_{|\beta}T + \psi_{|\beta}MTM, \tag{75}$$
$$\lambda\psi = \psi_{|\beta}M. \tag{76}$$

Proof. Just act with the element M on equation (70) to get equation (76). Next replace $\lambda\psi$ in equation (70) by $\psi_{|\beta}M$ to get equation (75). Similar arguments show the inverse direction of the Lemma. $\quad\square$

Lemma 6.4. *The function ψ solves equation (70) iff the function $\tilde{\psi} = \tilde{\psi}(z, \overline{\Gamma}g) =: \psi(z, \overline{\Gamma}gT)$ is a solution of the Lewis equation (74) with $\tilde{\phi}_{\overline{\Gamma}gM|\beta}M = \lambda\tilde{\phi}_{\overline{\Gamma}g}$ for $\tilde{\phi}_{\overline{\Gamma}g} = \tilde{\psi}(z, \overline{\Gamma}g)$.*

Proof. From Lemma 6.3 it follows that ψ solves equation (70) iff it solves equations (75) and (76). Explicitly equation (75) reads as follows

$$
\begin{aligned}
0 &= \phi_{\overline{T}gT} - \phi_{\overline{T}gTT^{-1}|\beta}T - \phi_{\overline{T}gTT^{-1}MT^{-1}MT|\beta}MTM \\
&= \phi_{\overline{T}gT} - \phi_{\overline{T}gT^{-1}T|\beta}T - \phi_{\overline{T}gMT^{-1}MT|\beta}MTM \\
&= \tilde{\phi}_{\overline{T}g} - \tilde{\phi}_{\overline{T}gT^{-1}|\beta}T - \tilde{\phi}_{\overline{T}gMT^{-1}M|\beta}MTM.
\end{aligned}
$$

But this is just equation (74) for the components of the function $\tilde{\psi}$. Equation (76) for the components of ψ on the other hand reads $\lambda\phi_{\overline{T}gT} = \phi_{\overline{T}gMT|\beta}M$, and hence $\tilde{\phi}_{\overline{T}gM|\beta}M = \lambda\tilde{\phi}_{\overline{T}g}$. The inverse direction of the Lemma is proved in a similar way. $\square$

The Lewis equation used by F. Martin hence also arises from the transfer operator $\mathcal{L}_\beta^{MM}$ for the geodesic flow $\Phi_t^\Gamma : SM_\Gamma \to SM_\Gamma$, it is written down only in a different basis in $\mathbb{C}^\mu$ compared to ours.

Consider now any Maass wave form $\phi = \phi(z)$ for a subgroup $\Gamma \subset \mathrm{SL}(2,\mathbb{Z})$ with $-\Delta_{LB}\phi(z) = \beta(1-\beta)\phi(z)$ and $\phi(gz) = \phi(z)$ for all $g \in \Gamma$. Following basically the arguments in [22] and of Lewis and Zagier [19] for the case of $\mathrm{SL}(2, \mathbb{Z})$ we can construct solutions of the Lewis equation for any modular group Γ.

For $\zeta \in \mathbb{C}$ and $z \in \mathbb{H}$ consider the function

$$
R_\zeta(z) = \frac{y}{(x-\zeta)^2 + y^2} \tag{77}
$$

with $z = x + iy$. A simple calculation shows that R_ζ^β is an eigenfunction of the Laplace-Beltrami operator $-\Delta_{LB}$ with eigenvalue $\lambda = \beta(1-\beta)$, namely

$$
-\Delta_{LB}R_\zeta^\beta(z) = \beta(1-\beta)R_\zeta^\beta(z). \tag{78}
$$

In [19] it is shown that the following transformation formula holds for $g = \begin{pmatrix} a & b \\ c & d \end{pmatrix} \in \mathrm{GL}(2,\mathbb{Z})$:

$$
R_{g\zeta}(gz) = \frac{(c\zeta + d)^2}{\det g} R_\zeta(z). \tag{79}
$$

Next one checks that for any two eigenfunctions u, $v : \mathbb{C} \to \mathbb{C}$ of $-\Delta_{LB}$ with identical eigenvalues the 1-form $\{u,v\}(z)$ with

$$
\{u,v\}(z) = (vu_y - uv_y)\,\mathrm{d}x + (uv_x - vu_x)\,\mathrm{d}y \tag{80}
$$

is closed and hence $\mathrm{d}\{u,v\}(z) = 0$. If $g : \mathbb{C} \to \mathbb{C}$ is any holomorphic change of coordinates then

$$
\{u \circ g, v \circ g\}(z) = \{u,v\}(gz). \tag{81}
$$

Given an eigenfunction ϕ of the Laplace Beltrami operator $-\Delta_{LB}$ with eigenvalue $\lambda = \beta(1-\beta)$ such that $\phi(gz) = \phi(z)$ for all $g \in \overline{\Gamma} \subset \mathrm{GL}(2, \mathbb{Z})$, consider for $\Re\zeta > 0$ the following function

$$\psi_\phi = \psi_\phi(\zeta, \overline{\varGamma}g) = \int_0^{i\infty} \{\phi \circ (gT^{-1}), R_\zeta^\beta\}(z). \tag{82}$$

Obviously the function ψ_ϕ depends only on the cosets $\overline{\varGamma}g$ since $\phi(hz) = \phi(z)$ for all $h \in \overline{\varGamma}$. Then one shows

Theorem 6.5. *For ϕ a Maass wave form of the modular group $\varGamma \subset \mathrm{SL}(2, \mathbb{Z})$ with eigenvalue $\lambda = \beta(1-\beta)$ the function ψ_ϕ in (82) solves the Lewis equation (75).*

Proof. By using the transformation properties of the function $R_\zeta(z)$ one finds for arbitrary $h = \begin{pmatrix} a & b \\ c & d \end{pmatrix} \in \mathrm{SL}(2, \mathbb{Z})$

$$(\psi_{\phi|\beta}h)(\zeta, \overline{\varGamma}g) = \int_{h(0)}^{h(i\infty)} \{\phi \circ (gT^{-1}), R_\zeta^\beta\}(z) \tag{83}$$

and hence

$$\psi_\phi - \psi_{\phi|\beta}T - \psi_{\phi|\beta}MTM =$$
$$\int_0^{i\infty} \{\phi \circ (gT^{-1}), R_\zeta^\beta\}(z) - \int_1^{i\infty} \{\phi \circ (gT^{-1}), R_\zeta^\beta\}(z) - \int_0^1 \{\phi \circ (gT^{-1}), R_\zeta^\beta\}(z)$$
$$= 0. \quad \square$$

Remark 6.6. Since the 1-form $\{\phi \circ gT^{-1}, R_\zeta^\beta\}(z)$ is closed one can move the path of integration from 0 to $i\infty$ in the complex z-plane and derives this way that the function $\psi_\phi(\zeta, \overline{\varGamma}g)$ is indeed holomorphic in the plane $\mathbb{C} \setminus [-\infty, 0]$.

In the special case $\varGamma = \mathrm{SL}(2, \mathbb{Z})$ formula (82) for the function ψ_ϕ simplifies to (see [19])

$$\psi_\phi(z) = \int_0^{i\infty} \{\phi, R_\zeta^\beta\}(z). \tag{84}$$

Obviously the function ψ_ϕ in (82) gives one direction of the Lewis-Zagier correspondence for arbitrary modular groups. One only has to check that the function has the correct asymptotic behaviour to define really an eigenfunction of the Manin-Marcolli transfer operator L_β^{MM}.

The converse correspondence, namely to attach a Maass wave form to an eigenfunction of the transfer operator is understood up to now only in the case of the full modular group $\mathrm{SL}(2, \mathbb{Z})$ [3]. One possibility here is the use of L-functions and the converse theorem of Maass. We follow again the paper by Lewis and Zagier [19].

Proposition 6.7. *For $\beta \in \mathbb{C}$ denote by $\sigma = \Re \beta$ its real part. There exists a 1–1 correspondence between the following classes of functions:*

---
[3] See, however, [11]

- *a solution u of $-\Delta_{LB}u = \beta(1-\beta)u$ in $\mathbb{H}$ with $u(z+1) = u(z)$ and $u(x+iy) = O(y^A)$ for some $A < \min(\sigma, 1-\sigma)$,*
- *a pair of L-series $L_\epsilon(\rho)$ ($\epsilon = 0,1$) convergent in some half plane,*
- *a periodic holomorphic function f in $\mathbb{C} \setminus \mathbb{R}$ with $f(z) = O(|\Im z|^{-A})$ for some A.*

Proof. We only sketch the main steps of the proof. Since u is a periodic eigenfunction of $-\Delta_{LB}$ it can be expanded in a Fourier series of the form

$$u(z) = \sqrt{y}\sum_{n\neq 0} A_n K_{\beta-\frac{1}{2}}(2\pi|n|y)e^{2\pi inx}, \tag{85}$$

where the A_n have at most polynomial growth in $|n|$.

Consider next the following two L-series

$$L_\epsilon(\rho) := \sum_{n=1}^\infty \frac{A_{n,\epsilon}}{n^\rho}, \quad \epsilon = 0,1, \tag{86}$$

with $A_{n,\epsilon} = A_n + (-1)^\epsilon A_{-n}$ and the holomorphic function

$$f(z) = \begin{cases} \sum_{n>0} n^{\beta-\frac{1}{2}} A_n e^{2\pi inz} & \text{for } \Im z > 0, \\ -\sum_{n<0} |n|^{\beta-\frac{1}{2}} A_n e^{2\pi inz} & \text{for } \Im z < 0. \end{cases} \tag{87}$$

On the other hand let be two Dirichlet series L_ϵ, $\epsilon = 0,1$, be given that are convergent in some half plane, or a periodic function holomorphic in $\mathbb{C}\setminus\mathbb{R}$ with $f(z) = O(|\Im z|^{-A})$ for some $A > 0$, then the expansion coefficients $\{A_n\}_{n\neq 0}$ are at most of polynomial growth in n. Define a function u on $\mathbb{H}$ as

$$u(z) = \sqrt{y}\sum_{n\neq 0} A_n K_{\beta-\frac{1}{2}}(2\pi|n|y)e^{2\pi inx}. \tag{88}$$

Then the functions $u_0(y) := \frac{1}{\sqrt{y}}u(iy)$, respectively $u_1(y) := \frac{\sqrt{y}}{2\pi i}\frac{\partial u}{\partial x}(iy)$, can be written as

$$u_\epsilon(y) = \sum_{n=1}^\infty (ny)^\epsilon A_{n\epsilon} K_{\beta-\frac{1}{2}}(2\pi ny), \quad \epsilon = 0,1, \tag{89}$$

such that

$$L_\epsilon^*(\rho) := \int_0^\infty u_\epsilon(y)y^{\rho-1}\,dy \tag{90}$$

fulfills

$$L_\epsilon^*(\rho) = \frac{1}{4\pi^{\rho+\epsilon}}\Gamma\left(\frac{\rho+\epsilon-\beta+\frac{1}{2}}{2}\right)\Gamma\left(\frac{\rho+\epsilon+\beta-\frac{1}{2}}{2}\right)L_\epsilon(\rho). \tag{91}$$

On the other hand one finds

$$(2\pi)^{-\rho}\Gamma(\rho)L_\epsilon\left(\rho-\beta+\tfrac{1}{2}\right) = \int_0^\infty \left(f(iy)-(-1)^\epsilon f(-iy)\right)y^{\rho-1}\,dy. \quad \square \tag{92}$$

Hence to any periodic eigenfunction u of $-\Delta_{LB}$ one can find a holomorphic function f on $\mathbb{C} \setminus \mathbb{R}$ which furthermore is periodic of period 1.

The following Proposition establishes a direct relation between the solutions ψ of the Lewis equation in the upper and lower half plane $\mathbb{H}$ and periodic holomorphic functions in the upper and lower half plane $\mathbb{H}$:

Proposition 6.8. *The function ψ is a holomorphic solution of the Lewis equation (73) in $\mathbb{H}$, respectively in $-\mathbb{H}$, iff the function $f = f(z) = \psi(z) + z^{-2\beta}\psi(\frac{-1}{z})$ is a periodic holomorphic function in $\mathbb{H}$, respectively $-\mathbb{H}$.*

Proof. If ψ solves equation (73) then one has

$$
\begin{aligned}
0 &= \psi(z+1) - \psi(z) + \left(\frac{1}{z+1}\right)^{2\beta}\psi\left(\frac{z}{z+1}\right) \\
&\quad -(z+1)^{-2\beta}\left[\psi\left(\frac{z}{z+1}\right) - \psi\left(\frac{-1}{z+1}\right) + \left(\frac{z+1}{z}\right)^{2\beta}\psi\left(\frac{-1}{z}\right)\right] \\
&= \left[\psi(z+1) + (z+1)^{-2\beta}\psi\left(\frac{-1}{z+1}\right)\right] - \left[\psi(z) + z^{-2\beta}\psi\left(\frac{-1}{z}\right)\right].
\end{aligned}
$$

Therefore the function $f(z) = \psi(z) + z^{-2\beta}\psi(\frac{-1}{z})$ is holomorphic in $\mathbb{H}$ and periodic with period 1. On the other hand if there is given a holomorphic function f on $\mathbb{H}$ with $f(z+1) = f(z)$ then the function $\psi(z) := f(z) - z^{-2\beta}f(\frac{-1}{z})$ is a solution of the Lewis equation for $SL(2,\mathbb{Z})$. The same kind of arguments can be applied for the lower half plane $z \in -\mathbb{H}$. □

In the next step one uses an old result by Maass [20].

Theorem 6.9. *There is a 1–1 correspondence between the Maass wave forms ϕ for $SL(2,\mathbb{Z})$ and pairs of L series $L_\epsilon^* = L_\epsilon^*(\rho)$, $\epsilon = 0,1$ with*

$$
L_\epsilon^*(1-\rho) = (-1)^\epsilon L_\epsilon^*(\rho). \tag{93}
$$

Proof. We sketch the main steps of the proof. We know already from the periodicity of the Maass form ϕ that there is a 1–1 correspondence with a pair of L-series L_ϵ^* convergent in some half plane. Since u is real analytic the condition $u(z) = u(\frac{-1}{z})$ however is equivalent to $\left(u(z) - u(\frac{-1}{z})\right)|_{z=i\rho} = 0$ and $\frac{d}{dz}\left(u(z) - u(\frac{-1}{z})\right)|_{z=i\rho} = 0$. But these two identities are equivalent to the equations $u_\epsilon(\frac{1}{y}) = (-1)^\epsilon y u_\epsilon(y)$ for $\epsilon = 0,1$. The last two equations are again equivalent to the L-series $L_\epsilon^* = (-1)^\epsilon \int_0^\infty u_\epsilon(y)y^{\rho-1}\,dy$ to fulfill the functional equations $L_\epsilon^*(1-\rho) = (-1)^\epsilon L_\epsilon^*(\rho)$. Thereby one uses the fact that the function $u_\epsilon(y)$ can be expressed as the Mellin transform $u_\epsilon(y) = \frac{1}{2\pi i}\int_{\Re\,\rho=C} L_\epsilon^*(\rho)y^{-\rho}\,d\rho$ of L_ϵ^* with $C \gg 0$. □

To finish the proof of Theorem 6.1 one has to show that the existence of the L-series L_ϵ^*, $\epsilon = 0,1$, fulfilling the functional equations (93) is equivalent to the fact that the function $\psi(z) = f(z) - z^{-2\beta}f(\frac{-1}{z})$, holomorphic in the upper and lower Poincaré plane, extends indeed across the positive real line to a holomorphic function on the cut plane $\mathbb{C} \setminus \mathbb{R}_-$ and fulfills the correct asymptotics for $x \to 0$ and $x \to \infty$. The details of this argument which are a little bit more cumbersome can be found in the Annals paper by Lewis and Zagier [19].

7 The Eichler-Shimura-Manin period polynomials

In the preceding sections we have discussed the eigenfunctions of the transfer operator $\mathcal{L}_\beta^{MM}$ with eigenvalue $\lambda = \pm 1$ for β-values corresponding to the so called spectral or nontrivial zeroes of the Selberg zeta function $Z_S(\beta)$ for the group $SL(2,\mathbb{Z})$. These eigenfunctions are the period functions of Lewis and Zagier. One can obviously ask if there is also a similar interpretation for the eigenfunctions of this transfer operator at β-values corresponding to the so called trivial zeroes $\zeta_R(2\beta) = 0$ of the Selberg function with $\Re \beta < 0$. Here ζ_R denotes Riemann's zeta function. For $\beta = -n$ it turns out that these eigenfunctions are either polynomials closely related to the holomorphic modular cusp forms or rational functions related to the holomorphic Eisenstein series for $SL(2,\mathbb{Z})$.

A function $f : \mathbb{H} \to \mathbb{C}$ holomorphic in $\mathbb{H}$ and at the cusp of $SL(2,\mathbb{Z})$ at $i\infty$ is called a modular form of weight $2k$ if

$$f_{|2k}g(z) = f(z) \tag{94}$$

for all $g \in SL(2,\mathbb{Z})$, where the slash action $_|$ is defined as

$$f_{|2k}g(z) = (cz+b)^{-2k} f\left(\frac{az+b}{cz+d}\right) \tag{95}$$

for $g = \begin{pmatrix} a & b \\ c & d \end{pmatrix}$.

Let $f(z) = \sum_{n=1}^\infty a_n q^n$ with $q = e^{2\pi i z}$ be the Fourier expansion of f. Then the period polynomial $r_f(z)$ attached to f can be defined as

$$r_f(z) = \tilde{f}(z) - z^{k-2}\tilde{f}\left(\frac{-1}{z}\right) \tag{96}$$

with $\tilde{f}(z) = \sum_{n=1}^\infty \frac{a_n}{n^{2k-1}} q^n$ the Eichler integral [36], respectively as

$$r_f(z) \sim \int_0^{i\infty} f(y)(y-z)^{2k-2}\,dy, \tag{97}$$

or respectively as

$$r_f(z) \sim \sum_{r=0}^{2k-2} \frac{(-2\pi i)^{-r}}{(2k-2-r)!} L_f(r+1) z^r \tag{98}$$

with $L_f(\rho) = \sum_{n=1}^{\infty} a_n n^{-\rho}$ the L-series of f.

It turns out that all three definitions are equivalent: namely, consider the differential operator

$$D := \frac{1}{2\pi i}\frac{d}{dz} = q\frac{d}{dq}. \tag{99}$$

Then one finds $D^{2k-1}(\tilde{f}) = f$. With Bol's identity

$$D^{2k-1}(F_{|2-2k}g) = (D^{2k-1}F)_{|2k}g \tag{100}$$

valid for any smooth function F in $\mathbb{C}$ one gets by a simple calculation

$$D^{2k-1}(r_f) = D^{2k-1}(\tilde{f} - \tilde{f}_{|2k}S) = f - f_{|2k}S = 0, \tag{101}$$

and hence r_f is a polynomial of degree $2k - 2$. But one easily verifies

$$\tilde{f}(z) = \int_z^{i\infty} (z - \tau)^{2k-2} f(\tau)\,d\tau$$

and hence, since f is a modular form of weight $2k$ for $SL(2, \mathbb{Z})$

$$z^{-2k}\tilde{f}\left(\frac{-1}{z}\right) = \int_z^0 (z - \tau)^{2k-2} f(\tau)\,d\tau.$$

Subtracting the two equations leads finally to

$$r_f(z) = \tilde{f}(z) - z^{-2k}\tilde{f}\left(\frac{-1}{z}\right) = \int_0^{i\infty} (z - \tau)^{2k-2} f(\tau)\,d\tau.$$

The representation of r_f in terms of the values of the L-series at the integer values $z = r + 1$ is obvious.

Consider then the space W_{2k-2} of polynomials P of degree less or equal to $2k - 2$

$$W_{2k-2} =: \{P \in \mathcal{P} \,:\, P_{|2-2k}(1 + Q) = P_{|2-2k}(1 + U + U^2) = 0\} \tag{102}$$

with Q and T the generators of $SL(2, \mathbb{Z})$ and $U = TQ$ fulfilling the relations $Q^2 = U^3 = 1$. Eichler showed in [12] that the space

$$\text{Span}\left\{ r_f(z), \overline{r_f(\bar{z})} \,:\right.$$
$$\left. f \text{ a holomorphic modular cusp form of } SL(2, \mathbb{Z}) \text{ of weight } 2k \right\}$$

has codimension 1 in W_{2k-2}. The result of Lewis, Zagier [36] and ourselves [7] is

Proposition 7.1. *Every $P \in W_{2k-2}$ is a solution of the Lewis equation for $\beta = 1-k$. Every polynomial solution of the Lewis equation for $\beta = 1-k$, $k > 1$ belongs to the space W_{2k-2}.*

This result can be extended also to the holomorphic Eisenstein series of weight $2k$ for the group $\mathrm{SL}(2,\mathbb{Z})$ which lead to rational solutions of the Lewis equation also at the β-values $1-k$ [7]. Obviously this is the reason why all the solutions of the Lewis equation are called "period functions". The solutions of the Lewis equation just discussed belong to β-values with $\zeta_R(2\beta) = 0$ corresponding to the trivial zeroes of the Riemann zeta function $\zeta_R(\beta) = \sum_{n=1}^{\infty} n^{-\beta}$. What about the non-trivial zeroes which according to Riemann's hypothesis are located on the critical line $\Re\beta = \frac{1}{2}$? Indeed there exist explicitly known solutions (see [7]) of the Lewis equation at β-values with $\zeta_R(2\beta) = 0$ and $\Re\beta > 0$ which correspond to the real analytic Eisenstein series determining the continuous spectrum of the hyperbolic Laplacian on the surface M [6].

What have the solutions of the Lewis equation to do with eigenfunctions of the transfer operator $\mathcal{L}_\beta$ for the full modular group $\mathrm{SL}(2,\mathbb{Z})$? We have seen already that any eigenfunction of the transfer operator with eigenvalue $\lambda = \pm 1$ determines a solution of the pair of equations (72) and (73). On the other hand given a solution ψ of equation (73) then also the function $\psi \pm \psi^\tau$ with $\psi^\tau(z) =: z^{-2\beta}\psi(\frac{1}{z})$ solves this equation. But $(\psi \pm \psi^\tau)^\tau = \pm(\psi \pm \psi^\tau)$. Hence if ψ is a solution of equation (73) and $\psi^\tau = \lambda\psi$ for $\lambda = \pm 1$ then ψ solves equation (70). The solutions of this last equation, holomorphic in the cut plane $\mathbb{C} \setminus R_-$, determine the eigenfunctions $f = f(z) = \psi(z+1)$ with eigenvalue $\lambda = \pm 1$ of the transfer operator $\mathcal{L}_\beta$ for $\mathrm{SL}(2,\mathbb{Z})$ as long as they fulfill the appropriate asymptotic behaviour valid in the corresponding region of the complex β-plane.

8 Transfer operators and Hecke operators for $\Gamma_0(n)$

The transfer operator for the Hecke congruence subgroup $\Gamma_0(n)$ as given by Manin and Marcolli has the form

$$(\mathcal{L}_\beta^{MM} \underline{f})(z) = \sum_{n=1}^{\infty} \frac{1}{(z+n)^{2\beta}} \chi_{\overline{\Gamma}_0^n} \left(\begin{pmatrix} -n & 1 \\ 1 & 0 \end{pmatrix} \right) \underline{f}(\frac{1}{z+n})$$

with $\overline{\Gamma}_0(n) = \Gamma_0(n) \cup \Gamma_0(n) \begin{pmatrix} 1 & 0 \\ 0 & -1 \end{pmatrix}$. The Lewis equation derived from this operator on the other hand reads

$$\underline{\phi} - \underline{\phi}_{|\beta}T - \underline{\phi}_{|\beta}MTM = 0$$

with $\underline{\phi} = (\phi_{\overline{\Gamma}_0(n)g})$ and $\overline{\Gamma}_0(n)g$ describing the different residue classes of $\Gamma_0(n) \setminus \mathrm{GL}(2,\mathbb{Z})$. Since $\mathrm{GL}(2,\mathbb{Z}) \supset \Gamma_0(m) \supset \Gamma_0(nm)$ there is a canonical projection

$$\sigma_{n,m} : I_{nm} \to I_m. \tag{103}$$

In [15] we showed that to any index $i \in I_n$ is attached a unique matrix $A_i = \begin{pmatrix} c & b \\ 0 & \frac{n}{c} \end{pmatrix}$ with $c \geq 1$, $c|n$ and $\gcd(c, \frac{n}{c}, b) = 1$. Hence, if

$$S_n = \left\{ \begin{pmatrix} a & b \\ c & d \end{pmatrix} : a > c \geq 0, \quad d > b \geq 0, \quad ad - bc = n \right\} \tag{104}$$

then $A_i \in S_n$. For $\begin{pmatrix} a & b \\ c & d \end{pmatrix} \in S_n$ with $b \neq 0$ consider the map

$$K\left(\begin{pmatrix} a & b \\ c & d \end{pmatrix} \right) = T^{\lceil \frac{d}{b} \rceil} Q \begin{pmatrix} a & b \\ c & d \end{pmatrix} \tag{105}$$

with $\lceil \frac{d}{b} \rceil = -\lfloor -\frac{d}{b} \rfloor$. Then $K\left(\begin{pmatrix} a & b \\ c & d \end{pmatrix} \right) \in S_n$. For $i \in I_n$ denote by k_i the number such that $K^j A_i \in S_n$ for all $0 \leq j < k_i$ and $K^{k_i} A_i = \begin{pmatrix} * & 0 \\ * & * \end{pmatrix}$. Then one shows (see [15])

Lemma 8.1. *To* $i \in I_{nm}$ *and* $0 \leq j \leq k_{\sigma_{n,m}(i)}$ *there exists exactly one index* $l = l(i,j) \in I_n$ *such that* $A_{l(i,j)}(K^j A_{\sigma_{n,m}(i)}) A_i^{-1} \in \mathrm{SL}(2,\mathbb{Z})$.

This then allowed us to prove in [15]

Theorem 8.2. *For any solution* $\underline{\phi} = (\phi_r)_{r \in I_n}$ *of the Lewis equation (70) for* $\Gamma_0(n)$ *with parameter* β *the function* $\underline{\psi} = (\psi_i)_{i \in I_{nm}}$ *with* $\psi_i(z) = \sum_{j=0}^{k_{\sigma_{n,m}(i)}} \phi_{l(i,j)|\beta} K^j A_{\sigma_{n,m}(i)}(z)$ *is a solution of the Lewis equation for the same* β-*value and the same* λ.

This shows that any solution of the Lewis equation for the group $\Gamma_0(n)$ determines a special nontrivial solution for the subgroup $\Gamma_0(nm)$. Besides this nontrivial solution $\underline{\phi}$ determines also a trivial solution $\underline{\psi}$ for the subgroup $\Gamma_0(nm)$ given by $\underline{\psi} = (\psi_j(z))_{j \in I_{nm}}$ with

$$\psi_j = \phi_{\sigma_{m,n}(j)}.$$

We call these solutions old solutions of the Lewis equation for the subgroup $\Gamma_0(nm)$. On the other hand one can show that any solution $\underline{\psi}$ of the Lewis equation for the subgroup $\Gamma_0(nm) \subset \Gamma_0(n)$ determines also a solution for $\Gamma_0(n)$:

Lemma 8.3. *For any solution* $\underline{\psi} = (\psi_i)_{i \in I_{nm}}$ *of the Lewis equation (70) for* $\Gamma_0(nm)$ *with parameter* β *the function* $\underline{\phi} = (\phi_l)_{l \in I_n}$ *with*

$$\phi_l(z) = \sum_{i \in \sigma_{m,n}^{-1}(l)} \psi_i(z)$$

is a solution of the Lewis equation for $\Gamma_0(n)$ *with the same parameter* β.

In the special case $n = 1$ one hence gets

Corollary 8.4. *Any solution ϕ of the Lewis equation for $\mathrm{SL}(2,\mathbb{Z})$ determines a nontrivial solution $\underline{\psi} = (\psi_i)_{i \in I_m}$ of this equation for the group $\Gamma_0(m)$ with*

$$\psi_i(z) = \sum_{j=0}^{k_i} \phi_{|2\beta} K^j A_i(z).$$

Remark 8.5. Since the entries of the matrices $K^j A_i$ are nonnegative the action $|\beta$ is well defined also for complex β.

Combining Theorem 8.2 and Lemma 8.3 one arrives at a family of linear operators $\tilde{T}_{n,m}$ mapping the space of solutions of the Lewis equation for $\Gamma_0(n)$ into itself:

$$(\tilde{T}_{n,m}\psi)_i = \sum_{r \in \sigma_{m,n}^{-1}(i)} \sum_{j=0}^{k_{\sigma_{n,m}}(r)} \psi_{l(i,j)} K^j A_{\sigma_{n,m}(r)}, \quad m = 1, 2, \ldots.$$

In the special case of the modular group $\mathrm{SL}(2,\mathbb{Z})$ one finds

$$\tilde{T}_m =: \tilde{T}_{1,m} = \sum_{A \in S_m^*} A \tag{106}$$

with

$$S_m^* = \left\{ \begin{pmatrix} a & b \\ c & d \end{pmatrix} \in S_m \ : \ \gcd(a,b,c,d) = 1 \right\}. \tag{107}$$

By using the explicit form of the Lewis correspondence for $\mathrm{SL}(2,\mathbb{Z})$ on the other hand T. Mühlenbruch recently transfered the Hecke operators T_m for the Maass wave forms to the space of period functions and derived this way the following explicit form for these operators when acting on the period functions for $\mathrm{SL}(2,\mathbb{Z})$

$$T_m = \sum_{A \in S_m} A.$$

Obviously our operators $\tilde{T}_m$ coincide for m prime with these Hecke operators T_m. For general m one finds

$$T_m = \sum_{d^2|m} \begin{pmatrix} d & 0 \\ 0 & d \end{pmatrix} \tilde{T}_{\frac{m}{d^2}}.$$

It is to be expected that our above operators $\tilde{T}_{n,m}$ for $m = 1, 2, \ldots$ are also closely related to the Hecke operators on the space of period functions for the modular groups $\Gamma_0(n)$ [4].

---

[4] see [13]

References

1. Atkin, A. O. L., Lehner, J.: Hecke operators on $\Gamma_0(m)$. Math. Ann. 185 (1970), 134–160.
2. Baladi, V.: Positive Transfer Operators and Decay of Correlations. Advanced Series in Nonlinear Dynamics 16. World Scientific, River Edge, NJ, (2000).
3. Bowen, R.: Equilibrium states and the ergodic theory of Anosov diffeomorphisms. Lecture Notes in Math. 470 (1975).
4. Chang, C.-H., Mayer, D.: Thermodynamic formalism and Selberg's zeta function for modular groups. Regul. Chaotic Dyn. 5 (2000), no. 3, 281–312.
5. Chang, C.-H., Mayer, D.: Eigenfunctions of the transfer operators and the period functions for modular groups. In "Dynamical, spectral, and arithmetic zeta functions" (San Antonio, TX, 1999). Contemp. Math. 290, pp. 1–40, Amer. Math. Soc., Providence, RI, 2001.
6. Chang, C.H., Mayer, D.: The period function of the nonholomorphic Eisenstein series for PSL(2, ℤ). Math. Phys. Elec. J. 4 (6) (1998).
7. Chang, C.H., Mayer, D.: The transfer operator approach to Selberg's zeta function and modular Maass wave forms for PSL(2, ℤ). In Emerging Applications of Number Theory, eds. D. Hejhal, M. Gutzwiller et al., IMA Volumes 109 (1999) 72–142.
8. Choie, Y., Zagier, D.: Rational period functions for PSL(2, ℤ). In "A tribute to Emil Grosswald: number theory and related analysis". Contemp. Math. 143, pp. 89–108 Amer. Math. Soc., Providence, RI, 1993.
9. Connes, A.: Formule de trace en geométrié non-commutative et hypothèse de Riemann. C. R. Acad. Sci. Paris 323 Serie I (1996) 1231–1236.
10. de Melo, W., van Strien, S.: One-Dimensional Dynamics. Springer Verlag Berlin (1993).
11. Deitmar, A., Hilgert, J.: A Lewis correspondence for submodular groups. Forum Math. 19 (2007) 1075–1099.
12. Eichler, M.: Eine Verallgemeinerung der Abelschen Integrale. Math. Zeitschrift 67 (1957) 267–298.
13. Fraczek, M., Mayer, D., Mühlenbruch, T.: A realization of the Hecke algebra on the space of period functions for $\Gamma_0(n)$. Crelle's J. Reine Angew. Math. 603 (2007) 133-166.
14. Grothendieck, A.: Produits tensoriels topologiques et espaces nucléaires. Mem. Am. Math. Soc. 16(1955).
15. Hilgert, J., Mayer, D., Movasati, H.: Transfer operators for $\Gamma_0(n)$ and the Hecke operators for the period functions of PSL(2, ℤ). Math. Proc. Camb. Phil. Soc. 139 (2005) 81–116.
16. Ising, E.: Beitrag zur Theorie des Ferromagnetismus. Zeitschrift für Physik 31 (1925) 253–258.
17. Lewis, J.B.: Spaces of holomorphic functions equivalent to the even Maass cusp forms. Invent. Math. 127 (1997), no. 2, 271–306.
18. Lewis, J.B.; Zagier, D.: Period functions and the Selberg zeta function for the modular group. In "The mathematical beauty of physics" (Saclay, 1996), Adv. Ser. Math. Phys. 24, pp. 83–97. World Sci. Publishing, River Edge, NJ, 1997.
19. Lewis, J.B.; Zagier, D.: Period functions for Maass wave forms. I. Ann. of Math. (2) 153 (2001), no. 1, 191–258.

20. Maass, H.:Über eine neue Art von nichtanalytischen automorphen Funktionen und die Bestimmung Dirichletscher Reihen durch Funktionalgleichungen. Math. Annalen 121 (1949) 141–183.

21. Manin, Yu.I., Marcolli M.: Continued fractions, modular symbols, and non-commutative geometry. Selecta Mathematica (New Series). 8, no. 3 (2002) 475–520.

22. Martin, F.: Périodes de formes modulaires de poids 1. Thèse, Université Paris 7 (2001).

23. Mayer, D.: The thermodynamic formalism approach to Selberg's zeta function for PSL(2, $\mathbb{Z}$). Bull. Amer. Math. Soc. (N.S.) 25 (1991), no. 1, 55–60.

24. Mayer, D.: The Ruelle-Araki transfer operator in classical statistical mechanics. Lecture Notes in Physics, 123. Springer-Verlag, Berlin-New York, 1980.

25. Mayer, D.: Continued fractions and related transformations. In: "Ergodic Theory, Symbolic Dynamics and Hyperbolic Spaces", pp. 175–222. Eds. T. Bedford et al., Oxford University Press, Oxford, 1991.

26. Mayer,D., Strömberg, F.: Symbolic dynamics for the geodesic flow on Hecke surfaces, to appear in Journal of Modern Dynamics 2 (4) (2008).

27. Merel, L.: Universal Fourier expansions of modular forms. In "On Artins conjecture for odd 2-dimensional representations", ch. IV. Ed.: G. Frey, Lecture Notes in Math. 1585, Springer Verlag, Berlin, 1994.

28. Mühlenbruch, T.: Systems of automorphic forms and period functions. Proefschrift Universiteit Utrecht, (2003).

29. Pollicott, M.: Distribution of closed geodesics on the modular surface and quadratic irrationals. Bull. Soc. Math. France 114 (1986) 431–446.

30. Ruelle, D.: Thermodynamic formalism. Addison-Wesley Publishing Co., Reading, Mass., 1978.

31. Ruelle, D.: Dynamical Zeta Functions for Piecewise Monotone Maps of the Interval. CRM Monograph Series 4 (1994) Am. Math. Soc..

32. Ruelle, D.: Zeta functions for expanding maps and Anosov flows. Invent. Math. 34 (1976) 231–242.

33. Sarnak, P.: Arithmetic chaos. Israel Math. Conf. Proc. 8 (1995) 183–236.

34. Series, C.: The modular surface and continued fractions. J. London Math. Soc. (2) 31 (1985) 69–80.

35. Venkov, A.B., Zograf, P.G.: Analogues of Artin's factorization formulas in the spectral theory of automorphic functions associated with induced representations of Fuchsian groups. (Russian) Izv. Akad. Nauk SSSR Ser. Mat. 46 (1982), no. 6, 1150–1158, 1343. Engl. tran.: Math. USSR-Izv. 21 (1983), no. 3, 435–443.

36. Zagier, D.: Periods of modular forms, traces of Hecke operators, and multiple zeta values. In "Research into automorphic forms and L functions" (Kyoto, 1992). Sūrikaisekikenkyūsho Kōkyūroku No. 843, (1993), 162–170.

37. Zagier, D.: Hecke operators and periods of modular forms. Israel Math. Conf. Proc. 3 (1990) 321–336.

38. Zagier, D.: Periods of modular forms and Jacobi theta functions. Invent. math. 104 (1991) 449–465.

V. On the Calculation of Maass Cusp Forms

Dennis A. Hejhal[1,2]

[1] School of Mathematics, University of Minnesota, Minneapolis, MN 55455, USA
[2] Department of Mathematics, Uppsala University, Box 480, S-75106 Uppsala, Sweden

1 Introduction

My mini-course at the Reisensburg meeting was intended mainly for students and focused on the general topic of "Maass Waveforms and Their Computation." The material that I spoke about was taken from a variety of sources, chief among them being

(for lecture 1) [8], [9, pp. 294(bottom)-295(top)], [14, p. 279 (para 1) and Fig. 8], [13, eqs. (5.1),(5.2)];

(for lecture 2) [8], [16, §8.2], [11, §3], [9], [5], [18], [14, §6], [15];

(for lecture 3) [11, §§4,5], [10, eqs. (9.6),(9.9),(9.13),(1.2)], [2].

As part of lecture 2, I gave a brief description of a general method – originating in an idea of Harold Stark – for calculating Maass cusp forms on cofinite Fuchsian groups $\Gamma \backslash H$ having one cusp.

(Recall that a *cusp form* is simply a waveform which vanishes exponentially fast in each cusp; cf. [8, p. 140].)

The method, which is partly heuristic, has a certain robustness (as well as generality) that was lacking in both [10] and [12].

Reference [9] contains an outline of how matters are implemented in the case of a Hecke triangle group $\mathbb{G}_N$ having signature $(0,3;\pi/2, \pi/N, \pi/\infty)$.

Though the underlying idea is simple enough, space limitations precluded saying more than just a few words about the algorithm's operational particulars in [9]. Good speaking style necessitated a similar abridgment in lecture 2.

Prompted in part by ongoing interest in utilizing Maass cusp forms as a "test basin" for a host of quantum chaos questions, we recently made available over the web a number of typical codes from [9]. See [7]. Beyond drawing attention to this fact, our primary goal here will be to give a relatively painless overview of the codes' structural specifics – with an eye towards *facilitating matters* should others (especially beginners!) wish to run their own experiments with these codes or with suitable revamps thereof.

See [2, 4, 20, 23] for a look at some of what has recently been achieved in the "revamp direction".

2 Recollection of a Few Preliminaries

The group $\mathbb{G}_N$ is generated by the transformations $E(z) = -1/z$ and $T(z) = z + \mathcal{L}$, where $\mathcal{L} = 2\cos\left(\frac{\pi}{N}\right)$ and $N \geq 3$. The set $\mathcal{F}_N = \{z \in H : |z| > 1, |x| < \mathcal{L}/2\}$ can be used as a fundamental region for $\mathbb{G}_N \backslash H$; its hyperbolic area $\mu(\mathcal{F}_N)$ is $\pi(1 - 2/N)$. Each Maass cusp form φ on $\mathbb{G}_N \backslash H$ can be expanded in the form

$$\varphi(x + iy) = \sum_{n=1}^{\infty} c_n \sqrt{y}\kappa(ny; R)cs(nx), \tag{1}$$

wherein $\kappa(u; R) \equiv K_{iR}(2\pi u/\mathcal{L})$ and $cs(v)$ stands for either $\cos(2\pi v/\mathcal{L})$ or $\sin(2\pi v/\mathcal{L})$ depending on whether φ is even or odd with respect to x.

The function φ is $\mathbb{G}_N$-invariant on H and has Laplace eigenvalue equal to $-\lambda$, where $\lambda = \frac{1}{4} + R^2$. One knows that $R > 0$; cf. [9, pp. 292-294 (top)] and the various references cited there. For each φ, one also knows that $c_n = O(n^{1/2})$ and that, at least in a *mean square* sense, $c_n = O(1)$. For this, cf. [8, p. 585], [16, eqs. (8.12), (8.15)], and [14, eq. (6.10)].

The function $K_{iR}(X)$ implicit in (1) is immediately seen, by virtue of its representation as a Fourier transform of $\exp(-X\cosh t)$, to be entire in R and nicely C^{∞} with respect to X for $X \geq X_0 > 0$. The function undergoes a kind of transition at the point $R = X$. For $R > X$, $K_{iR}(X)$ basically manifests a sinusoidal-like oscillatory behavior of local amplitude $\sqrt{2\pi}/\sqrt[4]{R^2 - X^2}$; for $0 \leq R < X$, a behavior akin to $c(X^2 - R^2)^{-1/4}\exp[R\cos^{-1}(R/X) - \sqrt{X^2 - R^2}]$. Cf. [5, 18] for fuller details; also [9, eqs. (24) and (25)]. For $R \gg X$, a calculation shows that the pertinent ϑ (i.e. argument in the sine) is about $R\log(2R/eX)$; the function $K_{iR}(X)$ thus has zeros approximately every $\pi/\log(2R/X)$ units in R. The smaller the X, the closer the zeros.

(These zeros are amenable to further study by interpreting the associated functions $K_{iR}(e^t)$ as eigenfunctions of a natural Sturm-Liouville problem on $[\log X, \infty)$. Cf. [24, §§4.15, 7.4, 2.21, 5.4 (para 2–4), 14.1, 14.5], [25, p. 237(bot)], and [18, p. 147(bot)]. For $0 \leq R \leq X$, use of [24, *loc. cit.*] readily shows that $K_{iR}(X) > 0$; also that $K_{iR}(X)$ is monotonic decreasing with respect to X.)

Both here and in everything that follows, we tacitly assume that the Bessel function $K_{iR}(X)$ has been pre-multiplied by $\exp(\frac{\pi}{2}R)$.

Already in connection with [10], we developed a fast numerical quadrature algorithm for computing $K_{iR}(X)$ using what amounts to paths of steepest descent. Keeping $X \geq \frac{1}{5}$ and $R \leq 10^5$ typically produced accuracy of between 10 and 12 significant digits.

It is this code that is employed in [7]. (As is apparent, for instance, from [19, eq. (11)], the restriction $X \geq \frac{1}{5}$ is mainly relevant for small R. For such R, a slight *restructuring* in the way several integration cut-off points are chosen suffices to allow use of the code for $X \in (0, \frac{1}{5}]$.)

One way the computation of $K_{iR}(X)$ can be speeded up for given X and *a set* of many tightly packed R-values is to exploit a Lagrange interpolation in R of some appropriately high degree D. The "nodal points" therein – at which K is evaluated by direct quadrature – should, at least for $R \gg X$, presumably be taken at a distance not exceeding some small fraction of $\pi/(\log(2R/X))$. Since $K_{iR}(X_0)$ tends to oscillate *more* rapidly than $K_{iR}(X)$, the correct "half-period" for less extreme values of R/X will necessarily exceed $\pi/(\log(2R/X_0))$ provided R is kept fairly big. Taking $X_0 = \frac{1}{5}$ as above, this gives values of .3411, .2729, .2274 when $R = 10^3, 10^4, 10^5$, respectively. Empirical tests can now be made – in each particular R regime – to check the efficacy of this procedure (over $\frac{1}{5} \leq X < \infty$) using a variety of D's and proposed fractions of the basic grid length $\pi/\log(10R)$. For R's of relatively modest proportions, one naturally replaces $\log(10R)$ by some suitable $\log(10R^*)$; for safety reasons, we opted to take $R^* = 10^3$. In this way, $D = 11$ and a fractional size of about $\frac{1}{11}$ were found to produce excellent accuracy for all $R \leq 10^5$ (and $X \geq \frac{1}{5}$) *so long as* each interpolatory polynomial was utilized only between nodes #5 and 6 (out of $\{0, 1, 2, \cdots, 11\}$). Compare: [10, p. 131(middle)] and [17, p. 386].

3 A Loose Description of the Algorithm

Given $\mathbb{G}_N$ and some R-interval $[R_1, R_2]$ that we wish to test for the presence of even/odd Maass cusp forms. There is no loss of generality in assuming that the "Weyl bound" for $[R_1, R_2]$ is no bigger than, say, 10; *i.e.*, that

$$\frac{\mu(\mathcal{F}_N)}{8\pi}(R_2^2 - R_1^2) \leq 10. \qquad (2)$$

The mode-of-attack outlined in [9, §4] has as its guiding philosophy the central tenet that Maass cusp forms on $\mathbb{G}_N \backslash H$ are *not* malicious. They "wish" to be found, and *can* be, provided that sensible precautions are taken in code design and in the subsequent assignment of any adjustable parameters within the code.

In this spirit, let φ be any $\mathbb{G}_N$-cusp form with $R \in [R_1, R_2]$. Let (1) be its Fourier expansion.

Depending on the multiplicity of $\lambda = \frac{1}{4} + R^2$, one can presumably normalize things by fixing the values of several early c_n in some simple $\{0, 1\}$ - way. Cf. (1). For simplicity, we assume that declaring $c_1 = 1$ already takes care of everything. (Compare [10, p. 10 (footnote)].)

One knows that $c_n = O(\sqrt{n})$. Depending on the implied constant C, it is certainly possible to choose some sensible (monotonic decreasing) function $M(y)$ so that, for each positive y,

$$\varphi(x + iy) = \sum_{n=1}^{M(y)} c_n \sqrt{y}\kappa(ny; R)cs(nx) + [\![10^{-16}]\!] , \qquad (3)$$

wherein $[\![10^{-16}]\!]$ signifies a quantity having absolute value no greater than 10^{-16}.

From a *practical* standpoint, one basically just selects some number $\hat{R} > R$ so that $K_{iR}(\hat{R}) < 10^{-\Delta}$ with an appropriately big Δ and then takes

$$M(y) = \frac{\mathcal{L}}{2\pi y}\hat{R}, \qquad (4)$$

at least for comparatively modest values of $\log^+(1/y)$. The number $\hat{R}$ can, if necessary, be determined with the aid of the machine. (Bear in mind here that $K_{iR}(X)$ is monotonic decreasing in X for $X \geq R$ and that eventually it must look like $c_R X^{-1/2}\exp(-X)$. The latter *format* is readily detected by the machine. Bear in mind too that the "exp" in §2 with $\cos^{-1}(R/X)$ is readily seen to be no bigger than $\exp[-\sqrt{8/9}X(1 - R/X)^{3/2}]$.)

The region $\mathcal{F}_N$ has $\exp(\pi i/N)$ as its lower right-hand vertex. Corresponding to this, let $M_0 = M(\sin\frac{\pi}{N})$. The interval $[R_1, R_2]$ is normally chosen so that the *same* M_0-value can be reasonably expected to hold for any "cuspidal" $R \in [R_1, R_2]$. Cf. (4) and the inherent flexibility therein in the choice of Δ. Condition (2) clearly facilitates matters when R_1 is large.

Machine-wise, $\varphi(x + iy)$ is now a finite Fourier series for each y. Take $Y < \sin(\frac{\pi}{N})$ and keep $1 \leqq n \leqq M(Y) < Q$. As a tautology, one finds that

$$c_n\sqrt{Y}\kappa(nY; R) = \frac{2}{Q}\sum_{j=1}^{Q}\varphi(z_j)cs(nx_j) + 2[\![10^{-16}]\!] , \qquad (5)$$

where $z_j = x_j + iY = (\mathcal{L}/2Q)(j - \frac{1}{2}) + iY$. (The proof of this fact is expedited by temporarily switching to complex exponentials.)

Let z_j^* be the pullback of z_j to $\bar{\mathcal{F}}_N$ under the action of $\mathbb{G}_N$. (If there are two, select one.) Since $Y < \sin(\frac{\pi}{N})$, the pullback process will necessarily involve at least one inversion for each j. [3]

By automorphy, however, $\varphi(z_j^*) = \varphi(z_j)$. Accordingly:

$$c_n\sqrt{Y}\kappa(nY; R) = \sum_{m=1}^{M_0} V_{nm}c_m + 4[\![10^{-16}]\!] \qquad (6)$$

for $1 \leqq n \leqq M(Y)$, wherein

$$V_{nm} = \frac{2}{Q}\sum_{j=1}^{Q}\sqrt{y_j^*}\kappa(my_j^*; R)cs(mx_j^*)cs(nx_j). \qquad (7)$$

---

[3] For a more general formulation of the familiar $\{E, T\}$ "flip-flop" used with $\mathbb{G}_N$, see [22].

The essential point here is that one gets M_0, *not* $M(Y)$, in (6).

Restricting (6) to $1 \leqq n \leqq M_0$, one can rewrite things as

$$\sum_{m=2}^{M_0} c_m \tilde{V}_{nm} = -\tilde{V}_{n1} + 4[\![10^{-16}]\!] \tag{8}$$

with $\tilde{V}_{nm} \equiv V_{nm} - \delta_{nm}\sqrt{Y}\kappa(nY;R)$. The terms $[\![10^{-16}]\!]$ are of course effectively 0 on the machine.

Since the pullback process is nontrivial for each z_j, system (8) is definitely not some tautology. Notice too that, formally, (8) makes perfectly good sense for *any* R – and that the numbers x_j, x_j^*, Y, y_j^* used in the computation of $\tilde{V}_{nm}$ are free of any outright R-dependence.

By throwing away one row, say $n = 1$, in (8), we now get a system [call it (8_1)] consisting of $M_0 - 1$ equations in $M_0 - 1$ unknowns.

If R is known, this system can presumably be solved to give $\{c_2, ..., c_{M_0}\}$. (The precision with which the c_k are obtained will naturally depend on both the conditioning-level of $[\tilde{V}_{nm}]$ as well as the inherent accuracy of the individual $\tilde{V}_{nm}$.)

If R is not known, one can proceed with two (or more) Y-values in parallel and simply try to *adjust* R so as to make the resulting c_k-vectors as nearly equal as possible. Substantial agreement in these vectors, particularly for sizable values of M_0, would be a strong indication that R is indeed an "eigenvalue" (at least for modest N; *cf.* §5 (paragraphs 2, 6)).

To circumvent any possible intervals of ill-conditioning, system (8_1) is typically solved using *several* sets of two (or more) Y-values once M_0 begins to become at all large.

This process of uncovering the "interesting" R-values in $[R_1, R_2]$ along with their solution vectors $(c_1, c_2, ..., c_{M_0})$ [out to, say, $9 \sim 11$ decimal places] comprises phase 1 of the overall algorithm.

Phase 2 refers to the subsequent process by which the phase 1 data-sets are inserted back into (6), *i.e.*

$$c_n = \frac{\sum_{m=1}^{M_0} V_{nm} c_m}{\sqrt{Y}\kappa(nY;R)}, \tag{9}$$

using a variety of successively smaller Y-values, in order to determine a multitude of "higher" c_n-values for each φ. (Observe that $M(Y)$ *does* approach ∞ as $Y \to 0^+$.)

Though (9) can be used for n up to $M(Y)$, cf. (4) and (6), underflow instabilities in the vicinity of $n = M(Y)$ are best avoided by keeping

$$n \leqq \frac{\mathcal{L}}{2\pi Y}\tilde{R}, \tag{10}$$

where $\tilde{R} \equiv \max(R, 4)$, say. Since Δ is large, the number $\tilde{R}$ lies well-inside $[R, \hat{R})$. A little machine calculation shows, in fact, that $K_{iR}(4) \in [.011, .884]$ when $0 \leqq R \leqq 4$ (the ".011" corresponding to $R = 0$).

For n satisfying (10), continued worries about underflow instabilities in (9) [stemming from sporadic, abnormally small denominators] can be mitigated by again running two or more Y-values in parallel and then appending the list of denominators to each respective output vector.

In this connection though, cf. [9, p. 300 items (B)(C)].

Finally, it pays to keep in mind that one excellent way of addressing the veracity of the *original* phase 1 data-sets is by simply watching the values $c_1 = 1, c_2, \cdots, c_{M_0}$ repeatedly reappear [to 9 $\sim$ 11 decimal places] as sets of Y-values significantly different than those utilized in phase 1 are substituted into (9) *à la phase 2*. The case of c_1 is perhaps the most striking.

4 Some Further Details

Consistent with the fact that neither $\mathcal{C}$, nor the number of "cuspidal" $R \in [R_1, R_2]$, nor the basic conditioning-level of $[\tilde{V}_{nm}]$ is typically known *a priori*, it is hardly surprising that our phase 1 code rests on any number of adjustable parameters and internal decision criteria.

The code's structure can be pretty well guessed from §§2,3 and [9, §4]. Cf. [7].

The main inputted quantities are: N, R_1, R_2, degree D, Y_1, Y_2 (we assume here that only *two* parallel "tracks" are used), M_0, Q (a single quantity presumed to exceed both $M(Y_1)$ and $M(Y_2)$) and, lastly, three length-scales H_1, H_2, H_3 (expressed most conveniently in the form $H_2 = H_1/MAG$, $H_3 = H_2/MAG2$).

The significance of the length scales is as follows.

H_1 is the coarse grid[4] over which the initial computation of all necessary K-Bessel values is carried out.

H_2 is the "once magnified" scale on which – by means of a rolling Lagrange interpolation of degree D – the matrices $[\tilde{V}_{nm}]$ are first assembled, then solved with Gaussian elimination, in an attempt to ascertain which intervals of length H_2 stand a reasonable chance of containing an "eigenvalue" (i.e., cuspidal R).

This is done by first looking for sign changes among the [piecewise linear] difference functions $c'_k - c''_k$ (in an obvious notation). To cut down on the number of possibilities, an internal decision procedure – or, "triggering mechanism" – is employed. Briefly put, the procedure checks that at least one $c'_k - c''_k$ manifests an appropriately transverse crossing (at, say, $R^\star$) and then seeks to ensure that the remaining differences $c'_\ell - c''_\ell$ ($\ell \leq 4$) manifest believable levels of smallness on the given H_2-interval and its closest neighbors. The calibration levels that go into the procedure are basically empirical.

H_3 is the "fully magnified" scale, reflecting the minimal accuracy with which we ultimately hope to get R. The scale is activated only near the numbers $R^\star$. For each such $R^\star$, numerical quadrature is used to obtain a

---

[4] in R

"pristine" set of K-Bessel values on a grid of $D + 1$ nodal points having gap H_2 and center $R^\star$. The matrices $[\tilde{V}_{nm}]$ are then prepared by Lagrange interpolation on a scale of H_3 and subsequently solved via Gaussian elimination. To determine the *optimal* value of R, an appropriate functional of the values $|c'_k/c''_k - 1|$, $k \leq M_0$, is minimized over the given H_3-grid.

The format of the functional used for this has a high degree of arbitrariness. We typically chose to work with only $2 \leq k \leq 4$ and then *displayed* a large batch of the "higher" differences $|c'_n - c''_n|$ $(n > 4)$ in our output. The extent to which these differences were nearly 0 served as a kind of "control" on the degree to which $R_{optimal}$ was [indeed] cuspidal.

Additional assurances came from running the same code at other values of Y_1, Y_2.

To strengthen the accuracy of any given $R_{optimal}$, it typically sufficed to "zero in on it" using smaller and smaller R-intervals $[R_1^{(\nu)}, R_2^{(\nu)}]$.

Before turning to phase 2, we need to make three important comments.

First, it virtually goes without saying that the parameters H_1, H_2, D need to be chosen in a way that ensures good accuracy in the various Lagrange interpolations. There are undoubtedly *many* ways in which this can be done. For the (empirically tested) set-up mentioned at the end of §2, wherein $D = 11$ and $R \leq 10^5$, it is basically necessary that one keep

$$\frac{1}{5} \leq \frac{2\pi}{\mathcal{L}} Y \leq \frac{2\pi}{\mathcal{L}} \sin(\frac{\pi}{N}) = \pi \tan(\frac{\pi}{N}).$$

This leads to the restriction $N \leq 49$; to allow sufficient room to vary Y, it seems wise to keep $N \leq 24$ (say). The parameter H_1 should be kept no bigger than about $\pi/11 \log(10R)$, with the same cutoff as in §2. We often took $H_1 = .025$ and $MAG = 25$ in our initial ["exploratory"] runs with $R \leq 10^4$.

To treat $N > 24$, the natural expectation of course is that it *suffices* to replace $\log(10R)$ with an appropriate $\log(AR)$. Cf. §2 paragraphs 7 and 8. (Verification would entail running a new battery of tests.)

Next, as explained in [9, p. 299], the presence of $\delta_{nm}\sqrt{Y}\kappa(nY; R)$ in $\tilde{V}_{nm}$ stands a reasonable chance of improving the conditioning of $[\tilde{V}_{nm}]$ to the extent that such terms help make the column norms all roughly of the *same* order-of-magnitude. By reviewing (7) and the reasoning for (10), it is evident that one should try to keep

$$\frac{2\pi}{\mathcal{L}} M_0 Y \leq \tilde{R} \quad \text{(or so).}$$

Cf. also here [5, 18]. Together with our earlier comment, this makes arranging things to have

$$\frac{1}{5} \leq \frac{2\pi}{\mathcal{L}} Y \leq \frac{2\pi}{\mathcal{L}} M_0 Y \leq \tilde{R} \tag{11}$$

seem optimal; we decided to adhere to (11) throughout our work with phase 1. To help gauge the *actual* conditioning-level of $[\tilde{V}_{nm}]$ near $R^\star$ (at level H_2), we found it helpful to append several crude indicators of coefficient velocity and acceleration (w.r.t. R) to the output data for each $R_{optimal}$. In this connection, cf. [10, pp. 52(8.8),(8.9) and 82 middle].

Finally, in regard to the choice of H_2, recall that perfect adherence to a Weyl law on $\mathbb{G}_N \backslash H$ (for either even/odd forms) carries with it a *mean R-gap* of $4\pi/R\mu(\mathcal{F}_N) = 4N/(N-2)R$ in the limit of large R. Cf. [10, p. 9]. Particularly in cases where the normalized spacing distribution is expected to be Poisson (i.e., $N = 3,4,6$), abnormally small gaps can*not* be summarily dismissed. (Cf. Eugene Bogomolny's lectures [3].)

This fact is clearly problematic for the phase 1 code. Indeed, it would appear almost certain that pairs of cusp forms "separated" by less than H_2 would simply be "missed" (by virtue of having never activated the triggering mechanism, depending as it does on the coefficient velocities being fairly stable). In our initial runs with $R \lesssim 10^3$, we generally took H_2 to be no bigger than about 5% of $4\pi/R\mu(\mathcal{F}_N)$ – and simply hoped for the best. ($H_2 = .001$ was a common choice.) Not too surprisingly perhaps, as R grew – especially beyond 10^3 – it became increasingly necessary to run phase 1 with *multiple* choices of $\{Y_1, Y_2\}$ in order to detect cuspidal R. The "culprit" for this appeared to be a mixture of 3 interrelated things: (a) H_2 being too big; (b) improperly adjusted control parameters in the triggering mechanism [at level H_2]; and (c) a gradual worsening in the average conditioning-level of $[\tilde{V}_{nm}]$.

Some experiments aimed at exploring the relative importance of a,b,c over, say, $\mathbb{G}_3$ and $\mathbb{G}_5$ could prove very illuminating.

Compared to phase 1, *phase 2* is virtually problem-free. To ensure that the denominators in (9) have a reasonable chance of staying large and that the K-Bessel calls "follow a well-trodden path accuracy-wise", it is natural to morph (10) into

$$\frac{1}{5} \lesssim \frac{2\pi}{\mathcal{L}} nY \lesssim \tilde{R}. \tag{12}$$

Treating $n \in [n_A, n_B]$ with $n_B/n_A < 5\tilde{R}$ thus necessitates keeping

$$\frac{\mathcal{L}}{10\pi n_A} \lesssim Y \lesssim \frac{\mathcal{L}\tilde{R}}{2\pi n_B}. \tag{13}$$

The case $n_A = 1$, $n_B = M_0$ agrees with (11); from (4) we see that $\mathcal{L}\tilde{R}/2\pi M_0 = (\sin \frac{\pi}{N})\tilde{R}/\hat{R}$.

5 Subtleties

To achieve success, many computational algorithms hinge on certain working hypotheses being fulfilled. The spectral codes considered here are no exception. There *are*, however, a number of subtleties involving phase 1 that we would be remiss not to at least draw attention to.

(i) The most important one concerns resonances. By mimicking the reasoning in [10, pp. 6(1.2), 97–101] (with Y therein taken large but fixed [5]) and concomitantly referring to [8, p. 155(12.2), 161(iv), 163(top), 200(bot)–201(13.15), 208(13.22)], it quickly becomes apparent that poles of the Eisenstein series $E(z; s; \mathbb{G}_N)$ [resonances] lying inside $\{Re(s) < \frac{1}{2}\}$ but *extremely close* to $\{Re(s) = \frac{1}{2}\}$ can produce "imposters" in phase 1. I.e., $\mathbb{G}_N$-invariant functions which are mistaken for genuine even cusp forms by the machine (as it runs through the phase 1 code).

The essential condition for this is that the pole, call it $\frac{1}{2} - \eta + i\gamma$, have a relatively modest γ ([6]) but an η-value which is effectively 0. The existence of such poles is a concern for *non*arithmetic $\mathbb{G}_N$ (i.e., $N \neq 3, 4, 6$) and, even more so (as algorithmic extensions are made), for 1-cusp groups Γ whose Teichmüller space is nontrivial. Cf. [2] and [6].

The "imposters" correspond to functions which, on $\mathcal{F}_N \cap \{z \in H : Im(z) < Y\}$, reduce to

$$\sqrt{\eta + \left(\frac{R - \gamma}{\sqrt{\eta}}\right)^2} \; E\left(z; \frac{1}{2} + iR\right) \qquad (14)$$

and have

$$\max\left(\sqrt{\eta}, \frac{|R - \gamma|}{\sqrt{\eta}}\right) < \text{something like (say) } 10^{-10}.$$

Cf. [8, p. 154(top) with $B = Y$]. The number η would thus need to be less than 10^{-20} or so; likewise for $|R - \gamma|$. On the machine, the latter condition would be indistinguishable from $R = \gamma$ (so, for R, we just declare $R \equiv \gamma$, analogously to [10, p. 6(1.2)]).

To the extent that the points z_j stay far away from the $\mathbb{G}_N$-image of $\{Im(z) \geq Y\}$ (in the hyperbolic metric), the phase 1 algebra for (14) will *look* essentially like that of (6). Bear in mind here that $|\phi(\frac{1}{2} + i\gamma; \mathbb{G}_N)| = 1$; the $n = 0$ term for (14) is thus very nearly zero. Cf. [8, pp. 65, 130]. (Also: [8, pp. 164, 165, 437(b), 587], [16, Theorem 3.2 (proof)], [21, Prop. 4.1].)

In situations where N and γ are modest, the existence of poles with $\eta < 10^{-20}$ seems very far-fetched, to put it mildly. For larger N, however, thanks to an old result of Selberg (see [8, p. 579]), poles of this kind *do* begin to appear; in fact near every point $\frac{1}{2} + i\tau$.

The cliché that comes to mind here is obvious: "how does one use a computer to check when a real number x is exactly zero?"

In cases like $\mathbb{G}_N$, where the Phillips-Sarnak conjecture predicts (for $N \neq 3, 4, 6$) that no even cusp forms exist, perhaps a *wiser* thing to do would be to carefully "unravel" the (overdetermined, parallel-track) linear algebra and concentrate *instead* on showing that "$|x|$" must necessarily exceed some

---

[5] do not confuse this Y with the one used earlier in eqs. (6) and (8)
[6] taken positive, without loss of generality

effectively computable $\varepsilon_0 > 0$ throughout certain γ-ranges. (This would refine [10, p. 26ff].)

Notice, too, that when the group Γ is nonarithmetic and has a nontrivial Teichmüller space, the matter of $\sqrt{\eta}$ dropping below 10^{-10} takes on an entirely different "feel"; namely, a *dynamic* one. Cf. [2] and [10, p. 7(top)]. Also: [8, pp. 231–232].

In the triangle group setting, there are 3 further subtleties of potential relevance in phase 1 (for modest values of N and R):

(ii) the possible occurrence of φ having an enormous $\mathcal{C}$; [7]

(iii) the possible occurrence of multiple eigenvalues;

(iv) the possible occurrence of a pair of cuspidal R having a microscopic gap.

The "short answer" to (ii) and (iii) [presupposing the guiding philosophy mentioned at the start of §3] is to "just switch the normalization."

Cases with a big $\mathcal{C}$ occasionally show themselves by appearing first with a substantially degraded accuracy in (8_1). Such R can often be "fixed" simply by tightening up the original parameters and then "zeroing in". Cf. §4 $[R_1^{(\nu)}, R_2^{(\nu)}]$.

In (iii), the *expectation* of course is that $\mathbb{G}_N$ has *no* multiple eigenvalues; cf. [10, p. 10]. The ε_0-strategy mentioned in connection with (i) has an immediate counterpart here; the pursuit of this might well prove worthwhile.

Subtlety (iv) has clear philosophical overlaps with both (i) and (iii). As such, rather than "just shrinking the H_j's and passing to double precision", it might be preferable to try to pursue an ε_0-strategy here as well. Cf. also here [23, pp. 369–370].

In cases where a Weyl law holds, a third option would be to examine the oscillations of the putative (i.e., "computed") counting-function remainder term $S(T)$ as T increases. If these appear to have the *wrong* mean-value, this can be taken as a hint that some eigenvalues are missing. Cf., e.g., [1, §8] and [2, §A.1]. The problem of course is that effective bounds on $\int_0^T S(t)dt$ are normally not known, particularly when the group Γ is nonarithmetic. Even *if* known, the matter of "where to look" would still remain.[8]

References

1. R. Aurich, F. Steiner, and H. Then, Numerical computation of Maass waveforms and an application to cosmology, this volume.
2. H. Avelin, On the deformation of cusp forms, Licentiat Thesis, Uppsala University, U.U.D.M. Report 2003:8, 2003, 85pp.

---

[7] (by [10, pp. 10(bottom), 127(c)], this is a concern only when $N \neq 3, 4, 6$)

[8] The availability of a fairly explicit Selberg-type trace formula would open the door to at least narrowing things down a bit.

3. E. Bogomolny, Quantum and arithmetical chaos, in *Frontiers in Number Theory, Physics, and Geometry I* (ed. by P. Cartier, B. Julia, et. al.), Springer-Verlag, 2006, pp. 3–106.

4. A. Booker, A. Strömbergsson, and A. Venkatesh, Effective computation of Maass cusp forms, Int. Math. Res. Not. (2006), Art. ID 71281, 34 pp.

5. A. Erdélyi, *Higher Transcendental Functions*, vol. 2, McGraw-Hill, 1953, especially pages 82(21), 86(7), 87(18), 88(19), 88(20, corrected).

6. D. Farmer and S. Lemurell, Deformations of Maass forms, Math. of Comp. 74 (2005), 1967–1982.

7. D. A. Hejhal, Spectral Codes, available at www.math.umn.edu/~hejhal

8. D. A. Hejhal, *The Selberg Trace Formula for $PSL(2,\mathbb{R})$*, vol. 2, Lecture Notes in Mathematics 1001, Springer-Verlag, 1983.

9. D. A. Hejhal, On eigenfunctions of the Laplacian for Hecke triangle groups, in *Emerging Applications of Number Theory* (ed. by D. Hejhal, et. al.), IMA Vol. 109, Springer-Verlag, 1999, pp. 291–315.

10. D. A. Hejhal, Eigenvalues of the Laplacian for Hecke Triangle Groups, Memoirs of the Amer. Math. Society No. 469, 1992.

11. D. A. Hejhal, On value distribution properties of automorphic functions along closed horocycles, in *XVIth Rolf Nevanlinna Colloquium* (ed. by I. Laine and O. Martio), de Gruyter, 1996, pp. 39–52.

12. D. A. Hejhal and S. Arno, On Fourier coefficients of Maass waveforms for $PSL(2,\mathbb{Z})$, Math. of Comp. 61 (1993), 245–267 and S11–S16.

13. D. A. Hejhal and H. Christianson, On correlations of CM-type Maass waveforms under the horocyclic flow, J. Physics A 36 (2003), 3467–3486.

14. D. A. Hejhal and B. Rackner, On the topography of Maass waveforms for $PSL(2,\mathbb{Z})$, Exper. Math. 1 (1992), 275–305.

15. D. A. Hejhal and A. Strömbergsson, On quantum chaos and Maass waveforms of CM-type, Found. of Physics 31 (2001), 519–533.

16. H. Iwaniec, *Spectral Methods of Automorphic Forms*, 2nd edition, GSM Vol. 53, American Math. Society and Revista Matemática Iberoamericana, 2002.

17. C. Jordan, *Calculus of Finite Differences*, 3rd edition, Chelsea Publishing, 1965.

18. W. Magnus, F. Oberhettinger, and R. Soni, *Formulas and Theorems for the Special Functions of Mathematical Physics*, Springer-Verlag, 1966, especially pages 85 (line 4), 139 (line 14), 141 (bottom) – 142 (line 11), 147 (bottom).

19. G. Pólya, Bemerkung über die Integraldarstellung der Riemannschen ξ-Funktion, Acta Math. 48 (1926), 305–317.

20. F. Strömberg, Maass waveforms on $(\Gamma_0(N),\chi)$ (computational aspects), this volume.

21. A. Strömbergsson, On the uniform equidistribution of long closed horocycles, Duke Math. J. 123 (2004), 507–547.

22. A. Strömbergsson, A pullback algorithm for general (cofinite) Fuchsian groups, 2000, available at www.math.uu.se/~astrombe

23. H. Then, Maass cusp forms for large eigenvalues, Math. of Comp. 74 (2005), 363-381.

24. E. C. Titchmarsh, *Eigenfunction Expansions associated with Second-order Differential Equations*, vols. 1 and 2, 2nd edition, Oxford University Press, 1962 and 1958.

25. E. C. Titchmarsh, *The Theory of the Riemann Zeta-Function*, Oxford University Press, 1951.

VI. Maass Waveforms on $(\Gamma_0(N), \chi)$ (Computational Aspects)

Fredrik Strömberg

TU Darmstadt, Fachbereich Mathematik, AG AGF, Schlossgartenstraße 7, D-64289 Darmstadt, Germany, `stroemberg@mathematik.tu-darmstadt.de`

Summary. The main topic of this paper is computational aspects of the theory of Maass waveforms, i.e. square-integrable eigenfunctions of the Laplace-Beltrami operator on certain Riemann surfaces with constant negative curvature and finite area. The surfaces under consideration correspond to quotients of the upper half-plane by certain discrete groups of isometries, the so called Hecke congruence subgroups.

It is known that such functions can also be regarded as wavefunctions corresponding to a quantum-mechanical system describing a particle moving freely on the surface. Since the classical counterpart of this motion is chaotic, the study of these wavefunctions is closely related to the study of quantum chaos on the surface. The presentation here, however, will be purely from a mathematical viewpoint.

Today, our best knowledge of generic Maass waveforms comes from numerical experiments, and these have previously been limited to the modular group, $PSL(2,\mathbb{Z})$, and certain triangle groups (cf. [13, 14, 16] and [39]). One of the primary goals of this lecture is to give the necessary theoretical background to generalize these numerical experiments to Hecke congruence subgroups and non-trivial characters. We will describe algorithms that can be used to locate eigenvalues and eigenfunctions, and we will also present some of the results obtained by those algorithms.

There are four parts, first elementary notations and definitions, mostly from the study of Fuchsian groups and hyperbolic geometry. Then more of the theoretical background needed to understand the rich structure of the space of Maass waveforms will be introduced. The third chapter deals with the computational aspects, and the final chapter contains some numerical results.

1 General Definitions and Notation

1.1 A Brief Introduction to Fuchsian Groups

For a more thorough (but still elementary) introduction to this subject see for example [21] or [10].

The following notation will be used; $M_2(K)$ is the ring of 2×2 matrices over some base ring K, usually $\mathbb{R}$ or $\mathbb{Z}$, $GL(2, K) \subseteq M_2(K)$ is the group of invertible 2×2 matrices and $SL(2, K) \subseteq M_2(K)$ is the group of matrices with determinant equal to 1. $\mathcal{H}$ denotes the Poincaré upper half-plane $\{z = x + iy \,|\, y > 0\}$ equipped with the hyperbolic metric and area measure

$$ds^2 = \frac{1}{y^2}(dx^2 + dy^2), \quad d\mu = \frac{1}{y^2}dxdy.$$

$PSL(2, \mathbb{R})$ is the group of Möbius transformations with real coefficients,

$$PSL(2, \mathbb{R}) = \left\{ z \to \frac{az + b}{cz + d} \,\middle|\, a, b, c, d \in \mathbb{R}, \ ad - bc = 1 \right\}.$$

It is clear that we can represent such mappings with matrices and there is an isomorphism $PSL(2, \mathbb{R}) \simeq SL(2, \mathbb{R})/\{\pm Id\}$, where Id is the identity element in $SL(2, \mathbb{R})$. We will use matrix and transformation notation interchangeably, all groups discussed are subgroups of $PSL(2, \mathbb{R})$ so they always contain $-Id$, hence there should be no confusion when we use the same notation for matrix groups as for transformation groups. Note that $\mathcal{H}$, the hyperbolic metric and the hyperbolic measure are invariant under $PSL(2, \mathbb{R})$. The only subgroups of $PSL(2, \mathbb{R})$ that we are interested in are the discrete subgroups.

Definition 1.1. *A* Fuchsian group *is a discrete subgroup of* $PSL(2, \mathbb{R})$.

The basic example of a Fuchsian group, for us, is the *modular group*,

$$PSL(2, \mathbb{Z}) = SL(2, \mathbb{Z})/\{\pm Id\} \subseteq PSL(2, \mathbb{R}).$$

This is just the subgroup of Möbius transformations in $PSL(2, \mathbb{R})$ with integer coefficients. This is also an example of what is called an *arithmetic group* (cf. [21, ch. 5]) and is of interest for number theorists among others.

If Γ is a Fuchsian group it is a standard result that the space of Γ-orbits, $\Gamma \backslash \mathcal{H} = \{\Gamma z \,|\, z \in \mathcal{H}\}$, can be given the analytical structure of a Riemann surface with marked points (this is also called an *orbifold*) (cf. [25, II.F.], and for more details [32]). On the other hand, a classical result, the Klein-Poincaré uniformization theorem, asserts that any Riemann surface M with constant negative curvature equal to -1 can be realized as $\Gamma \backslash \mathcal{H}$ for some Fuchsian group Γ (see [22] for more details on uniformization). We can visualize the Riemann surface $\Gamma \backslash \mathcal{H}$ via a fundamental domain F for Γ.

Definition 1.2. *A closed set* $F \subseteq \mathcal{H}$ *is a* fundamental domain *for* Γ *if*

a) $\cup_{T \in \Gamma} T(F) = \mathcal{H}$, *and*
b) *if* F° *denotes the interior of* F *then* $T_1(F^\circ) \cap T_2(F^\circ) = \emptyset$ *if* $T_1 \neq T_2 \in \Gamma$.

Note that we use *closed* fundamental domains in line with for example [21], so they will in general contain some equivalent boundary points. There are two important examples of fundamental domains.

Definition 1.3. *Let Γ be a Fuchsian group and let $p \in \mathcal{H}$ be a point that is not fixed by any element in Γ. Let $d(z, w)$ denote the hyperbolic distance between z and w and define*

$$D_\Gamma(p) = \{z \in \mathcal{H} \mid d(z, p) \leq d(z, Tp), \forall T \in \Gamma\}.$$

The set $D_\Gamma(p)$ is called a Dirichlet fundamental domain *for Γ, with center p.*

Let $\Gamma_z = \{T \in \Gamma \mid Tz = z\}$ denote the stabilizer in Γ of the point $z \in \mathcal{H} \cup \mathbb{R} \cup \{\infty\}$. Another important type of fundamental domain, F_Γ, the *Ford fundamental domain*, can be described in terms of the exterior of isometric circles. Let $D_T = \{z \in \mathcal{H} \mid |T'(z)| \geq 1\}, \forall T \in PSL(2, \mathbb{R})$. Then, if $\Gamma_\infty = \{T \in \Gamma \mid T\infty = \infty\}$ is non-empty and generated by $S : z \mapsto z + 1$, we define

$$F_\Gamma = \bigcap_{T \in \Gamma \backslash \Gamma_\infty} D_T \cap \left\{z \in \mathcal{H} \mid |\Re z| \leq \frac{1}{2}\right\}.$$

The boundary (in $\mathcal{H} \cup \mathbb{R} \cup \{\infty\}$) of a fundamental domain F might contain vertices that are fixed-points of elements of Γ; these points are called elliptic or parabolic vertices respectively if the transformations fixing them are either elliptic or parabolic. A parabolic vertex is usually referred to as a cusp, and is viewed as either a point removed from the surface $\Gamma \backslash \mathcal{H}$, or a point at infinity.

1.2 Hecke Congruence Groups

Let N be any positive integer. We define the *principal congruence subgroup* of level N, $\Gamma(N) \subseteq PSL(2, \mathbb{Z})$, by

$$\Gamma(N) = \left\{\begin{pmatrix} a & b \\ c & d \end{pmatrix} \in PSL(2, \mathbb{Z}) \,\middle|\, \begin{pmatrix} a & b \\ c & d \end{pmatrix} \equiv \begin{pmatrix} 1 & 0 \\ 0 & 1 \end{pmatrix} \mod N\right\}.$$

This is a subgroup of finite index in $PSL(2, \mathbb{Z})$, and any subgroup of $PSL(2, \mathbb{Z})$ containing some $\Gamma(N)$ is called a *congruence subgroup*. The *Hecke congruence subgroup*, $\Gamma_0(N)$, is defined by

$$\Gamma_0(N) = \left\{\begin{pmatrix} a & b \\ c & d \end{pmatrix} \in PSL(2, \mathbb{Z}) \,\middle|\, c \equiv 0 \mod N\right\}.$$

It is obvious that $\Gamma(N) \subseteq \Gamma_0(N)$ so $\Gamma_0(N)$ is a congruence subgroup.

The *standard fundamental domain* for $PSL(2, \mathbb{Z}) = \Gamma_0(1)$ is the set

$$\mathcal{F}_1 = \left\{z \in \mathcal{H} \mid |z| \geq 1, |\Re z| \leq \frac{1}{2}\right\},$$

which is also a Ford fundamental domain. Suppose from now on that we have fixed a set of right coset representatives $\{V_k\}_{k=1}^{\upsilon_N}$ for $\Gamma_0(N)$ in $\Gamma_0(1)$. (See p. 217) on how to construct such maps V_k). One knows that

$$v_N = [\Gamma_0(1) : \Gamma_0(N)] = N \prod_{p|N} \left(1 + p^{-1}\right).$$

We can now fix a fundamental domain for $\Gamma_0(N)$ corresponding to these coset representatives (cf. [21, thm. 3.1.2])

$$\mathcal{F}_N = \cup_{k=1}^{v_N} V_k(\mathcal{F}_1). \tag{1}$$

Note that $\mathcal{F}_N$ need not be normal in the sense of [12], but it will be bounded by finitely many geodesics and it also has some other nice properties as we will see (Theorem 1.6). Denote by κ_0 the total number of parabolic vertices of $\mathcal{F}_N$, and by κ the number of inequivalent cusps of $\mathcal{F}_N$ (note $\kappa \leq \kappa_0$). It is known that

$$\kappa = \sum_{d|N,d>0} \phi\left(\left(d, \frac{N}{d}\right)\right),$$

where ϕ is Eulers totient function,

$$\phi(n) = \#\left\{j \in \{1, 2, 3, \ldots, n\} | (j, n) = 1\right\}$$

(see [20, 35]). Let $v_1, \ldots, v_{\kappa_0}$ be the set of parabolic vertices of $\mathcal{F}_N$ and $p_1, \ldots, p_\kappa$ a set of inequivalent cusps. Without loss of generality we set $p_1 = \infty$.

Definition 1.4. *A cusp normalizing map* associated to a cusp p_j, *is a map* $\sigma_j \in PSL(2,\mathbb{R})$ *satisfying*

i) $\sigma_j(\infty) = p_j$, and
ii) $\sigma_j S \sigma_j^{-1} = S_j$,

where $Sz = z+1$ is a generator of Γ_∞, and S_j is a generator of Γ_{p_j}, the (cyclic infinite) stabilizer group of the cusp p_j. Then σ_j is uniquely determined up to a translation.

For the groups $\Gamma_0(N)$ we will always choose the cusp normalizing maps σ_j of the form $\sigma_j = A_j \rho_j$, where $A_j \in SL(2,\mathbb{Z})$ maps ∞ to p_j and ρ_j is a scaling by h_j, the so-called *width* of the cusp p_j (cf. [20, pp. 36-37]). It is important to note that $h_j|N$. For the cusp $p_1 = \infty$ we will always take σ_1 as the identity.

We will now provide an important fact about the fundamental domain for $\Gamma_0(N)$ that will be of use to us later. First, for each parabolic vertex v_ℓ of $\mathcal{F}_N$ we choose a map $U_\ell \in \Gamma_0(N)$ which maps v_ℓ to its cusp representative in $\{p_1, \ldots, p_\kappa\}$; say $U_\ell(v_\ell) = p_{j(\ell)}$. If v_ℓ is already a cusp representative we take $U_\ell = Id$. This will give us a finite collection of maps $\mathcal{U} = \{U_\ell\}_{1 \leq \ell \leq \kappa_0}$. Now, to any point $w \in \mathcal{F}_N$ we associate the "closest" (see Remark 1.5) parabolic vertex, v_w, and a corresponding map $U_w = U_{\ell(v_w)} \in \mathcal{U}$.

Remark 1.5. By the "closest" vertex to a point $w \in \mathcal{F}_N$ we mean the vertex v_ℓ with respect to which the point w has the *greatest* height, i.e. that v_ℓ for which $\Im(\sigma_{j(\ell)}^{-1} U_\ell w)$ is maximal. We use $I(w)$ to denote the index of the cusp representative corresponding to the vertex closest to w.

Define the *height* of $w \in \mathcal{F}_N$, $h(w)$, by

$$h(w) = \Im\left(\sigma_{I(w)}^{-1} U_w\left(w\right)\right),$$

and the *minimal height* of the fundamental domain $\mathcal{F}_N$, Y_0, by

$$Y_0 = Y_0(N) = \inf_{w \in \mathcal{F}_N} h(w) = \inf_{w \in \mathcal{F}_N} \Im\left(\sigma_{I(w)}^{-1} U_w\left(w\right)\right). \tag{2}$$

It is clear that Y_0 depends only on the fundamental domain $\mathcal{F}_N$ (i.e. the choice of the coset representatives $\{V_k\}$) and not on the choice of the cusp representatives $\{p_j\}$. A standard compactness argument also implies that Y_0 is strictly positive and larger than some fixed quantity only depending on the fundamental domain, and the following theorem actually gives us an explicit bound.

Theorem 1.6. *For any positive integer N, the minimal height of $\mathcal{F}_N$ satisfies the following inequality*

$$Y_0(N) \geq \frac{\sqrt{3}}{2N}.$$

Proof. Let $\{V_k\}$ be the fixed set of coset representatives of $\Gamma_0(N)$ in $\Gamma_0(1)$ and $\mathcal{F}_N$ as above.

Let $w \in \mathcal{F}_N$. We want to show that $\Im\left(\sigma_{I(w)}^{-1} U_w w\right) \geq \frac{\sqrt{3}}{2N}$, and according to Remark 1.5 it is enough to show that $\Im\left(\sigma_{j(\ell)}^{-1} U_\ell w\right) \geq \frac{\sqrt{3}}{2N}$ for some ℓ. Since $w \in \mathcal{F}_N = \cup V_k(\mathcal{F}_1)$ there is an index k such that $w \in V_k(\mathcal{F}_1) = R_k$. Let $v_{\ell(k)} = V_k(\infty)$ be the vertex at infinity of R_k, and $p_{j(k)}$ the cusp representative equivalent to $v_{\ell(k)}$, i.e. $U_{\ell(k)} v_{\ell(k)} = p_{j(k)}$.

Note that we can write $\sigma_j^{-1} = l_j s_j \omega_j$ where l_j is a translation, $\omega_j \in PSL(2, \mathbb{Z})$ is such that $\omega_j S_j \omega_j^{-1} = S^{h_j}$, (where $S^h z = z + h$) and s_j is a scaling with $\frac{1}{h_j}$, where the positive integer h_j is called the *width* of the cusp p_j (cf. [20, p. 36-37]). The width, h_j is independent of the choice of cusp-representative in the class of p_j (cf. [24, p. 59]). Now $\omega_{j(k)} U_{\ell(k)} V_k(\mathcal{F}_1)$ is simply a horizontal translate of $\mathcal{F}_1$, so $\Im(\omega_{j(k)} U_{\ell(k)} w) \geq \frac{\sqrt{3}}{2}$ and thus $\Im(\sigma_{j(k)}^{-1} U_{\ell(k)} w) \geq \frac{\sqrt{3}}{2h_{j(k)}}$. From [20, p. 36-37] it is clear that $1 \leq h_j \leq N$ for all j, hence $\Im(\sigma_{j(k)}^{-1} U_{\ell(k)} w) \geq \frac{\sqrt{3}}{2N}$, and accordingly $Y_0 \geq \frac{\sqrt{3}}{2N}$. $\square$

Remark 1.7. If P is a prime ≥ 3 we will see that for a natural choice (which we use in the numerical work) of fundamental domain we have $Y_0(P) = \frac{\sqrt{3}}{2P}$.

Let $Tz = \frac{-1}{z}$ and $Sz = z + 1$. We know that the maps $V_1 = Id$ and $V_k = TS^{\left(k - \frac{P+3}{2}\right)}$ for $k = 2, \ldots P + 1$ is a set of right coset representatives for $\Gamma_0(P)$ in $\Gamma_0(1)$ (cf. [30, p. 135]). The corresponding fundamental domain is $\mathcal{F}_P = \cup_{k=1}^{P+1} V_k(\mathcal{F}_1)$, it has one cusp of width 1 at $p_1 = \infty$ and one cusp of width P at $p_2 = 0$. The cusp normalizing maps are $\sigma_1 = Id$ and $\sigma_2 z = \frac{-1}{Pz}$. Let $e =$

$\frac{1}{2}+i\frac{\sqrt{3}}{2}$. Then we see that $\Im(V_{P+1}(e)) = \frac{\sqrt{3}}{2}\left|\frac{1}{2}+i\frac{\sqrt{3}}{2}+\frac{P-1}{2}\right|^{-2} = \frac{\sqrt{3}}{2}\frac{4}{P^2+3} \leq \frac{\sqrt{3}}{2P}$ (since $P \geq 3$), and clearly $\Im(\sigma_2^{-1}V_{P+1}(e)) = \Im\left(\frac{e}{P}+\frac{1}{2}-\frac{1}{2P}\right) = \frac{\sqrt{3}}{2P}$. Hence the height of $V_{P+1}(e)$ is equal to $\frac{\sqrt{3}}{2P}$, which means that $Y_0 \leq \frac{\sqrt{3}}{2P}$. The conclusion is that for a prime level $P \geq 3$ and for this particular choice of fundamental domain we have $Y_0 = \frac{\sqrt{3}}{2P}$.

1.3 Introduction to Dirichlet Characters

First we fix $m \in \mathbb{Z}^+$. A Dirichlet character, χ mod m, is a group homomorphism from $(\mathbb{Z}/m\mathbb{Z})^*$ to the unit circle S^1, which is viewed as a function on $\mathbb{Z}$ (with period m) by assigning $\chi(n) = 0$ if $(n, m) > 1$. We then define the map $\chi : SL(2, \mathbb{Z}) \to S^1$ by

$$\chi\left(\begin{pmatrix} a & b \\ c & d \end{pmatrix}\right) = \chi(d).$$

Note that $\chi \neq 0$ on $\Gamma_0(m)$ since $ad - bc = 1$ and $c \equiv 0$ mod m imply that $\chi(ad) = \chi(a)\chi(d) = 1$. Observe also that χ is a group homomorphism from $\Gamma_0(m)$ to S^1.

If χ has a period less than m (for values of n restricted by $(n, m) = 1$) then χ is said to be *imprimitive,* otherwise χ is said to be a *primitive* character. (Note that the *trivial* character, $\chi(n) \equiv 1$, is imprimitive for $m \geq 2$). If q is the smallest period of χ, then q is called the *conductor* of χ.

If χ has period m and conductor q there is a unique way to define a character χ' mod m', for all modulus m' which are multiples of q or multiples of m.

We say that the character χ is *even* if $\chi(-1) = \chi(1) = 1$, and *odd* if $\chi(-1) = -\chi(1) = -1$.

Since we are actually concerned with $PSL(2, \mathbb{Z})$ we need $\chi(A) = \chi(-A)$ to hold for any $A \in SL(2, \mathbb{Z})$, hence we will only consider even characters.

It is known (see [9]) that primitive *real* characters exist only for moduli of the following types:

$$m = N_1, 4N_2, \text{ or } 4N_3,$$

where $N_1 \equiv 1$ mod 4, $N_2 \equiv 2$ mod 4 and $N_3 \equiv 3$ mod 4 and N_1, N_2 and N_3 are square-free.

One should note that these are precisely the fundamental discriminants of real quadratic fields, and it is also a fact that all real characters are given by quadratic residue symbols (Kronecker's extension of the Legendre symbol) cf., e.g. [8, p. 37].

Since characters are multiplicative it is fairly easy to evaluate them at products of integers, but when it comes to linear combinations of integers it is harder. A useful trick for evaluating characters at certain linear combinations is to factor the character instead of the integer. Here we will give the simplest case, which will be of use for the proof of Proposition 2.22 in Subsect. 2.8.

Fact 1.8. Suppose that χ is a real primitive character mod N and that $m \in \mathbb{Z}^+$ is such that $m|N$ and $\left(m, \frac{N}{m}\right) = 1$. Then it is known (cf. [27, (3.1.4)]) that we can factor χ as $\chi = \chi_m \chi_{\frac{N}{m}}$, where χ_m and $\chi_{\frac{N}{m}}$ are real primitive characters mod m and $\frac{N}{m}$ respectively.

Example 1.9. Let χ, N and m be as above, and let $t = qA + \frac{N}{q}B$ for some $A, B \in \mathbb{Z}$. Then

$$\chi(t) = \chi_m(t)\chi_{\frac{N}{m}}(t) = \chi_m\left(mA + \frac{N}{m}B\right)\chi_{\frac{N}{m}}\left(mA + \frac{N}{m}B\right)$$

$$= \chi_m\left(\frac{N}{m}B\right)\chi_{\frac{N}{m}}(mA).$$

Definition 1.10. *A character χ defined on $\Gamma_0(N)$ is said to leave the cusp p_j open if it is trivial on the stabilizer subgroup Γ_{p_j}, i.e. if $\chi(S_j) = 1$. If a character leaves all cusps open we say that the character is* regular *for $\Gamma_0(N)$.*

If a character does not leave the cusp open we say that the cusp is closed, *and it is known that there is no contribution to the continuous part of the spectrum from a closed cusp (cf. [12, pp. 91-99]).*

Observe that the cusp at ∞ is fixed by $S : z \mapsto z + 1$ so $\chi(S) = 1$ for any Dirichlet character. Since we know that $S_j = \sigma_j S \sigma_j^{-1}$ we will also have $\chi(S_j) = 1$ whenever σ_j is a $(\Gamma_0(N), \chi)$-normalizer as described in Definition 2.4 below. In particular, for square free $N \equiv 1 \bmod 4$ all Dirichlet characters are regular (cf. [5]).

The basic assumption from now on (unless anything else is explicitly stated) is that any character χ is a real and even Dirichlet character.

1.4 A Brief Introduction to Maass Waveforms

For a Fuchsian group Γ, a function f defined on the upper half-plane is called Γ-automorphic if it is invariant or transforms in a chosen manner (i.e. with respect to characters or general multipliers) under the action of Γ by the so-called slash operator of weight $k \in \mathbb{Z}$ defined by

$$f_{|[k,A]}(z) = \left(\frac{cz + d}{|cz + d|}\right)^{-k} f(Az), \ A \in \Gamma.$$

Such functions can be viewed as "living" on the Riemann surface $\Gamma \backslash \mathcal{H}$ (or coverings of it in the case of characters). In this paper we will only consider certain non-holomorphic L^2-functions, the so called Maass waveforms (to be defined below) and only at weight $k = 0$ (note that $f_{|[0,A]}(z) = f(Az)$).

In the hyperbolic metric on $\mathcal{H}$ the Laplace-Beltrami operator, Δ, takes the form

$$\Delta = y^2\left(\frac{\partial^2}{\partial x^2} + \frac{\partial^2}{\partial y^2}\right).$$

Definition 1.11. *If Γ is a Fuchsian group, a* Maass waveform *for Γ is a function f defined on $\mathcal{H}$ with the following properties*

i) $\Delta f + \lambda f = 0$, $\lambda \geq 0$;
ii) $f(Az) = f(z)$, for all $A \in \Gamma$;
iii) $\int_{\Gamma \backslash \mathcal{H}} |f|^2 d\mu < \infty$.

We usually write the eigenvalue as $\lambda = \frac{1}{4} + R^2$. It is conjectured (Selberg) that for congruence subgroups, if $\lambda \neq 0$, then we can take $R \in [0, \infty)$ ($\lambda \geq \frac{1}{4}$); any eigenvalues $\lambda \in (0, \frac{1}{4}]$ are usually called *exceptional*. We denote the space of Maass waveforms for Γ by $\mathcal{M}(\Gamma)$, and the space corresponding to a particular eigenvalue by $\mathcal{M}(\Gamma, \lambda)$. If χ is a group homomorphism from Γ to S^1 (cf. Subsection 1.3) we can replace condition ii) by

ii') $f(Az) = \chi(A)f(z)$, for all $A \in \Gamma$.

We then say that f is a Maass waveform for (Γ, χ), the space of which we denote by $\mathcal{M}(\Gamma, \chi)$ or $\mathcal{M}(\Gamma, \chi, \lambda)$. It is also known that each space $\mathcal{M}(\Gamma, \chi, \lambda)$ is finite dimensional (see for example [12, p. 140, thm. 11.11]), that Maass waveforms span the discrete part of the spectra of Δ, and that the continuous part is spanned by the Eisenstein series (see for example [19,38] or [12]).

It is an important fact that if Γ is a congruence subgroup then every Maass waveform f in $\mathcal{M}(\Gamma, \chi, \lambda)$ with $\lambda > 0$ is a *cusp form*, meaning that it tends rapidly to 0 in each cusp. This fact follows from [12, pp. 327(note 15), 78(claim 9.6), 284(line 1), 328(line -2)] since the scattering matrix of $\Gamma(N)$ can be computed explicitly and shown not to have any poles for $\frac{1}{2} < s < 1$; cf., e.g., [28] or [18], and [27, p. 114].

We can equip the space $\mathcal{M}(\Gamma, \chi, \lambda)$ with the *Petersson inner-product*,

$$< f, g > = \int_{\Gamma \backslash \mathcal{H}} f \bar{g} d\mu,$$

where the integration can be taken over any fundamental domain for Γ. With this inner-product $\mathcal{M}(\Gamma, \chi, \lambda)$ is now a finite dimensional Hilbert space.

If Γ is a congruence subgroup and χ is trivial on $\Gamma(N)$, the eigenvalues $0 = \lambda_0 < \lambda_1 \leq \lambda_2 \leq \cdots$, counted with multiplicity, form a discrete sequence and a general Weyl's law is known to hold. In particular, for Hecke congruence subgroups and the trivial character, we have (see [33, thm. 2])

$$N_{\Gamma_0(N)}(T) = \frac{\mu(\mathcal{F}_N)}{4\pi} T - \frac{2\kappa}{\pi} \sqrt{T} \ln \sqrt{T} + A\sqrt{T} + O\left(\frac{\sqrt{T}}{\ln \sqrt{T}}\right), \qquad (3)$$

$$\text{as } T \to \infty,$$

where A is a certain constant depending on the level N and κ is the number of cusps of $\Gamma_0(N)$.

However it is widely believed (by the Sarnak-Phillips philosophy, see for example [29]) that for a "generic" non-cocompact but cofinite Fuchsian group

Γ, unless there are some arithmetic or geometric symmetries present, there should only be at most a finite discrete spectrum (generic here is in the sense that in some appropriate deformation space the exceptional groups have measure 0).

The Maass waveforms are also more mysterious than the holomorphic automorphic functions since there are only very few examples of explicit formulas for constructing Maass waveforms, whereas there are numerous examples of explicit formulas for holomorphic modular functions.

2 Some Structural Theory of $\mathcal{M}(\Gamma_0(N), \chi, \lambda)$

For the rest of this section we put $\Gamma = \Gamma_0(N)$ for brevity and we also assume that χ is a Dirichlet character. To facilitate the computation of Maass waveforms we need as much information as possible about the various symmetries that can be used. First we will see that in the case of a real character we can assume that our functions are real-valued. Then we will consider the obvious translational symmetry which makes the Fourier expansion possible, and the reflectional symmetry which simplifies the said Fourier expansion. Then we will describe the less obvious symmetries, the Hecke operators and the involutions, which further refine the spectral eigenspaces.

2.1 The Conjugation Operator

Let $f \in \mathcal{M}(\Gamma, \chi, \lambda)$ and consider the conjugation operator K, $Kf = \overline{f}$. It is clear that K commutes with the Laplacian so Kf is also an eigenfunction of Δ with the same eigenvalue as f. We also see that for $A \in \Gamma$ we get

$$\begin{aligned}
(Kf)_{|A}(z) = Kf(Az) &= \overline{f(Az)} \\
&= \overline{\chi(A)f(z)} \\
&= \overline{\chi(A)}(Kf)(z),
\end{aligned}$$

so Kf is automorphic with respect to the conjugate character, i.e. $Kf \in \mathcal{M}(\Gamma, \overline{\chi}, \lambda)$. In particular, if χ is a real character, then $\overline{f} = Kf \in \mathcal{M}(\Gamma, \chi, \lambda)$. By considering the functions $\frac{1}{2}(f + Kf)$ and $\frac{1}{2i}(f - Kf)$ (which are real-valued) we get the following proposition.

Proposition 2.1. *If χ is a real-valued character then $\mathcal{M}(\Gamma, \chi, \lambda)$ has a $\mathbb{C}$-linear basis consisting of real-valued functions.*

2.2 The Fourier Series

Notation 2.2. In connection with Fourier series of functions in $\mathcal{M}(\Gamma, \chi, \lambda)$ we will use the following notation:

$$e(x) = e^{2\pi i x},$$

and

$$\kappa_n(y) = \kappa_n(R, y) = \sqrt{y} K_{iR}(2\pi|n|y),$$

where $\lambda = \frac{1}{4} + R^2$. Observe that $\kappa_{-n}(y) = \kappa_n(y)$.

If $f \in \mathcal{M}(\Gamma, \chi, \lambda)$, we define functions $f_j = f_{|\sigma_j}$ and we know that $f_j(z + 1) = f(\sigma_j S z) = f(S_j \sigma_j z) = \chi(S_j) f_j(z) = f_j(z)$ if the character leaves the cusp p_j open.

Let $\lambda > 0$. By separating variables in the Laplace equation it follows that f_j has a Fourier expansion of the form $\sum_{n \neq 0} c_j(n) \kappa_n(y) e(nx)$ (cf. [12], and recall that f is necessarily a cusp form, cf. Sect. 1.4).

Proposition 2.3. *If χ is regular for Γ then a function $f \in \mathcal{M}(\Gamma, \chi, \lambda)$ admits a Fourier expansion at each cusp of Γ, and these expansions are given by the functions f_j. Explicitly we have the following expansion at the cusp p_j :*

$$f_j = f_{|\sigma_j} = \sum_{n \neq 0} c_j(n) \kappa_n(y) e(nx), \ j = 1, \ldots, \kappa.$$

2.3 Involutions and Normalizers

Definition 2.4. *For $\Gamma \subseteq PSL(2, \mathbb{R})$ we say that $g \in PGL(2, \mathbb{R})$ is a normalizer of Γ in $PGL(2, \mathbb{R})$ if*

$$g \Gamma g^{-1} = \Gamma.$$

This means in particular that $A \in \Gamma \Rightarrow \exists B \in \Gamma$ such that $gA = Bg$. But we need a stronger definition if a character is present.

We say that g is a normalizer *of (Γ, χ) if g is a normalizer of Γ and for all $A \in \Gamma$:*

$$\chi(g A g^{-1}) = \chi(A).$$

The set of all normalizers of Γ in $PGL(2, \mathbb{R})$ forms a group, the so-called normalizer subgroup *of Γ in $PGL(2, \mathbb{R})$.*

In the above definition we have $PGL(2, \mathbb{R}) = GL(2, \mathbb{R})/C$, where C is the center in $GL(2, \mathbb{R})$ and consists of all diagonal matrices of the form $\left(\begin{smallmatrix} a & 0 \\ 0 & a \end{smallmatrix} \right)$ with $a \neq 0$.

We define an action of $GL(2, \mathbb{R})$ on $\mathcal{H}$ as follows, for $g = \left(\begin{smallmatrix} a & b \\ c & d \end{smallmatrix} \right)$ and $z \in \mathcal{H}$:

$$g(z) = \begin{cases} \frac{az+b}{cz+d}, & \text{if } ad - bc > 0, \\[2mm] \frac{a\bar{z}+b}{c\bar{z}+d}, & \text{if } ad - bc < 0. \end{cases}$$

Note that all of C acts trivially on $\mathcal{H}$, and hence we also obtain a well-defined action of $PGL(2, \mathbb{R})$ on $\mathcal{H}$. And if g is a normalizer of (Γ, χ) and $f \in \mathcal{M}(\Gamma, \chi, \lambda)$ then $f_{|g} \in \mathcal{M}(\Gamma, \chi, \lambda)$.

The usual definition of a (linear) *involution* of a vector space S is a linear operator $T : S \to S$ such that $T^2 = Id$ (or equivalently $T^{-1} = T$).

Definition 2.5. *A linear operator* $T : \mathcal{M}(\Gamma, \chi, \lambda) \to \mathcal{M}(\Gamma, \chi, \lambda)$ *will be called a* (Γ, χ)-*involution if* T^2 *is the identity operator, that is*

$$T^2 f = f$$

for any function $f \in \mathcal{M}(\Gamma, \chi)$.

Example 2.6. There are three typical examples of (Γ, χ)-involutions that we will use.

- All $W \in PSL(2, \mathbb{R})$ which are normalizers of (Γ, χ) and satisfy $W^2 \in \Gamma$ and $\chi(W^2) = 1$ are (Γ, χ)-involutions when acting via $f \mapsto f_{|W}$. In particular we will use $\omega_N : z \mapsto \frac{-1}{Nz}$, which can be represented by $\begin{pmatrix} 0 & -1/\sqrt{N} \\ \sqrt{N} & 0 \end{pmatrix} \in PSL(2, \mathbb{R})$.
- If $J = \begin{pmatrix} 1 & 0 \\ 0 & -1 \end{pmatrix}$ is a normalizer of (Γ, χ) then $f \mapsto f_{|J}$ is a (Γ, χ)-involution. Note that this is just a reflection in the imaginary axis, $J(z) = -\bar{z}$.

It is important not to confuse the cusp normalizing map defined in the beginning with the normalizer of (Γ, χ) defined above. But some cusp normalizing maps might be normalizers of (Γ, χ) and even involutions of $\mathcal{M}(\Gamma, \chi)$. In general, if it is possible, this is a very good choice of cusp normalizing map as we will see.

2.4 The Reflection Operator

By looking at the fundamental domain of $PSL(2, \mathbb{Z})$ (or the Ford fundamental domain of any congruence subgroup) we see that there is an obvious symmetry; reflection in the imaginary axis. This symmetry provides us with a partitioning of the spectrum into even and odd functions and amounts actually to looking separately at Dirichlet or Neumann boundary conditions when we view the spectral problem on the fundamental domain itself (which has finite boundary points) and not on the corresponding Riemann surface $PSL(2, \mathbb{Z}) \backslash \mathcal{H}$ (with the only boundary point at the cusp).

As remarked in Example 2.6, the reflection operator is represented by the matrix $J = \begin{pmatrix} 1 & 0 \\ 0 & -1 \end{pmatrix}$, and for any matrix $T = \begin{pmatrix} a & b \\ c & d \end{pmatrix}$ we define

$$T^* = JTJ^{-1} = \begin{pmatrix} 1 & 0 \\ 0 & -1 \end{pmatrix} T \begin{pmatrix} 1 & 0 \\ 0 & -1 \end{pmatrix} = \begin{pmatrix} a & -b \\ -c & d \end{pmatrix}. \qquad (4)$$

Then J is a (Γ, χ)-involution whenever $T^* \in \Gamma$ and $\chi(T^*) = \chi(T)$ for all $T \in \Gamma$. In particular this is clearly true for $\Gamma_0(N)$ when χ is a Dirichlet character since then $\chi(T^*) = \chi(d) = \chi(T)$.

Suppose now that J is a (Γ, χ)-involution. We can diagonalize $\mathcal{M}(\Gamma, \chi, \lambda)$ with respect to J and the eigenvalues will be either 1 or -1. $f \in \mathcal{M}(\Gamma, \chi, \lambda)$ is said to be *even* or *odd*, respectively, if $f_{|J} = f$ or $f_{|J} = -f$. Every even or odd function f in $\mathcal{M}(\Gamma,\chi,\lambda)$ has a cosine resp. sine Fourier series. This can be seen as follows. Let

$$f(z) = \sum_{n\neq 0} c(n)\kappa_n(y)e(nx)$$

with $\kappa_n(y)$ as in Notation 2.2 on page 195 (remember that $\kappa_{-n}(y) = \kappa_n(y)$). Then

$$f_{|J}(z) = \sum_{n\neq 0} c(n)\kappa_n(y)e(-nx)$$

$$= \sum_{n\neq 0} c(-n)\kappa_n(y)e(nx).$$

If $f_{|J} = f(-\bar{z}) = \varepsilon f$, then we get $c(-n) = \varepsilon c(n)$ so

$$f(z) = \sum_{n\neq 0} c(n)\kappa_n(y)e(nx)$$

$$= \sum_{n=1}^{\infty} \kappa_n(y)\left(c(n)e(nx) + c(-n)e(-nx)\right)$$

$$= \sum_{n=1}^{\infty} c(n)\kappa_n(y)\left(e(nx) + \varepsilon e(-nx)\right)$$

$$= \sum_{n=1}^{\infty} c(n)\kappa_n(y)\begin{cases} 2\cos(2\pi nx), & \varepsilon = 1, \\ 2i\sin(2\pi nx), & \varepsilon = -1. \end{cases}$$

Set $a(n) = 2c(n)$ and $b(n) = 2ic(n)$, then f will have a Fourier cosine or sine series with coefficients $a(n)$ or $b(n)$ respectively, i.e.

$$f(z) = \sum_{n=1}^{\infty} a(n)\kappa_n(y)\cos(2\pi nx), \text{ or}$$

$$f(z) = \sum_{n=1}^{\infty} b(n)\kappa_n(y)\sin(2\pi nx).$$

So we know that $f = f_0$ can always be taken as an eigenfunction of J, but we should also note that this need *not* be the case with the other Fourier series, f_j. In the next section we give a condition for the simultaneous diagonalization of all f_j with respect to J.

2.5 Complete Symmetrization

Since it is usually preferable to work with symmetrized Fourier series containing only cosines or sines instead of the exponential functions we make a comment on when this is possible.

Theorem 2.7. *Let $\frac{u}{v}$ be any cusp satisfying $(u, v) = 1$, $v|N$ and $\left(v, \frac{N}{v}\right) = 1$ or 2. Then if we take $p_j = \frac{u}{v}$ as a cusp representative, there is a choice of cusp normalizing map σ_j such that if $f \in \mathcal{M}(\Gamma, \chi, \lambda)$ is an eigenfunction of J with eigenvalue ϵ then f_j is also an eigenfunction of J with the eigenvalue $\chi(d)\epsilon$, where d is unique modulo N determined by $d \equiv 1 \bmod v$ and $d \equiv -1 \bmod \frac{N}{v}$.*

Proof. (We refer to [20, p. 36] for the basic facts).

We know that every cusp for Γ has a representative of the form $\frac{u}{v}$ where $v|N$ and $(u, v) = 1$, and $\frac{u}{v}$ and $\frac{u'}{v'}$ are equivalent cusps if and only if $v = v'$ and $u \equiv u' \bmod \left(v, \frac{N}{v}\right)$. Hence $\frac{u}{v}$ is equivalent to $-\frac{u}{v}$ if and only if $2u \equiv 0 \bmod \left(v, \frac{N}{v}\right)$ and this is possible if and only if $\left(v, \frac{N}{v}\right) = 1$ or 2. Suppose now that this is the case, and that we have $p_j = \frac{u}{v}$ for such choice of u, and that $A = \left(\begin{smallmatrix} a & b \\ Nc & d \end{smallmatrix}\right) \in \Gamma_0(N)$ is such that $Ap_j = -p_j$. One verifies by a straightforward computation that up to a sign change $(A \longleftrightarrow -A)$, the most general form of A is given by taking c as an arbitrary integer satisfying $\left(\frac{N}{v}\right)cu \equiv -2 \bmod v$ (thus c is uniquely determined modulo v if $\left(v, \frac{N}{v}\right) = 1$, and uniquely determined modulo $\frac{v}{2}$ if $\left(v, \frac{N}{v}\right) = 2$), and then letting

$$d = a = -1 - \frac{N}{v} \cdot cu, \qquad b = \frac{u(1 - d)}{v}.$$

In particular, note that $d \equiv -1 \bmod \left(\frac{N}{v}\right)$ and $d \equiv 1 \bmod v$.

Observe now that $f_{|T|J} = f_{|J|T^*}$ (where T^* is as in (4)). This means in particular that if f has J-eigenvalue ϵ, then $f_{j|J} = f_{|\sigma_j|J} = f_{|J|\sigma_j^*} = \epsilon f_{|\sigma_j^*}$. We can choose the cusp normalizing map of p_j as

$$\sigma_j = \begin{pmatrix} u & x \\ v & y \end{pmatrix} \begin{pmatrix} \sqrt{m} & 0 \\ 0 & \sqrt{m}^{-1} \end{pmatrix}, \tag{5}$$

for some integers x, y satisfying $uy - vx = 1$ and $m = \frac{N}{(N, v^2)}$ is the width of the cusp p_j (note that $m = \frac{N}{v}$ or $\frac{N}{2v}$ if $\left(v, \frac{N}{v}\right) = 1$ or 2). This choice of σ_j is unique up to right multiplication with S^r for some $r \in \mathbb{R}$. Then we will have

$$\sigma_j^* = J\sigma_j J^{-1} = \begin{pmatrix} u & -x \\ -v & y \end{pmatrix} \begin{pmatrix} \sqrt{m} & 0 \\ 0 & \sqrt{m}^{-1} \end{pmatrix},$$

which is clearly a cusp normalizing map for $-p_j$. Since $Ap_j = -p_j$, then another cusp normalizing map of $-p_j$ is the map $A\sigma_j$, and hence we have $\sigma_j^* = A\sigma_j S^t$ for some $t \in \mathbb{R}$.

It follows that $f_{j|J} = \epsilon f_{|\sigma_j^*} = \epsilon f_{|A\sigma_j S^t} = \chi(d)\epsilon f_{j|S^t}$, and hence if $t \in \mathbb{Z}$ we have $f_{j|J} = \chi(d)\epsilon f_j$, as desired.

Hence it only remains to make sure that we can obtain $t \in \mathbb{Z}$ above. In general this can always be achieved by replacing σ_j by $\tilde{\sigma}_j$ where $\tilde{\sigma}_j = \sigma_j S^{\frac{t}{2}}$, since then $\tilde{\sigma}_j^* = \sigma_j^* S^{-\frac{t}{2}} = A\sigma_j S^t S^{-\frac{t}{2}} = A\sigma_j S^{\frac{t}{2}} = A\tilde{\sigma}_j$. However, if either $\left(v, \frac{N}{v}\right) = 1$ or m is odd, then we can take x and y to be *integers* in (5). To see this, note that by writing out the upper right element in the matrix relation $\sigma_j^* = A\sigma_j S^t$ (and using $au + bv = u$) one obtains $-x = umt + ax + by$. Using the formulas for a and b this implies that $mt = \frac{N}{v}cx + \frac{d-1}{v}y$. Hence $t \in \mathbb{Z}$ holds if and only if

$$\frac{N}{v}cx + \frac{d-1}{v}y \equiv 0 \bmod m. \tag{6}$$

Recall that $m = \frac{N}{v}$ or $\frac{N}{2v}$, so that (6) is equivalent to $\frac{1-d}{v}y \equiv 0 \bmod m$. Hence we achieve our goal if we let x, y be any integers satisfying $uy - vx = 1$ and $y \equiv 0 \pmod{m}$. This is clearly possible if either $\left(v, \frac{N}{v}\right) = 1$ or m is odd, since then $(v, m) = 1$. (In the remaining case, i.e. $\left(v, \frac{N}{v}\right) = 2$ and $m = \frac{N}{2v}$ is even, one can check that one must allow x or y to be non-integral to obtain $t \in \mathbb{Z}$.)
□

Corollary 2.8. *Suppose that N can be written as $N = 2^r p_1 \cdots p_m$, with $0 \leq r \leq 3$, and if $m > 0$ then $p_1, \ldots, p_m$ are distinct odd primes, and suppose also that in every character decomposition $\chi = \chi_v \chi_{\frac{N}{v}}$, $v|N$, the character $\chi_{\frac{N}{v}}$ is even. Then we can take all f_j as simultaneous eigenfunctions of J with the same eigenvalue.*

Note that if we do *not* have simultaneous eigenfunctions of J we have to work (numerically) with exponentials instead of sines and cosines, and complex instead of real Fourier coefficients.

2.6 Hecke Operators

It is well-known that the classical theory of Hecke operators (cf., e.g. [1], [11]) can be carried over from the case of holomorphic modular forms to the case of Maass waveforms. This section provides an outline of the theory for those not familiar with Hecke theory, and it will also serve as a recollection of fundamental facts and a common ground of notation for the more experienced.

The Hecke operators considered here are operators acting on spaces of modular functions (i.e. functions automorphic with respect to congruence subgroups).

We will define the Hecke operators in a similar way as in Atkin-Lehner [1]. Let $N \in \mathbb{Z}^+$ be given. For a prime p there is a subgroup of finite index in $\Gamma_0(N)$,

$$\Gamma_0(N, p) = \left\{ \begin{pmatrix} a & b \\ c & d \end{pmatrix} \in \Gamma_0(N) \,\middle|\, b \equiv 0 \bmod p \right\}.$$

For any $d \in \mathbb{Z}^+$, we define the map $A_d : z \mapsto dz$ in $PSL(2, \mathbb{R})$. For $d = p$ we then have $A_p^{-1}\Gamma_0(N, p)A_p = \Gamma_0(Np) \subseteq \Gamma_0(N)$ and it is easy to see that $\chi(A_p^{-1}BA_p) = \chi(B)$ so that A_p gives a map between the spaces of Maass waveforms

$$A_p^{-1} : \mathcal{M}(\Gamma_0(N), \chi) \to \mathcal{M}(\Gamma_0(N, p), \chi), \tag{7}$$

$$f \mapsto f_p = f_{|A_p^{-1}}.$$

We then want to map f_p to a function once again in $\mathcal{M}(\Gamma_0(N), \chi)$ but in a non-trivial way (i.e. not via A_p). Let $\{R_j\}_{j=1}^\mu$ be the set of right coset representatives of $\Gamma_0(N, p)$ in $\Gamma_0(N)$ used in [1, lemma 5]. R_j will then have the lower right entry $d \equiv 1 \bmod N$. If V is any element in $\Gamma_0(N)$, then for each j there exists a unique i such that $R_j V = g R_i$ for some $g \in \Gamma_0(N, p)$, and different j's give different i's. Using our special choice of R_j one also checks that $\chi(V) = \chi(g)$ in this relation. Thus for any function $h \in \mathcal{M}(\Gamma_0(N, p), \chi)$ it is clear that the function $\sum_{j=1}^\mu h_{|R_j}$ belongs to $\mathcal{M}(\Gamma_0(N), \chi)$. Hence we may define the operator $T_p : \mathcal{M}(\Gamma_0(N), \chi) \to \mathcal{M}(\Gamma_0(N), \chi)$ as follows:

$$
T_p f = \frac{1}{\sqrt{p}} \sum_{j=1}^\mu f_{p|R_j}.
$$

The factor of $\frac{1}{\sqrt{p}}$ is a convenient normalization. Working out the coset representatives explicitly (see [1, lemma 5]) gives the following formula for a prime number p with $(N, p) = 1$

$$
T_p f(z) = \frac{1}{\sqrt{p}} \sum_{j=0}^{p-1} f\left(\frac{z+j}{p}\right) + \frac{1}{\sqrt{p}} \chi(p) f(pz),
$$

and for prime q with $q|N$ (compare, e.g., [27, p. 142 (4.5.26)]):

$$
T_q f(z) = U_q f(z) = \frac{1}{\sqrt{q}} \sum_{j=0}^{q-1} f\left(\frac{z+j}{q}\right).
$$

We will follow the convention to call T_q with $q|N$ an *exceptional* Hecke operator and denote it by U_q instead of T_q. In principle, due to the multiplicative nature of the Hecke operators (cf. Theorem 2.10 below) it is sufficient to define Hecke operators for primes, and then use the multiplicativity (cf. (9) below and [27, lemma 4.5.12]) to recursively define T_n for any positive integer n with $(n, N) = 1$. To use the above construction for a composite number n is quite elaborate and it is not done this way in the literature. Hecke (cf. [11, nos. 32, 35, 36, 37 and 41, p. 859]) does not explain how he arrives at the exact definition of the operators T_n. To get a motivation for the formula (8) below (including the character) it is easiest to use the approach of double cosets as in Shimura ([35]) or Miyake ([27]). Using either of these constructions yields the following definition.

Definition 2.9. *For any $n \in \mathbb{Z}^+$ and any $f \in \mathcal{M}(\Gamma_0(N), \chi, \lambda)$ we define the Hecke operator T_n by the formula*

$$
T_n f(z) = \frac{1}{\sqrt{n}} \sum_{\substack{ad=n \\ d>0}} \chi(a) \sum_{j=0}^{d-1} f\left(\frac{az+j}{d}\right). \tag{8}
$$

Theorem 2.10. *The Hecke operators T_n with $(n, N) = 1$ are endomorphisms of the space $\mathcal{M}(\Gamma_0(N), \chi, \lambda)$ which have the following properties:*

$$T_n T_m = \sum_{\substack{d | (m,n) \\ d > 0}} \chi(d) T_{\frac{mn}{d^2}}, \tag{9}$$

for any integers n and m with $(n, N) = (m, N) = 1$, and (for the adjoint operator)

$$T_n^* = \overline{\chi(n)} T_n. \tag{10}$$

In particular $\{T_n | (n, N) = 1\}$ forms a commutative family of normal operators which also commutes with the Laplacian and the reflection J.

Proof. To prove the multiplicative property cf. [35, eq. (3.5.10)], look at the action of T_n on the Fourier coefficients of a Maass waveform f. Suppose that $c(k)$ and $b(k)$ are the respective Fourier coefficients of f and $T_n f$, then the following relation holds:

$$b(k) = \sum_{ad=n, a, d > 0} \chi(a) c\left(\frac{kd}{a}\right), \tag{11}$$

with the usual convention that $c(r) = 0$ if $r \notin \mathbb{Z}$. We prove the equality (9) by comparing the action on the Fourier coefficients of the left hand side and the right hand side.

Next, to prove (10), a quick computation using (9) shows that it suffices to treat the case $n = p$ a prime. To prove that $T_p^* = \overline{\chi(p)} T_p$ one uses the following relation between the Petersson inner products on the two groups:

$$< T_p f, g >_{\Gamma_0(N)} = \frac{1}{\sqrt{p}} < f_p, g >_{\Gamma_0(N,p)}.$$

This relation is an easy extension of [1, lemma 12] to the case of non trivial character using the fact that $\chi(R_j) = 1$ for all right coset representatives R_j of $\Gamma_0(N, p)$ in $\Gamma_0(N)$ as given in [1, lemma 5]. Alternatively: see [20, thm. 6.20] and [27, thm. 4.5.4(1)]. It is easy to show that T_n commutes with Δ and J. $\square$

Remark 2.11. Actually ([27, thm. 4.5.13]) we also have

$$U_q T_n = T_n U_q,$$

for $q | N$ and $(n, N) = 1$ but in general the operators U_q are not normal so they are put aside for a while, until the next section, where we will introduce the space of newforms.

Theorem 2.12. *There exists an orthogonal basis $\{\phi_j\}$ in $\mathcal{M}(\Gamma_0(N), \chi, \lambda)$ consisting of eigenfunctions to all Hecke operators T_n with $(n, N) = 1$.*

Proof. Observe that $\mathcal{M}(\Gamma_0(N), \chi, \lambda)$ is a finite dimensional vector space (see [12, pp. 140, 298]) so we can choose an ON basis $\{f_j\}_{j=1}^m$ and represent the operators T_n by matrices $\Lambda(n)$ in this basis. From Theorem 2.10 above it is clear that $\{\Lambda(n) \,|\, (n, N) = 1\}$ is a family of commuting normal $m \times m$ matrices. From elementary linear algebra arguments (the spectral theorem or [31, thm. 9.4.1]) we know that there exists a basis $\{\phi_j\}_{j=1}^m$ where each ϕ_j is an eigenfunction of all $\Lambda(n)$, and this is exactly what we want. Compare: [20, thm. 6.21] and [27, thm. 4.5.4(3)]. $\square$

2.7 Oldforms and Newforms

From (11) it is clear that if f has Fourier coefficients $c(n)$ and $T_n f = \lambda(n) f$, then

$$c(n) = \lambda(n) c(1).$$

We would like to make sure that $c(1) \neq 0$ here, but unfortunately as we will see below, it may happen that $c(1) = 0$.

Suppose that χ is an imprimitive character mod N of conductor q. If $q|M$ and $Md|N$ for some positive integers M, d, let χ_M be the character induced from χ on $\Gamma_0(M)$. Then

$$f(z) \in \mathcal{M}(\Gamma_0(M), \chi_M, \lambda) \Rightarrow f(z), f_{|A_d}(z) \in \mathcal{M}(\Gamma_0(N), \chi, \lambda)$$

(with A_d as in Sect. 2.6, cf. (7)), which is clear from the fact that $A_d \Gamma_0(N) A_d^{-1} = \Gamma_0\left(\frac{N}{d}, d\right) \subseteq \Gamma_0\left(\frac{N}{d}\right) \subseteq \Gamma_0(M)$, and $\chi(A_d B A_d^{-1}) = \chi(B)$.

Definition 2.13. *Let $f \in \mathcal{M}(\Gamma_0(N), \chi, \lambda)$. We say that f is an* oldform *if there exist positive integers M and d with $Md|N$, a character χ_M mod M and a function $f_1 \in \mathcal{M}(\Gamma_0(M), \chi_M, \lambda)$ such that we can write $f = f_{1|A_d}$. The smallest such integer M is called the* conductor *(or level) of f. Compare: [20, p. 107 (bottom)] and [27, p. 162 (bottom)].*

Suppose now that f is an oldform with Fourier series given by

$$f(z) = \sum_{m \neq 0} c(m) \sqrt{y} K_{iR}(2\pi|m|y) e(mx).$$

Then

$$f_{|A_d}(z) = f(dz) = \sum_{m \neq 0} c(m) \sqrt{dy} K_{iR}(2\pi|m|dy) e(mdx)$$

$$= \sqrt{d} \sum_{n \equiv 0 \bmod d} c\left(\frac{n}{d}\right) \sqrt{y} K_{iR}(2\pi|n|y) e(nx).$$

Hence if $d > 1$ then the first Fourier coefficient of $f(dz)$ is zero. This is an example of one of the inconveniences of oldforms.

Let $\mathcal{M}^{old}(\Gamma_0(N), \chi, \lambda)$ be the linear space spanned by the oldforms and then define $\mathcal{M}^{new}(\Gamma_0(N), \chi, \lambda)$ as the orthogonal complement with respect to the Petersson inner product. We then have

$$\mathcal{M}(\Gamma_0(N), \chi, \lambda) = \mathcal{M}^{old}(\Gamma_0(N), \chi, \lambda) \oplus \mathcal{M}^{new}(\Gamma_0(N), \chi, \lambda).$$

It is easy to see that both subspaces are stable under the Hecke operators T_n with $(n, N) = 1$, since $T_n^* = \chi(n)T_n$.

Definition 2.14. *A function $f \in \mathcal{M}^{new}(\Gamma_0(N), \chi, \lambda)$ is called a normalized newform of level N if f is an eigenfunction of J and all Hecke-operators T_n, $(n, N) = 1$, and the first Fourier coefficient of f is 1, i.e. $c(1) = 1$.*

Proposition 2.15. *If χ is a primitive character mod N and $q|N$, then the operator U_q is unitary on $\mathcal{M}(\Gamma_0(N), \chi, \lambda)$; i.e.*

$$U_q^* = U_q^{-1}.$$

Proof. The proof is elementary (at least for prime indices N) and goes by first proving that $U_q^* = \omega_N U_q \omega_N^{-1}$ and then using properties of the Petersson inner-product. Here ω_N is defined as in Example 2.6. (For a sketch of the proof see [5]). $\square$

The following theorem is crucial to the theory of oldforms/newforms. See Atkin-Lehner [1] and Miyake [27] for similar results in the classical setting. For a more general adelic version, see Miyake [26]. The following version for Maass waveforms is in principle a direct adaptation of [36, thm. 4.6] to the case of non-trivial character.

Theorem 2.16. *Let χ be a Dirichlet character with conductor q. The spaces of newforms have the following properties.*

(a) For each multiple N of q, the normalized newforms of level N and character χ, $\{F_j\}_{j=1}^{m_N}$, form an orthogonal basis in $\mathcal{M}^{new}(\Gamma_0(N), \chi, \lambda)$. If we want to stress the level and/or character we write $F_j = F_j^{(N, \chi)}$. This is called the Hecke-(eigen)basis of $\mathcal{M}^{new}(\Gamma_0(N), \chi, \lambda)$.
(b) Each F_j is an eigenfunction of all T_n with $n \in \mathbb{Z}^+$.
(c) A basis in the space $\mathcal{M}(\Gamma_0(N), \chi, \lambda)$ is given by

$$\left\{ F_i^{(M, \chi)}(dz) \,\middle|\, d, M \in \mathbb{Z}^+, q|M, dM|N, 1 \leq i \leq m_M \right\}.$$

(d) Let χ' be a Dirichlet character mod M. If $f \in \mathcal{M}^{new}(\Gamma_0(N), \chi, \lambda)$ and $g \in \mathcal{M}^{new}(\Gamma_0(M), \chi', \mu)$ with $\lambda, \mu > 0$, then either there exists an infinite number of primes p for which the Hecke eigenvalues of T_p are different for f and g, or else $N = M$, $\lambda = \mu$, $\chi' = \chi$ and $f \equiv g$.

Proof. Observe that the only thing that differs from [36, thm. 4.6] is that we have a non-trivial character.

The proof of (a)-(c) can be done exactly as in the proof of [36, thm. 4.6]. The proof of "multiplicity one", (d) is a bit more complicated to extend. The main difference from the case of trivial character is the fact that ω_N (cf. Example 2.6) and T_n no longer commute and we need the Euler products of both f and $\omega_N f$ for the proof of Lemma 4.12 in [36], however it is a simple matter to do the same thing as in the proof of [27, thm. 4.6.19], noting that on the newspace both T_p and U_q are normal so any eigenform of all Hecke operators T_n is also an eigenform of all T_n^* and U_q^* (i.e $T_n^* = \omega_N^{-1} T_n \omega_N$ and $U_q^* = U_q^{-1}$ on the newspace). $\square$

Remark 2.17. Note that U_q is not normal on the full space $\mathcal{M}(\Gamma_0(N), \chi, \lambda)$ (unless, of course, if χ is primitive, in which case there are only newforms) so therefore we would have no multiplicity one theorem if we tried to diagonalize the full space with respect to all Hecke operators.

Theorem 2.18. *Let f be a normalized newform with $T_n f = \lambda(n) f$ and $f_{|J} = \epsilon f$. If the Fourier expansion of f is*

$$f(z) = \sum_{|m| \geq 1} c(m) \sqrt{y} K_{iR}(2\pi m y) e(mx)$$

(where $c(1) = 1$) then

$$c(m) = \lambda(m), \text{ and}$$
$$c(-m) = \epsilon \lambda(m),$$

for all $m \in \mathbb{Z}^+$. We also have the following multiplicativity relations

$$c(m)c(n) = \sum_{d|(m,n), d>0} \chi(d) c\left(\frac{mn}{d^2}\right), \, (n, N) = 1, \, m \in \mathbb{Z}, \qquad (12)$$
$$c(m)c(p) = c(mp), \, p|N, \, m \in \mathbb{Z}.$$

Proof. The first part follows by adding f and $f_{|J}$ and reordering the sum. The second part follows at once from the multiplicative relation in Theorem 2.10 which then apply to the Hecke eigenvalues $\lambda(n)$. By setting $c(1) = 1$ in the relation $c(n) = \lambda(n)c(1)$ we are done. $\square$

Proposition 2.19. *Suppose that f is a normalized newform in $\mathcal{M}^{new}(\Gamma_0(N), \chi, \lambda)$, and that q is a prime such that $q|N$, then the following holds.*

(a) If χ is primitive then $|c(q)| = 1$.

(b) If χ is trivial and $q^2 \nmid N$ then $c(q) = \frac{\lambda_q}{\sqrt{q}}$, where $\lambda_q = \pm 1$. In fact λ_q is the eigenvalue of $-W_q$, cf. Subsect. 2.8 below.

(c) If χ is trivial and $q^2|N$, then $c(q) = 0$.

Proof. For (a), use Prop. 2.15. The assertions (b) and (c) are the analogs of [1, thm. 3]. Compare: [27, thm. 4.6.17 and cor. 4.6.18(2)], and [31, thm. 9.4.8]. □

Recall that by Theorem 2.16 we can always choose a Hecke eigenbasis in the newspace $\mathcal{M}^{new}(\Gamma_0(N), \chi, \lambda)$ where the functions are normalized by $c(1) = 1$.

We stress that if the character χ is non-trivial, then we can *not* in general assume that the Hecke basis is real valued. For by Theorem 2.10 we have $T_n^* = \overline{\chi(n)}T_n$ whenever $(n, N) = 1$, and thus

$$\overline{\lambda(n)} = \overline{\chi(n)}\lambda(n) \tag{13}$$

and $\overline{c(n)} = \overline{\chi(n)}c(n)$. Hence, in particular, if $\chi(n) = -1$ then $c(n)$ is purely imaginary, whereas if $\chi(n) = 1$ then $c(n)$ is real.

Remark 2.20. Suppose that χ is real and that f is a normalized newform in the space $\mathcal{M}^{new}(\Gamma_0(N), \chi, \lambda)$ with $f_{|J} = \epsilon f$ and Hecke eigenvalues $\lambda(n)$. Then $\epsilon \bar{f}$ is also a normalized newform in $\mathcal{M}^{new}(\Gamma_0(N), \chi, \lambda)$, but with Hecke eigenvalues $\overline{\lambda(n)}$. Now f and $\epsilon \bar{f}$ are in general linearly independent since $\lambda(n) \neq \overline{\lambda(n)}$ whenever $\chi(n) \neq 1$ and $\lambda(n) \neq 0$. *Thus the newspace $\mathcal{M}^{new}(\Gamma_0(N), \chi, \lambda)$ is in general multidimensional.*

An exception to the above fact are the CM-type forms à la [17] (originally constructed by Maass in [23]). These forms are real and (hence) they have $c(n) = 0$ for all $n \in \mathbb{Z}^+$ with $(n, N) = 1$ and $\chi(n) \neq 1$.

2.8 The Cusp Normalizing Maps as Normalizers of $(\Gamma_0(N), \chi)$

Suppose that $Q|N$ and $\left(Q, \frac{N}{Q}\right) = 1$. Then following [1] and [2] we can attach a normalizer of $\Gamma_0(N)$ to Q, call it W_Q. This can be taken of the form

$$W_Q = \begin{pmatrix} Qx & y \\ Nz & Qw \end{pmatrix},$$

where $x, y, z, w \in \mathbb{Z}$, $x \equiv 1 \bmod \frac{N}{Q}$, $y \equiv 1 \bmod Q$ and $Q^2 xw - Nzy = Q$. One can also verify that W_Q is a normalizer of $(\Gamma_0(N), \chi)$ for certain χ (for examples see below). A remarkable fact about W_Q is that as an operator on $\mathcal{M}(\Gamma_0(N), \chi, \lambda)$ it is independent of the choices of x, y, z and w. In fact, if $W_Q' = \begin{pmatrix} Qx' & y' \\ Nz' & Qw' \end{pmatrix}$ for any $x', y', z', w' \in \mathbb{Z}$ and $\det W_Q' = Q$, then (as operators) $W_Q' = \overline{\chi_Q(y')\chi_{\frac{N}{Q}}(x')}W_Q$ (see [2, prop. 1.1]). In particular, this implies that for any $f \in \mathcal{M}(\Gamma_0(N), \chi, \lambda)$

$$f_{|J|W_Q} = \overline{\chi_Q(-1)}f_{|W_Q|J}.$$

For us the normalizer W_Q will be most useful for real characters because of the following proposition (cf. [2, prop. 1.1]).

Proposition 2.21. *Suppose that $Q|N$ and $\left(Q, \frac{N}{Q}\right) = 1$, and let $\chi = \chi_Q \chi_{\frac{N}{Q}}$. Then for any $f \in \mathcal{M}(\Gamma_0(N), \chi, \lambda)$ we have $W_Q f = f_{|W_Q} \in \mathcal{M}(\Gamma_0(N), \overline{\chi_Q} \chi_{\frac{N}{Q}}, \lambda)$, and $W_Q^2 f = \chi_Q(-1)\overline{\chi_{\frac{N}{Q}}(Q)}f$. In particular, if χ_Q is real and $\chi_Q(-1)\overline{\chi_{\frac{N}{Q}}(Q)} = 1$ then W_Q maps $\mathcal{M}(\Gamma_0(N), \chi, \lambda)$ into itself and W_Q is a (Γ, χ)-involution.*

The normalizer W_Q can in many cases be taken to be a cusp normalizing map, as explained in the following proposition.

Proposition 2.22. *Suppose that $Q|N$ and $\left(Q, \frac{N}{Q}\right) = 1$. Then we can choose W_Q as a cusp normalizing map of the cusp $\frac{Q}{N}$.*

In reference to Definition 1.4, note that W_Q is equal to $Q^{-\frac{1}{2}}W_Q \in PSL(2, \mathbb{R})$.

Proof. We may choose $x = z = 1$ in the formula for W_Q, that is

$$W_Q = \begin{pmatrix} Q & y \\ N & wQ \end{pmatrix},$$

where $y, w \in \mathbb{Z}$, and $Qw - \frac{N}{Q}y = 1$. Note that the width of the cusp $\frac{Q}{N}$ is $\frac{N}{\left(N, \left(\frac{N}{Q}\right)^2\right)} = \frac{N}{\frac{N}{Q}\left(Q, \frac{N}{Q}\right)} = Q$ (cf. [20, p. 37]). We know that the cusp normalizing map of $\frac{Q}{N}$ can be written uniquely up to a translation as $\sigma = A\rho$, where $A \in SL(2, \mathbb{Z})$ maps ∞ to $\frac{Q}{N}$, and ρ is a scaling by the width of the cusp. Hence, writing W_Q as

$$W_Q = \begin{pmatrix} 1 & y \\ \frac{N}{Q} & wQ \end{pmatrix}\begin{pmatrix} Q & 0 \\ 0 & 1 \end{pmatrix},$$

we see that W_Q is a cusp normalizing map for $\frac{Q}{N}$. $\square$

Another interesting property of W_Q is the following (cf. [2, prop. 1.2, 1.3]).

Proposition 2.23. *Let $Q \in \mathbb{Z}^+$ be such that $Q|N$ and $\left(Q, \frac{N}{Q}\right) = 1$, and let q and p be primes with $(p, N) = 1$, $q|N$ and $(q, Q) = 1$. Then*

$$T_p W_Q f = \overline{\chi_Q(p)}W_Q T_p f, \text{ and}$$
$$U_q W_Q f = \overline{\chi_Q(q)}W_Q U_q f,$$

for $f \in \mathcal{M}(\Gamma_0(N), \chi, \lambda)$, where χ_Q is the character mod Q induced by χ (i.e. $\chi = \chi_Q \chi_{\frac{N}{Q}}$).

Observe that for a Hecke eigenfunction f we get $T_p(W_Q f) = \overline{\chi_Q(p)}W_Q T_p f = \overline{\chi_Q(p)}\lambda_p W_Q f$ where λ_p is the Hecke eigenvalue of f for T_p. In other words $W_Q f$

is also an eigenfunction of T_p with eigenvalue $\lambda_p \overline{\chi_Q(p)}$. In view of "multiplicity one" (cf. Thm. 2.16 (d)), if $\chi_Q \equiv 1$ and f is a newform, then $W_Q f$ must be a multiple of f, i.e. $W_Q f = \mu_Q f$, where $\mu_Q^2 = \chi_{\frac{N}{Q}}(Q)$ (cf. Prop. 2.21). Now if we choose the cusp normalizer σ_j of $\frac{Q}{N}$ as $\sigma_j = W_Q$ (cf. Prop. 2.22), then we have $f_j = \mu_j f_1$, and in terms of the different Fourier expansions, this means that (for all k)

$$c_j(k) = \mu_j c_1(k), \tag{14}$$

where $\mu_j^2 = \chi_{\frac{N}{Q}}(Q)$. This relation can either be used as a means to test the accuracy of our program or to reduce the size of the linear system in Sect. 3.2 (cf. the normalization (B) on p. 212). However, it is important to observe that (14) does *not* in general allow us to reduce the Fourier coefficients at *all* cusps to the ones at $i\infty$. Cf. Prop. 2.22.

Our comment in Prop. 2.19(b) about $-W_q$ is justified by combining the foregoing remarks with the three references cited in the proof of Prop. 2.19. Cf. [2, §1,2], [20, pp. 113–118] and [27, Cor. 4.6.18] for some related ideas.

3 Computational Aspects

3.1 Introduction

We know that a Maass waveform $f \in \mathcal{M}(\Gamma_0(N), \chi, \lambda)$, with $\lambda > 0$ is completely described by its Fourier series at ∞, but unfortunately to assure stability of the numerical method we need knowledge of the Fourier series at *all* cusps of $\mathcal{F}$, i.e. for $1 \le j \le \kappa$,

$$f_j(z) = \sum_{|n| \ge 1} c_j(n) \kappa_n(y) e(nx).$$

The aim here is to *compute* such functions f, that is, we wish to find an eigenvalue $\lambda = \frac{1}{4} + R^2$ (for convenience we will usually refer to R as the eigenvalue) and a set of Fourier coefficients $\{c_j(n) \,|\, 1 \le j \le \kappa, \ |n| \ge 1\}$. In general (the exceptions are the CM-type forms, cf., e.g., [17]) there are no known formulas for neither the eigenvalues nor the coefficients and we have to be content with numerical approximations.

There are two steps (Phase 1 and Phase 2) in the algorithm. First we locate R (up to the desired accuracy) and an initial set of Fourier coefficients, and then, if we want to, we can use this initial set to generate a larger set of coefficients.

3.2 Phase 1

The method here is a generalization of Hejhal's algorithm in [14] to groups with several cusps and non-trivial characters, and some of the ideas are also inspired by the algorithm of Selander and Strömbergsson in [34]. The idea is that given a real number R we use linear algebra to compute a set of numbers that are likely to be close to "true" Fourier coefficients if R is close to a "true" eigenvalue of a Maass waveform.

Preliminary Numerical Remarks

First we introduce an "effective zero". In standard double precision arithmetic we know that $x + 10^{-16}x \approx x$, so if we fix $\epsilon < 10^{-16}$ then a truncation error of ϵ can in principle be neglected (computationally). We will use $[[\epsilon]]$ to denote a quantity with absolute value less than ϵ.

Suppose that $f \in \mathcal{M}(\Gamma_0(N), \chi, \lambda)$ with Fourier expansions at the cusps

$$f_j(z) = \sum_{|n| \geq 1} c_j(n) \kappa_n(y) e(nx), \ 1 \leq j \leq \kappa. \tag{15}$$

There is a trivial (a priori) bound on the coefficients: $c_j(n) = O(\sqrt{n})$ (cf. [12, p. 585] and [19, thm. 3.2]). Combining this with $\kappa_n(y) = \sqrt{y} K_{iR}(2\pi|n|y)$ and the fact that $K_{iR}(u) \sim \sqrt{\frac{\pi}{2u}} e^{-u}$, as $u \to \infty$ (for fixed R), we see that the tail of the sum (15) satisfies $\sum_{|n| \geq M} = O\left(e^{-2\pi y M}\right)$ as $M \to \infty$, with the implied constant depending on R and y. Hence for any fixed y (and R) we can take $M = M(y)$ such that

$$f_j(z) = \sum_{1 \leq |n| \leq M(y)} c_j(n) \kappa_n(y) e(nx) + [[\epsilon]], \tag{16}$$

for $1 \leq j \leq \kappa$. Using similar observations as in [14, p. 311] it is clear that, for R large, we can take $M(y)$ as

$$M(y) = \left\lfloor \frac{R + AR^{\frac{1}{3}}}{2\pi y} \right\rfloor, \tag{17}$$

for some constant A. In practice it turns out that $A = 12$ or 15 is good enough. We will now see how to use the automorphy condition ((19) below) to get a good linear system that can be solved for the $c_j(n)$.

The Linear System

In order to make the algorithm stable we need to use the Fourier coefficients at all cusps, which means that we have to use different expansions at different regions in a clever way as follows (see also [34]). Recall the notation from Subsect. 1.2.

We will view $f \in \mathcal{M}(\Gamma_0(N), \chi, \lambda)$ as pieced together of the Fourier series at all cusps, meaning that we use the f_j that is most convenient at each point. Hence, given $w \in \mathcal{F}_N$ we will use the identity

$$f(w) = \chi(U_w^{-1}) f_{I(w)}(\sigma_{I(w)}^{-1} U_w w). \tag{18}$$

Cf. Remark 1.5 on p. 190.

Given any point $z \in \mathcal{H}$ and any $j \in \{1, \ldots \kappa\}$, we let $z_j = \sigma_j z$ (which may or may not be inside $\mathcal{F}_N$), and let w_j be the pullback of z_j into $\mathcal{F}_N$, i.e. $w_j = T_j(z_j) \in \mathcal{F}_N$ with $T_j \in \Gamma_0(N)$, and let $z_j^* = \sigma_{I(j)}^{-1} U_{w_j} w_j$ (here $I(j) = I(w_j)$), then $z_j^* = \sigma_{I(j)}^{-1} U_{w_j} T_j \sigma_j z$, so

$$\begin{aligned} f_j(z) = f(\sigma_j z) &= f(T_j^{-1} U_{w_j}^{-1} \sigma_{I(j)} z_j^*) \\ &= \chi(T_j^{-1} U_{w_j}^{-1}) f_{I(j)}(z_j^*). \end{aligned} \tag{19}$$

This relation (implicit automorphy, cf. [14, p. 298 (15)]) between $f_j(z)$ and $f_{I(j)}(z_j^*)$ is what enables us to get hold of all Fourier series. Consider the truncated Fourier series

$$\hat{f}_j(z) = \sum_{1 \le |n| \le M(y)} c_j(n) \kappa_n(y) e(nx).$$

One way to view this series is as a Discrete Fourier Transform, and we can perform an inverse transform over the following set of sampling points along a horocycle:

$$\left\{ z_m = x_m + iY \,\middle|\, x_m = \frac{1}{2Q}\left(m - \frac{1}{2}\right), 1 - Q \le m \le Q \right\},$$

for some $Y < Y_0$ (recall (2) on p. 191), and $Q > M(Y)$. The inverse transform gives us that for $1 \le |n| \le M(Y) < Q$ we have

$$\begin{aligned} c_j(n)\kappa_n(Y) &= \frac{1}{2Q} \sum_{m=1-Q}^{Q} \hat{f}_j(z_m) e(-n x_m) \\ &= \frac{1}{2Q} \sum_{m=1-Q}^{Q} f_j(z_m) e(-n x_m) + [[\epsilon]], \end{aligned}$$

where in the last step we used (16). This system is now almost a tautology, but we can use the implicit automorphy (19) to get a good "mix" of the Fourier coefficients and a far from trivial system.

Mimicking the discussion leading to (18) we let $T_{mj} \in \Gamma_0(N)$ be the pullback map of $\sigma_j z_m$ into $\mathcal{F}_N$, and let $w_{mj} = T_{mj}(\sigma_j z_m) \in \mathcal{F}_N$, $I(m,j) = I(w_{mj})$, $U_{mj} = U_{w_{mj}} \in \Gamma_0(N)$, $z_{mj}^* = \sigma_{I(m,j)}^{-1} U_{mj} w_{mj}$. Furthermore we let $\chi_{mj} = \chi(T_{mj}^{-1} U_{mj}^{-1})$. Using (the analog of) (19) we then obtain:

$$c_j(n)\kappa_n(Y) = \frac{1}{2Q} \sum_{m=1-Q}^{Q} \chi_{mj} f_{I(j,m)}(z_{mj}^*) e(-n x_m) + [[\epsilon]],$$

and we now want to substitute the truncated Fourier expansion for $f_{I(j,m)}$. Since we know that $\Im(z^*_{mj}) = \Im(\sigma^{-1}_{I(j,m)} U_{mj} T_{mj} \sigma_j(z_{mj})) \geq Y_0$ we can actually use the same truncation point, $M_0 = M(Y_0)$ for all series, and in interchanging the order of summation we get the following expression, valid for $1 \leq |n| \leq M(Y) < Q$ and $1 \leq j \leq \kappa$:

$$c_j(n)\kappa_n(Y) = \sum_{i=1}^{\kappa} \sum_{1 \leq |k| \leq M_0} c_i(k) V^{ji}_{nk} + 2[[\epsilon]], \qquad (20)$$

where

$$V^{ji}_{nk} = \frac{1}{2Q} \sum_{\substack{m=1-Q \\ I(j,m)=i}}^{Q} \chi_{mj} \kappa_k(y^*_{mj}) e(kx^*_{mj}) e(-nx_m). \qquad (21)$$

Remark 3.1. It seems clear that in order for the system (20) to be useful, we need to use the automorphy relation non-trivially sufficiently often. In practice we found that a necessary (and sufficient!) condition for the numerics to behave well is that $z^*_{mj} \neq z_m$ for all j, m. This condition is ensured if we keep $\Im(z^*_{mj}) > Y$ for all j, m. This is of course *automatically* fulfilled if we choose $Y < Y_0$ as above, but in many cases it is possible to verify numerically that the same inequality also holds for certain choices of Y(slightly) larger than Y_0. This means that we may be able to use a $Y > Y_0$ and a corresponding $M(Y) < M_0$. The drawback is that such a Y can not be used for Phase 2 later since there the pullbacks z^*_{mj} of the horocyclic points z_m will eventually fill out the whole fundamental domain (as $Q \to \infty$ and $Y \to 0$ simultaneously).

We now have a linear system that can be used to obtain the coefficients. Note that the V^{ji}_{nk} can be small due to either the K-Bessel decay or "bad mixing" in the sense that the numbers of m such that $I(j,m) = i$ might differ much (meaning that the pullback of the horocycle does not encircle each cusp equally much). Bad mixing is avoided basically through increasing the length of the horocycle by means of decreasing Y. The system (20) can be expressed as

$$0 = \sum_{i=1}^{\kappa} \sum_{|k| \leq M_Y} c_i(k) \tilde{V}^{ji}_{nk} + 2[[\epsilon]], \qquad (22)$$

where $\tilde{V}^{ji}_{nk} = V^{ji}_{nk} - \delta_{nk}\delta_{ji}c_j\kappa_n(Y)$. The term $-\kappa_n(Y)$ occurring in all diagonal entries gives us good reason to hope that this system should turn out to be well-conditioned, just as in [14, p. 298 (19) and second paragraph of p. 299]). We now have a linear system which is satisfied by (linear combinations of) Fourier coefficients of Maass waveforms $f \in \mathcal{M}(\Gamma_0(N), \chi, \lambda)$ (up to the error $2[[\epsilon]]$). If we let V denote the $(\kappa M_0 \times \kappa M_0)$-matrix $\tilde{V}^{ji}_{nk}$ and C denote the κM_0-vector of Fourier coefficients $c_j(k)$, we can write the system as

$$VC = 0. \qquad (23)$$

Observe that the solution space of this system for a true eigenvalue R is at least a 1-dimensional linear space, so in order to get a *unique* solution we need to use some sort of normalization.

Of course, if f and all the various f_j are all Fourier cosine series, there is an immediate "cosine" counterpart of (23). Similarly for sine series. This remark is important in all cases where Cor. 2.8 applies. The matrix coefficients in these analogs of (23) are all real!

Normalization

Depending on the dimension of the particular space $\mathcal{M}(\Gamma_0(N), \chi, \lambda)$ we are investigating, we have to employ different types of normalizations algorithmically. In order to test the accuracy of our coefficients (cf. p. 215 below) we will always try to get eigenfunctions of Hecke operators and/or involutions whenever possible. We will now discuss two examples (A) and (B) of possible normalizations.

(A) *Newforms, trivial character.* According to a standard conjecture (and experimental findings) the space of newforms with trivial character should have dimension 1. So, by Theorem 2.16 we can search for a normalized newform by setting $c_1(1) = 1$ *and dropping* the first equation from V. If we find a true R-value under this assumption, the coefficients will automatically satisfy all multiplicative relations (cf. Theorem 2.18).

(B) *N and real primitive χ satisfying the hypotheses of Cor. 2.8.* To discuss this case some preliminaries are necessary. We present them in outline form.

1) Let $\mathcal{MR}(\Gamma_0(N), \chi, \lambda) \subseteq \mathcal{M}(\Gamma_0(N), \chi, \lambda)$ be the space of all real-valued Maass waveforms. By simple use of the operator K, one sees that

$$\mathcal{M}(\Gamma_0(N), \chi, \lambda) = \mathcal{MR}(\Gamma_0(N), \chi, \lambda) + i\mathcal{MR}(\Gamma_0(N), \chi, \lambda)$$

as vector spaces. The operators W_Q and T_n are all endomorphisms on $\mathcal{M}(\Gamma_0(N), \chi, \lambda)$ and $\mathcal{MR}(\Gamma_0(N), \chi, \lambda)$ (see Prop. 2.21 and Thm. 2.10). One knows that $W_Q^2 = \pm Id$, and that $T_n W_Q = \overline{\chi_Q(n)} W_Q T_n$ for $(n, N) = 1$ (can be proved by induction on Prop. 2.23).

3) When χ is primitive, there are no oldforms. Cf. Thm. 2.16. Thm. 2.16(d) assures us of multiplicity one for any newforms.

4) Apply Thm. 2.12 to $\mathcal{M}(\Gamma_0(N), \chi, \lambda)$ and get an orthogonal basis $F_1, \ldots, F_m$ consisting of normalized newforms. (Do not confuse things here with (15)). By Thm. 2.18 and Cor. 2.8 each F_j is either a cosine or sine series at infinity and remains of the same type at the other cusps.

5) It is now natural to apply Remark 2.20 to the basis functions F_j. If we form the normalized newform $\epsilon \overline{F_j}$ there are two possibilities.

a) We arrive back at F_j. This is equivalent to saying $\overline{\lambda(n)} = \lambda(n)$, for all n. Since $\overline{\lambda(n)} = \chi(n)\lambda(n)$, for $(n, N) = 1$, by (13), this would imply $\lambda(n) = 0$ whenever $\chi(n) = -1$.

b) We arrive at a form independent of F_j. By mult. 1, we thus have that the sequences $\overline{\lambda(n)}$ and $\lambda(n)$ are essentially different as $n \to \infty$. Accordingly, here, via (13), we have that $\lambda(n)$ is nonzero and pure imaginary for infinitely many primes with $\chi(n) = -1$. While, $\lambda(1) = 1$, of course.

6) We apply 5) to each F_j, and thus deduce a natural splitting of the basis $\{F_1, \dots, F_m\}$ into singlets and pairs.

7) In pursuing the numerics on the machine, we hypothesize (at least initially) that either $m = 1$ (singlet) or $m = 2$ (one pair) for our given space $\mathcal{M}(\Gamma_0(N), \chi, \lambda)$. In the following comments, we assume $\epsilon = 1$. The treatment of $\epsilon = -1$ is similar.

8) Take $m = 1$ first. Look at $\frac{1}{2}F_1$. This is a unit normalized *real* cosine series. Its expansions at the other cusps are again *real* cosine series. Take the cosine analog of (23) and declare $\lambda(1) = 1$. Solve the system in the usual way with two parallel Y. This case should have a unique solution.

Cases where $\chi(n) = -1$ should yield coefficients that are exactly 0. This will be very compelling to see! (Cf. Table 5, right column).

9) Now assume $m = 2$. The (complex) basis is $\{F_1, F_2\}$, where $F_2 = \epsilon \overline{F_1} = \overline{F_1}$ is the "Remark 2.20 flip" of F_1. The idea here is to construct a real basis from F_1 and F_2. To this end we construct $F_+ = \frac{1}{4}(F_1 + F_2)$ and $F_- = \frac{1}{4i}(F_1 - F_2)$. These functions are now *real* cosine series at *all* cusps. F_+ is unit normalized, while the $n = 1$ term for F_- is zero. Observe that the Fourier coefficients of F_+ and F_- are $\Re(\lambda(n))$ and $\Im(\lambda(n))$ respectively (here $\lambda(n)$ are the coefficients of F_1).

The "cosine" analog of (23) holds. To single out F_+, we impose the condition that $\lambda(n) = 0$ for $\chi(n) = -1$. We reduce the size of the "cosine" system (23) accordingly, and set $\lambda(1) = 1$, and use our two parallel Y. This process determines λ (the eigenvalue) and F_+. If we instead impose the condition $\lambda(n) = 0$ for $\chi(n) = 1$, and set $\lambda(p_0) = 1$ where $\chi(p_0) = -1$ and use the appropriately reduced "cosine" system with parallel Y, we obtain aF_- with some nonzero $a \in \mathbb{R}$. To get F_1 and F_2, we form $2F_+ \pm 2i\left(\frac{1}{a}\right)aF_-$ and determine a either by Prop. 2.19(a) or by use of relation (12). If desired, λ can again be solved for at this stage of the game.

10) An alternative reduction in the size of the "cosine" system (23) can be achieved by looking at the action of W_Q on $\mathcal{M}(\Gamma_0(N), \chi, \lambda)$. Cf. Prop. 2.23 and the discussion afterwards. (This discussion can be generalized to $(n, N) = 1$ by induction; notice too that there are no oldforms when χ is primitive). Each $W_Q(F_j)$ must be a *nonzero* multiple of some F_k by virtue of its Hecke action. Depending on whether $m = 1$ or 2, the resulting permutation structure must either be (1), (12), or (1)(2). In each instance, the relation $W_Q^2 = \pm Id$ affords one some *a priori* control on any "non zero constants" that arise.

Let $\{W_{Q_1}, \dots, W_{Q_r}\}$ be any commutative family of W_Q's satisfying $W_Q^2 = Id$ and the hypotheses of Prop. 2.21. Cf. [2, Prop. 1.4]. Since each W_Q acts unitarily, it is natural to seek a basis of $\mathcal{MR}(\Gamma_0(N), \chi, \lambda)$ consisting of simultaneous eigenfunctions of the W_{Q_j}. By Prop. 2.22, the Fourier expansion of any such basis function at the cusp $\frac{Q_j}{N}$ will be "redundant"; there is thus a corresponding reduction in the size of the "cosine" analog of (23).

Remark 3.2. If Cor. 2.8 applies, the reduction with respect to the W_Q's, as mentioned above, can be applied also to the case of the trivial character in a similar manner.

Remark 3.3. Note that in the case of $N = 4$ and trivial character, using Thm. 2.7 and the fact that the cusp normalizing map of the cusp $p_3 = -\frac{1}{2}$, $\sigma_3 : z \mapsto \frac{z}{2z+1}$ is in the normalizer of $\Gamma_0(4)$ we can actually use pure sine/cosine series and real coefficients at all cusps in this case too. And we can also use reduction with respect to $\sigma_2 = \omega_N$ and σ_3. This is *not* true in general for non-square-free levels, e.g. $\Gamma_0(9)$. Then we *must* use complex arithmetic.

Example 3.4. As an illustration of (B) consider the case of prime level $N \equiv 1 \bmod 4$ and $\chi = \left(\frac{N}{\cdot}\right)$. With the notation as in (B), for the sake of simplicity, suppose that $m = 2$ and $\epsilon_1 = \epsilon_2 = 1$. Let F_1 and $F_2 = \overline{F_1}$ be the Hecke eigenbasis of normalized newforms in $\mathcal{M}(\Gamma_0(N), \chi, \lambda)$. There is only one W_Q, i.e. ω_N, which is now also a cusp normalizing map, and $\omega_N^2 = Id$. We will see how to diagonalize with respect to ω_N to get a reduction of the linear system. Observe that $T_n (\omega_N F_1) = \chi(n) \omega_N T_n F_1 = \chi(n) \lambda(n) \omega_N F_1$, and hence by multiplicity one we must either have $\omega_N F_1 = \pm F_1$ (since $\omega_N^2 = Id$) or we have $\omega_N F_1 = \mu F_2$ and $\omega_N F_2 = \mu^{-1} F_1$, for some constant $\mu \in S^1$ (again since $\omega_N^2 = Id$). In case that $\omega_N F_1 = \pm F_1$ we already have an eigenfunction of ω_N in F_1 and we don't have to do anything more. If $\mu = \pm 1$ then the functions $F_\pm$ (defined under (B)) are also invariant under ω_N, i.e. $\omega_N F_\pm = \pm F_\pm$ if $\mu = 1$ and $\omega_N F_\pm = \mp F_\pm$ if $\mu = -1$. If $\mu \neq \pm 1$, then we can form

$$f_\pm = \frac{1}{1 \pm \mu} (F_1 \pm \mu F_2).$$

It is easily verified that $f_\pm$ are both unit normalized (have first Fourier coefficient equal to 1), real-valued on $\mathcal{H}$, and satisfy $\omega_N f_\pm = \pm f_\pm$. Denote the Fourier coefficients of f_+ and f_- by $c^+(n)$ and $c^-(n)$, respectively. For $(n, N) = 1$, one immediately checks that

$$c^+(n) = c^-(n) = \lambda(n) \quad \text{for} \quad \chi(n) = +1$$

and

$$c^+(n)c^-(n) = \lambda(n)^2 \leq 0 \quad \text{for} \quad \chi(n) = -1.$$

To find f_+, we use the "cosine" analog of (23). We reduce the size of the system by using the fact that ω_N is a cusp normalizing map, i.e. we set

$$c_2(k) = c_1(k), \ k = 1, \dots, M_0$$

together with $c_1(1) = 1$. And then we drop corresponding equations. Similarly for f_-. In this case, when we run with the parallel Y, we have a lot of extra tests; Hecke relations coming from the $\lambda(n)$, $c^+(n) = c^-(n)$ if $\chi(n) = 1$ and $c^+(n)c^-(n) \leq 0$ if $\chi(n) = -1$. All these can be used as independent tests for the accuracy of our program.

By running through the search of λ for both f_+ and f_- we get further insurance.

Locating the Eigenvalues

We will see how, with the use of the normalized linear system introduced above, we can locate an eigenvalue R. Cf. [15, p. 5]

Consider the linear system (23). The entries of the coefficient matrix $V = V(R, Y)$ are clearly real-analytic functions of R, but as functions of Y they might be only piecewise continuous. For a value of R corresponding to a Maass waveform f we expect the solution vector $C = C(R, Y)$ to give us the actual Fourier coefficients if the appropriate normalization is used (e.g. $c(1) = 1$), and Y is kept less than Y_0 for the sake of well-conditioning.

This fact is used to determine whether a given R is close to a true eigenvalue or not by solving for $C(R, Y)$ and looking for known properties of Fourier coefficients of a Maass waveform. The three most important properties (here) are:

a) *Invariance under change of Y*. As described above, for a true eigenvalue the solution vector will be invariant when we change Y (keeping $Y < Y_0$).
b) *Hecke relations.* We can look for Hecke eigenfunctions by seeking solutions which satisfy multiplicative relations (cf. (12)), for example the relation $c_1(2)c_1(3) = c_1(6)$.
c) *Involutions of $(\Gamma_0(N), \chi)$.* We use cusp normalizing maps which are involutions with eigenvalues, $\mu_j = \pm 1$, and seek solution vectors which satisfy $c_j(k) = \mu_j c_1(k)$ (cf. (14), Example 3.4, and B item 10).

For a given interval $I = [R_1, R_2]$, we want to find all Maass waveforms with eigenvalues R in this interval. The idea here is to do this by solving the linear system (23) for two different values of Y in parallel and then try to find $R \in I$ such that the solution vectors $C(R, Y_i)$ match (usually we only require the first few coefficients to match, and use the rest only as an insurance).

In practice we first divide I into a number of equally sized small chunks:

$$R_1 = x_0 < \cdots < x_j < \cdots < x_{N_1} = R_2,$$

where the number N_1 is chosen in a way that it is probable that any interval $I_j = [x_j, x_{j+1}]$ contains at most one eigenvalue, using e.g. Weyl's law (3) and some assumption on the behavior of the nearest-neighbour spacings (e.g. that they are exponentially distributed).

We then look at each small I_j to see if there is a change of sign in any of the differences $c_k - c'_k$, $k = 2, 3, 4$, where $c_k = c_k(R, Y_1)$ and $c'_k = c'_k(R, Y_2)$ are entries in $C(R, Y_1)$ and $C(R, Y_2)$ respectively. If there are sign changes for all considered differences in an interval I_{j_0} we go to the next stage of the investigation and "zoom in" into this interval. At this point we usually form a functional

$$h(R) = \omega_2(c_2 - c_2') + \omega_3(c_3 - c_3') + \omega_4(c_4 - c_4'),$$

and try to minimize it over this interval. We use $\omega_j \in \{\pm 1\}$ to "align" the differences so that $h(R)$ changes sign where all three differences change sign. The minimization can be done in a number of ways. One very efficient approach is to use the method of false position to get successively better approximations to the location of the minimum (which if it exists is near a point where h changes sign).

When a value of R that approximates a zero of h is found it is listed as a candidate for a true eigenvalue. These candidates are stored and subjected to further examination.

Remark 3.5. Zeros of the K-Bessel functions in the left hand side of (20) (or "random noise") can trigger false indications of zeros in the intervals I_{j_0}. Keeping track of how fast $c_j - c_j'$ changes over each interval, most of these "false" intervals are immediately discarded.

If we only use two Y-values and a few coefficients, it can not be excluded that the located minimum of h is not close to a real eigenvalue, but such mistakes will be spotted either when we try to refine the eigenvalue or when we look closer at properties a)-c) for a larger set of coefficients.

Eventually we declare that the R which we have found is close to an eigenvalue of the Laplacian, and that our $\{c_j(n)\}$ are close to the Fourier coefficients of the corresponding Maass waveform. It is worth stressing that, strictly speaking, these assertions are never proved *rigorously* through our computations — although the excellent agreement seen when testing properties a) - c) on a large number[1] of coefficients on top of those used in the functional $h(R)$ clearly gives a very strong heuristic justification. The question of giving rigorous proofs is dealt with in [6].

Remark 3.6. To speed up the process of looking for sign changes over the large number of small intervals I_j we use Lagrange interpolation to evaluate the matrix V at this point (usually interpolation of degree 14 is good enough). Cf. [15, p. 6].

It should be remarked that the algorithm reproduces both the known old-forms from $\Gamma_0(1)$ (see [16] or [13]), the newforms on $\Gamma_0(5)$ as in [4], as well as the explicitly known CM-forms (cf. [17]).

There is obviously no guarantee that we find all eigenvalues in a specified interval in this manner, but comparing the results with the detailed version of Weyl's law (3) might give us an indication of missing eigenvalues (cf. [4, §A.1] and [3, §8]).

---

[1] e.g. 50-10000 produced by the Phase 2 algorithm, cf. Sect. 3.3

The Pullback Algorithm

Since the implicit automorphy (19) plays an important part in the algorithm, it is crucial to have an efficient means to compute the pullback $z^* \in \mathcal{F}_N$ of a point $z \notin \mathcal{F}_N$ (observe that the notation z^* differs from the one used earlier). We recall that in the case of the modular group $\Gamma_0(1)$, it is easy to make a pullback to the standard fundamental domain, $\mathcal{F}_1 = \{z = x + iy \in \mathcal{H} \mid |x| \leq \frac{1}{2}, |z| \geq 1\}$ using a sequence of "flip-flops" through the generators $E : z \mapsto -\frac{1}{z}$ and $S : z \mapsto z + 1$ (cf. [31, p. 51] or [4, pp. 44-46]).

Instead of extending this algorithm to the case of $\Gamma_0(N)$ by using side pairing generators of a suitable fundamental domain, we will use the facts that $\Gamma_0(N)$ is a subgroup of finite index in $PSL(2, \mathbb{Z})$, and that it is easy to find a set of coset representatives.

Let $\{V_j\}_{j=1}^{v_N}$ and $\mathcal{F}_N = \cup V_j(\mathcal{F}_1)$ be as on p. 189. Given $z \notin \mathcal{F}_N$ we make a pullback into $\mathcal{F}_1$; $\tilde{z} = T(z) \in \mathcal{F}_1$ with $T \in \Gamma_0(1)$. Then we find the index j such that $T^{-1} \in \Gamma_0(N)V_j$, and note that $V_j T \in \Gamma_0(N)$. Hence the $\Gamma_0(N)$ pullback of z is given by

$$z^* = V_j T(z) \in \mathcal{F}_N.$$

This gives a pullback algorithm for any N, but we need the coset representatives $\{V_j\}_{j=1}^{v_N}$ to apply it. The observation that $\{V_j\}_{j=1}^{v_N}$ is by definition a maximal set of $v_N = N \prod_{p|N} (1 + p^{-1})$ maps in $PSL(2, \mathbb{Z})$, all independent over $\Gamma_0(N)$ allows us to use the following simple recursive algorithm:

> Traverse $\Gamma_0(1)$ as a tree in S, E and S^{-1} and collect maps independent over $\Gamma_0(N)$ until exactly v_N independent maps have been found.

This calculation is done once and for all for each group and the resulting representatives are stored. A possible alternative here is to use explicit formulas for the coset representatives, but the advantage of the above algorithm is that it can be extended to other congruence subgroups, such as $\Gamma_1(N)$, $\Gamma^0(N)$ or $\Gamma(N)$ as well. Actually it can be extended to any finite-index subgroup of $PSL(2, \mathbb{Z})$ that possesses some precise characteristic which we can use to see if a given map belongs to the group or not.

3.3 Phase 2

After Phase 1 is completed we have an (approximate) eigenvalue R and a corresponding set of (approximate) Fourier coefficients

$$\{c_j(k) \mid 1 \leq j \leq \kappa, 1 \leq |k| \leq M_0\}.$$

Suppose now that we want to compute $c_j(n)$ for $N_A \leq n \leq N_B$. Going back to the identity (20),

$$c_j(n)\kappa_n(Y) = \sum_{i=1}^{\kappa} \sum_{|k|\leq M_0} c_i(k)V_{nk}^{ji} + 2[[\epsilon]]$$

(valid for $Y < Y_0$, $1 \leq |n| \leq M(Y) < Q$ and $1 \leq j \leq \kappa$), we see that in order
to compute $c_j(n)$ for $n \in [N_A, N_B]$ (where $N_B > M_0$), we take $Y < Y_0$ such
that $M(Y) > N_B$. To avoid cancellation we also have to make sure that the
particular choice of Y does not make $\kappa_n(Y)$ too small. We can now formulate
the core of Phase 2 as a theorem.

Theorem 3.7. *Suppose that $f \in \mathcal{M}(\Gamma_0(N), \chi, \frac{1}{4}+R^2)$ has Fourier coefficients
$\{c_j(n)\}$. Let $Y < Y_0$, and let n, Q be integers such that $Q > M(Y)$ and
$1 \leq |n| \leq M(Y)$. We then have*

$$c_j(n) = \frac{\sum_{i=1}^{\kappa} \sum_{|k|\leq M_0} c_i(k)V_{nk}^{ji}}{\kappa_n(Y)} + Err_n(Y), \qquad (24)$$

where V_{nk}^{ji} is given by (21) on p. 211, and the error $Err_n(Y)$ is given by

$$Err_n(Y) = \frac{[[\epsilon]]}{\kappa_n(Y)}.$$

In order to use Theorem 3.7 to successfully compute the Fourier coefficients
we have to adjust the value of Y (and accordingly $M(Y)$ and Q also) to keep
the error $Err_n(Y)$ from growing.

If we just want to confirm the existence of an eigenvalue in the neighbour-
hood of R, or if we want to make a picture of an eigenfunction, the first set
of coefficients from Phase 1 is usually sufficient. But, even for those purposes,
the method of Phase 2 is a very cheap way to improve the last part of the
already obtained coefficients, as well as obtaining a lot more.

3.4 Remarks on the Performance of the Algorithm

The main difference (with regard to performance issues) between groups with
one cusp, e.g. the modular group or Hecke triangle groups, and the groups
$\Gamma_0(N)$ with $N > 1$ should now be clear. The presence of extra cusps introduces
more sets of coefficients, thus increasing the size of the linear system used in
Phase 1. We will give some examples of how the relevant factors scale with
respect to N (and R).

Timing for a Single R

Suppose we are given a level N and a potential eigenvalue R, and that we want
to compute the corresponding (minimal) set of coefficients, $C = \{c_j(n) \mid 1 \leq |n| \leq M_0, 1 \leq j \leq \kappa\}$, i.e. we have to solve the system (22) under some nor-
malization assumption. Let $T(R, N)$ be the time required to perform this task.

It turns out in practice that the time required to solve the system is negligible compared to the time required to compute the coefficient matrix, and the main contribution to $T(R, N)$ in fact comes from the K-Bessel computations. Observe that in the evaluation of V_{nk}^{ji} the only K-Bessel terms we need are $K_{iR}(2\pi|k|y_{mj}^*)$ for $1 - Q \leq m \leq Q$, these are independent of n and i, hence the total number of K-Bessel computations is $2Q(M_0\kappa) \sim 2\kappa M_0^2$ (as we can take $Q \sim M_0$). Using $Y_0 \geq \frac{\sqrt{3}}{2N}$ we see that, roughly $M_0 = [\text{const}]N(R + 1)$, and

$$T(R, N) \ll \kappa N^2 (R + 1)^2. \tag{25}$$

Remember that κ is dependent on the number of prime factors of N, and for a prime number $\kappa = 2$. From this formula we see that the time to compute even a single eigenfunction with a small eigenvalue increases drastically as the level N grows. No matter how much we improve the speed of computing the individual K-Bessel functions, attaining levels like $N \approx 1000$ seems to be out of reach at the present time. In Table 1 we give some examples of $T(R, N)$ for different (prime) N, with $R = 9.5336\ldots$ (the first eigenvalue on $PSL(2, \mathbb{Z})$).

Table 1. Time to compute the coefficient vector for $R = 9.5336\ldots$ on a 3.2GHz CPU

N	M_0	$T(R, N)$
2	13	0.03s
5	32	0.14s
13	90	1.17s
17	118	2.20s
41	307	23.9s
101	757	287.4s

Recall steps a)-c) on p. 215. To find eigenvalues we need to compute the matrix $V(R, Y)$ at least $14 \times 2 = 28$ times for step 1 (using Lagrange interpolation of degree 14 for the function $R \mapsto V(R, Y)$, cf. Remark 3.6) and then at least $6 \times 2 = 12$ times for the step 3 (assuming that it takes approximately 6 iterations with the method of false position to obtain the required precision).

The K-Bessel Function

The K-Bessel function is computed using a number of different algorithms, each with different range of applicability (cf. [13], [39] and [4]).

One possible way to speed things up is to use an adaptive Lagrange interpolation method for the function $x \mapsto K_{iR}(x)$. However, due to the oscillatory nature of $K_{iR}(x)$ for small x we need a large number of interpolation points when N is large (i.e. when Y_0 is small) so this type of interpolation is in general only efficient in Phase 2 or when R is large.

Timing for Phase 2

Since we only use the first few V_{nk}^{ji} with $|k| \leq M_0$ in Theorem 3.7 we need about $2QM_0\kappa$ K-Bessel calls to compute $c_j(n)$. Remember here that for (24) to hold we need $Q > M(Y) > |n|$, and we also need $Y < $ const $(1 + R)/n$, hence, if we take $Y = [\text{const}] \cdot (1 + R)/n$, then we have $M(Y) = [\text{const}] \cdot (1 + R)/Y = [\text{const}] \cdot n$. The time to compute one coefficient $c_j(n)$ is thus roughly proportional to

$$n\kappa M_0 = \kappa N(R + 1)n,$$

and hence the time to compute the first A coefficients is approximately proportional to

$$\kappa N(R + 1)A^2, \text{ as } A \to \infty.$$

Observe that the K-Bessel functions in V_{nk}^{ji} do *not* depend on n or i. Hence to avoid unnecessary computations we can use a vector to store all $4Q\kappa M_0$ K-Bessel values used in (24). We will then only need to perform these computations when we change Y, but unfortunately we will have to deal with memory issues instead, and these can degrade the performance as much as the computations themselves. To compute a large number of coefficients we have to keep a balance between the number of computations and the size of the allocated memory. In practice this is a rather difficult task.

4 Results

The focus of the current project has been to provide a robust and efficient algorithm for computing Maass waveforms on Hecke congruence subgroups $\Gamma_0(N)$ for any integer N. The following are our main results so far.

- Extensive lists of eigenvalues for small and prime $N = 5, 7, 13$, with either trivial character or the real Dirichlet character mod N given by the symbol $\left(\frac{N}{\cdot}\right)$.
- Shorter lists of eigenvalues for *all* $N \leq 30$. More precisely, we believe our lists contain *all* eigenvalues with $R \leq 10$ for $N = 1, 2, 3, 4, 5, 6, 7, 8, 9, 10$, 11, 12, 13, 14, 15, 17, 19, 21, 22, 23, 25, 26, 29, $R \leq 5$ for $N = 16, 20, 24$ $R \leq 3$, for $N = 30$ and $R \leq 2$ for $N = 18, 27$ and 28.
- The first eigenvalue, $\lambda_1 = \frac{1}{4} + R_1^2$, for all prime $N \leq 131$.
- Large sets of Fourier coefficients for a few Maass waveforms, e.g. for $N = 5$ the following coefficients have been computed: all $c(n)$ with $|n| \leq 70209$ for $R = 3.2642\ldots$; all $c(n)$ with $|n| \leq 89263$ for $R = 4.8937\ldots$; all $c(n)$ with $|n| \leq 56420$ for $R = 4.1032\ldots$ (the first two eigenvalues correspond to the character $\left(\frac{N}{\cdot}\right)$ and the last one corresponds to the trivial character).

Work on extending these data, as well as carrying out more extensive statistical tests, is currently being pursued. In this section we will give some examples and comments on each item in the above list.

Further data than that which is presented in this section is available from the author upon request.

4.1 Eigenvalues

Tables 3, 2 and 6 provide some examples of eigenvalues for different groups. In Tables 2 and 6 the eigenvalues of the operators J and ω_N are indicated in the respective columns. The eigenvalues are denoted by $-$ for -1, $+$ for 1, and $*$ is used to indicate that both eigenvalues are present. In all tables we use $H_1 = |c_1(2)c_1(3) - c_1(6)|$ as one indication of the accuracy of the program (we based our search on property a) on p. 215; hence no Hecke relation like H_1 was built in to our algorithm).

It should be remarked that in the case of Table 3, $\Gamma_0(4)$, all computed newforms are eigenfunctions with eigenvalue -1 of both non-trivial cusp normalizers, i.e. $\sigma_2 = \omega_4 : z \mapsto -1/4z$, and $\sigma_3 : z \mapsto z/(2z+1)$, which are both $\Gamma_0(4)$-involutions.

4.2 Lowest Eigenvalues

We have used the Phase 1 algorithm to compute small eigenvalues for $\mathcal{M}(\Gamma_0(N), \chi)$ as N ranges through the primes up to 131 and χ is either trivial or the real Dirichlet character $\left(\frac{N}{\cdot}\right)$. The first located eigenvalues are listed in Table 4.

Recall that all eigenvalues fall into distinct classes with respect to the involutions J and ω_N. We denote the classes by $\{++, +-, -+, --\}$.

In the case of a non-trivial character, CM-forms are present, and in all cases considered such a form was found to occur as the lowest eigenvalue. All the CM-forms occurring here are of the type considered in [7] (i.e. the narrow class number of $\mathbb{Q}[\sqrt{N}]$ is 1), and by the explicit formula (cf. [7, p. 112]) the CM-forms with lowest eigenvalues are of the type $--$. In Table 4, H_1 and H_2 are two parameters indicating the error. H_1 is always defined by $H_1 = |c_1(2)c_1(3) - c_1(6)|$, and H_2 is defined as either the true error in R for CM-forms, or as $H_2 = \left||c_1(N)| - \frac{1}{\sqrt{N}}\right|$ (cf. Prop. 2.19) otherwise.

4.3 Fourier Coefficients

Table 5 gives some examples of Fourier coefficients and Hecke relations obtained by the Phase 2 algorithm. Figure 1 (a) is a histogram of the coefficients $c_1(p)$ of the CM-form with eigenvalue $3.26\ldots$ where the first 3407 primes for which $\chi(p) = \left(\frac{5}{p}\right) = 1$ are used. Figure 1 (b) is a histogram of the $\tilde{c}_1(p)$ for one of the Maass waveforms on $(\Gamma_0(N), \chi_5)$ with eigenvalue $R = 4.8937\ldots$. Here $\tilde{c}_1(p) = c_1(p)$ if $\chi(p) = \left(\frac{5}{p}\right) = 1$ and $-ic_1(p)$ if $\chi(p) = \left(\frac{5}{p}\right) = -1$ (to get real coefficients), and the first 7462 primes are used. The experimentally

222   Fredrik Strömberg

obtained density functions (curves) in Figs. 1 (a) and (b) should be compared with the known density function $\frac{1}{\pi}\left(4 - x^2\right)^{-\frac{1}{2}}$ in the CM-case (a) and the conjectured density function $\frac{1}{2\pi}\sqrt{4 - x^2}$ in the non-CM case (b) (cf. [16, 17]).

Fig. 1. Histogram of Fourier coefficients for Maass forms on $\Gamma_0(5)$, $\left(\frac{5}{\cdot}\right)$

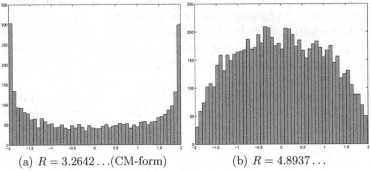

(a) $R = 3.2642\ldots$(CM-form)   (b) $R = 4.8937\ldots$

Table 2. Eigenvalues for $N = 29$, $\chi = \left(\frac{29}{\cdot}\right)$

R	J	ω_N	H_1	H_2[a]
0.9535979998	-	-	0.9E-10	0.5E-10 (CM)
1.804894697229	-	*	0.2E-13	0.2E-12
1.907196000095	+	+	0.3E-13	0.6E-12 (CM)
1.9491483792	+	*	0.9E-14	0.2E-10
2.1682948312	-	*	0.7E-13	0.3E-10
2.8084567797	-	*	0.8E-12	0.4e-10
2.8252375274	+	*	0.2E-13	0.8E-10
2.8607940001	-	-	0.5E-13	0.1E-11 (CM)

[a] H_2 is the true error in R for CM-forms and otherwise $|c_+(4) - c_-(4)|$. Cf. Remark 3.4.

Table 3. Eigenvalues for $\Gamma_0(4)$

Sine			Cosine		
R	H_1	$H_2{}^c$	R	H_1	$H_2{}^c$
3.70330780123	0.5E-11	0.7E-11	5.879354157759	0.3E-12	0.6E-12
[b] 5.41733480684	0.2E-11	0.1E-11	8.042477591693	0.7E-11	0.4E-12
6.62042287384	0.1E-11	0.5E-12	[b] 8.92287648699	0.5E-11	0.2E-11
[b] 7.22087197596	0.2E-12	0.4E-11	9.859896162239	0.7E-13	0.9E-13
[b] 8.27366588959	0.4E-11	0.1E-11	[b] 10.92039200294	0.4E-11	0.3E-11
8.52250301688	0.8E-12	0.2E-12	[b] 12.09299487508	0.1E-13	0.3E-10
[a] 9.53369526135	0.3E-11	0.4E-11	12.87761656411	0.1E-11	0.2E-12
9.93491995937	0.3E-12	0.7E-12	13.17207496748	0.5E-11	0.7E-11
[b] 10.71270690070	0.2E-13	0.3E-09	[a] 13.77975135189	0.2E-11	0.1E-09
10.76471068319	0.3E-12	0.2E-12	14.47780273166	0.1E-12	0.7E-12
[b] 11.31767970147	0.4E-12	0.5E-12	[b] 14.68501595060	0.2E-09	0.7E-11
11.97277669404	0.1E-11	0.4E-11	[b] 15.31419658418	0.9E-11	0.2E-11
[a] 12.17300832468	0.2E-11	0.2E-11	15.7434456453	0.2E-10	0.6E-10
[b] 12.82198816197	0.8E-12	0.4E-12	16.09359237739	0.7E-11	0.6E-11
[b] 13.31016428347	0.9E-13	0.2E-11	[b] 16.4041087751	0.5E-10	0.7E-10
[b] 14.09720373392	0.9E-11	0.2E-11	[b] 16.94026094636	0.1E-10	0.3E-11
14.1438850658	0.2E-10	0.1E-10	17.19512721203	0.8E-11	0.2E-11
[a] 14.3585095183	0.2E-10	0.4E-10	[a] 17.73856338106	0.8E-11	0.1E-10
14.97031306857	0.2E-11	0.2E-11	[b] 17.8780025394	0.2E-10	0.2E-10
[b] 15.27402248481	0.2E-11	0.3E-12	18.24637789203	0.1E-11	0.3E-11

[a] Oldform from $\Gamma_0(1)$. This eigenspace is 3-dimensional.

[b] Oldform from $\Gamma_0(2)$. This eigenspace is 2-dimensional.

[c] From Proposition 2.19(c) we know that $|c(4)| = 0$ for newforms, and it is easy to see that in an oldspace $f_{\pm}(z) = f(z) \pm \varepsilon f(dz)$, $d|4$, $\varepsilon \in \{\pm 1\}$ (cf. notation in Remark 3.4, and hence $c_+(n) = c_-(n)$ for $(n, 4) = 1$. Accordingly we set $H_2 = |c(4)|$ for newforms and $H_2 = |c_+(5) - c_-(5)|$ for oldforms.

Table 4: Lowest eigenvalues for $\Gamma_0(N)$

N	χ	J	ω_N	R	H_1	H_2
2		-	-	5.4173348068447	0.3E-13	0.5E-12
3		-	-	4.38805356322	0.1E-10	0.8E-11
4		-	+	3.70330780123	0.2E-11	0.7E-11
5		-	-	3.02837629306	0.4E-11	0.7E-12
5	$\left(\frac{5}{\cdot}\right)$	-	-	3.264251302636	0.7E-11	0.2E-12 ^{CM}
6		-	+	2.592379771762	0.8E-12	0.3E-10
7		-	-	1.92464430511	0.2E-11	0.8E-11
8		-	-	2.425056827087	0.3E-11	0.2E-11
9		-	+	3.536002092938	0.3E-11	0.9E-13
10		-	+	1.446399133415	0.1E-11	0.1E-10

^{CM} This is a CM-form. For definition of H_2 see the end of the table

N	χ	J	ω_N	R	H_1	H_2
11		-	-	2.03309099385	0.3E-12	0.1E-11
12		-	-	1.69106559844	0.3E-12	0.2E-10
13		-	-	0.9708154174	0.7E-10	0.2E-10
13	$\left(\frac{13}{\cdot}\right)$	-	-	1.31473442107	0.1E-10	0.7E-12 ^{CM}
14		-	-	1.794165360458	0.6E-12	0.3E-10
15		-	-	1.518429336032	0.9E-11	0.9E-12
17		-	-	1.4414285450	0.5E-10	0.3E-10
17	$\left(\frac{17}{\cdot}\right)$	-	-	0.7498863406	0.5E-10	0.4E-10 ^{CM}
19		-	-	1.0919915598	0.2E-13	0.3E-10
23		-	+	1.3933371415	0.6E-10	0.2E-10
29		+	+	1.017266551	0.5E-09	0.8E-09
29	$\left(\frac{29}{\cdot}\right)$	-	-	0.953598000	0.3E-11	0.5E-10 ^{CM}
31		+	+	0.789356178	0.8E-10	0.4E-08
37		-	-	0.64230596	0.4E-08	0.6E-08
37	$\left(\frac{37}{\cdot}\right)$	-	-	0.630391294	0.7E-09	0.7E-09 ^{CM}
41		+	+	0.66572483	0.9E-08	0.3E-08
41	$\left(\frac{41}{\cdot}\right)$	-	-	0.37767450	0.2E-08	0.1E-07 ^{CM}
43		-	-	0.65545239	0.5E-09	0.7E-08
47		-	+	0.5854522	0.5E-07	0.1E-08
53		-	-	0.8039894582	0.2E-12	0.4E-08
53	$\left(\frac{53}{\cdot}\right)$	-	-	0.799094454	0.5E-10	0.8E-09 ^{CM}
59		+	+	0.595829688	0.4E-10	0.3E-08
61		-	-	0.4180624	0.4E-07	0.9E-07
61	$\left(\frac{61}{\cdot}\right)$	-	-	0.42868522	0.2E-08	0.3E-08 ^{CM}
67		-	-	0.67759021	0.3E-07	0.1E-08
71		+	+	0.35745048	0.8E-08	0.5E-08
73		-	-	0.517887505	0.1E-08	0.2E-09
73	$\left(\frac{73}{\cdot}\right)$	-	-	0.20488585	0.5E-07	0.7E-08 ^{CM}
79		+	+	0.55177485	0.2E-06	0.9E-12
83		-	+	0.640287578	0.3E-09	0.3E-09
89		+	+	0.4894360	0.1E-07	0.1E-07
89	$\left(\frac{89}{\cdot}\right)$	-	-	0.2273961	0.3E-08	0.6E-07 ^{CM}
97		-	-	0.41693167	0.1E-07	0.8E-08
97	$\left(\frac{97}{\cdot}\right)$	-	-	0.16846	0.4E-05	0.1E-05 ^{CM}
101		+	+	0.45375925	0.6E-09	0.3E-08
103		-	-	0.56540670	0.9E-08	0.5E-10
107		-	+	0.5816789	0.2E-07	0.8E-07
109		-	-	0.423655	0.2E-05	0.9E-07
109	$\left(\frac{109}{\cdot}\right)$	-	-	0.2822870	0.8E-09	0.3E-07 ^{CM}
113		-	-	0.46200265	0.4E-09	0.3E-07
113	$\left(\frac{113}{\cdot}\right)$	-	-	0.21379233	0.2E-08	0.8E-08 ^{CM}
127		+	+	0.373385109	0.5E-07	0.4E-10
131		-	+	0.261072758	0.7E-09	0.4E-11

^{CM} This is a CM-form

For a CM-form H_2 is the true error and for a non-CM form we know by Proposition 2.19(a) that $|c(N)| = 1$, and we put $H_2 = ||c(N)| - 1|$.

Table 5. Fourier coefficients for $N = 5$ and $\chi = \left(\frac{5}{\cdot}\right)$

$R = 4.893781291438$		$R = 3.264251302636$ (CM) [a]	
$c(2)$	1.217161411799i	$c(2)$	0 0.1E-12
$c(3)$	0.295119713347i	$c(3)$	0 0.1E-12
$c(5)$ [b]	$\exp(1.157414657530i)$	$c(5)$	-1 0.1E-12
$c(7)$	-1.138873755146i	$c(7)$	0 0.1E-12
$c(11)$	-0.041396292578	$c(11)$	-1.11318397521973 0.2E-07
$c(13)$	-0.558344591841i	$c(13)$	0 0.4E-11
$c(17)$	-0.212576664102i	$c(17)$	0 0.1E-10
$c(19)$	0.608670097807	$c(19)$	-0.17157173156738 0.6E-08
$c(23)$	-1.205831908853i	$c(23)$	0 0.5E-09
$c(29)$	0.162328579004	$c(29)$	0.44610100984573 0.1E-07
$c(31)$	-0.556019364974	$c(31)$	-0.16668809652328 0.6E-08
$c(37)$	0.411889174623i	$c(37)$	0 0.4E-11
$c(41)$	-0.835153489179	$c(41)$	0.84569376707077 0.1E-07
$c(43)$	-1.240299728133i	$c(43)$	0 0.1E-10
$c(47)$	-0.605458209042i	$c(47)$	0 0.6E-10
$c(53)$	1.458187614411i	$c(53)$	0 0.1E-09
$c(59)$	0.922374897976	$c(59)$	-0.59840279817581 0.5E-08
$c(61)$	-1.247703551349	$c(61)$	-1.82808148860931 0.4E-07
$c(67)$	0.33772433772i	$c(67)$	0 0.1E-08
$c(71)$	-0.74219625846	$c(71)$	1.29922688007355 0.5E-07
$c(73)$	-0.44062515256i	$c(73)$	0 0.1E-12
$c(79)$	0.71467631335	$c(79)$	-1.48637688159943 0.2E-07
$c(83)$	0.99829038428i	$c(83)$	0 0.1E-12
$c(89)$	-1.42635989529	$c(89)$	1.43317067623138 0.5E-07
$c(97)$	0.07295604310i	$c(97)$	0 0.1E-12
$c(1553)$	-1.3885884029i		
$c(72977)$	-0.9912088597i		

Some Corresponding Hecke relations (for $R = 4.893781291438$)

$\lvert c(2)c(3) - c(6) \rvert$	0.2E-12	$\lvert c(895) - c(5)c(179) \rvert$	0.1E-11
$\lvert\, \lvert c(5) \rvert - 1 \rvert$	0.1E-12	$\lvert c(15625) - c(5)^6 \rvert$	0.1E-10
$\lvert c(93) - c(3)c(31) \rvert$	0.3E-11		
$\lvert c(195) - c(3)c(5)c(13) \rvert$	0.2E-11		
$\lvert c(9763) - c(13)c(751) \rvert$	0.3E-10		
$\lvert c(72991) - c(47)c(1553) \rvert$	0.2E-09		

[a] Since coefficients of CM-forms can be computed explicitly (cf. [17, p. 5]), the first column contains the coefficients computed according to the explicit formula, and the second column is the difference between this value and the value obtained with the methods presented in this paper.

[b] By Proposition 2.19(a) we know that $\lvert c(5) \rvert = 1$.

Table 6. Eigenvalues for $N = 23$, $\chi = 1$

R	J	ω_N	H_1	$H_2 = \|\|c(23)\| - \frac{1}{\sqrt{23}}\|$[a]
1.3933371415	-	+	0.8E-11	0.4E-11
1.5061266371	-	-	0.3E-11	0.1E-10
1.5795892401	+	+	0.7E-12	0.3E-11
1.9352504926	-	-	0.3E-11	0.3E-10
2.4307774166	-	-	0.1E-10	0.9E-10
2.4643338705	+	+	0.7E-11	0.9E-09
2.4936237868	-	+	0.2E-11	0.7E-11
2.5321343295	+	+	0.1E-11	0.1E-09
2.6109599620	+	-	0.4E-12	0.3E-11
2.7362777132	-	+	0.1E-10	0.1E-09
2.8079149467	-	-	0.6E-13	0.1E-10
3.0166209534	+	-	0.3E-11	0.5E-11
3.2357419725	+	+	0.2E-12	0.2E-10
3.3723339187	-	-	0.2E-10	0.8E-11
3.5202562438	+	-	0.1E-11	0.8E-13
3.5245436383	-	+	0.5E-10	0.1E-09
3.5882850865	-	+	0.9E-12	0.3E-11
3.7062705391	-	-	0.6E-10	0.2E-12
3.8167552363	+	-	0.3E-12	0.4E-12
3.8189248762	-	-	0.5E-10	0.2E-09

[a] From Proposition 2.19(b) we know that $|c(23)| = \frac{1}{\sqrt{23}}$.

5 Acknowledgment

These lectures are based on Chapter 1 of my Ph.D. thesis [37] and I am grateful to my advisor Dennis Hejhal for valuable comments and suggestions.

References

1. A. O. L. Atkin and J. Lehner, *Hecke operators on $\Gamma_0(M)$*, Math. Ann. **185** (1970), 134-160.
2. A. O. L. Atkin and W.-C. W. Li, *Twists of newforms and pseudo-eigenvalues of W-operators*, Invent. Math. **48** (1978), no. 3, 221–243.
3. R. Aurich, F. Steiner, and H. Then, *Numerical computation of Maass waveforms and an application to cosmology*, this volume.
4. H. Avelin, *On the Deformation of Cusp Forms*, 2003, Filosofie Licentiat avhandling.
5. E. Balslev and A. Venkov, *Spectral theory of Laplacians for Hecke goups with primitive character*, Acta Math. **186** (2001), 155–217.

6. A. Booker, A. Strömbergsson, and A. Venkatesh, *Effective Computation of Cusp Forms*, Int. Math. Res. Not. (2006), Art. ID 71281, 34 pp.
7. D. Bump, *Automorphic Forms and Representations*, Cambridge Studies in Advanced Mathematics, no. 55, Cambridge University Press, 1996.
8. H. Cohn, *Advanced Number Theory*, Dover, 1980.
9. D. Davenport, *Multiplicative Number Theory*, 2 ed., Graduate Texts in Mathematics, no. 74, Springer-Verlag, 1980.
10. L. R. Ford, *Automorphic Functions*, 2nd ed., Chelsea Publishing, 1972.
11. E. Hecke, *Mathematische Werke*, Vandhoeck & Ruprecht, Göttingen, 1970.
12. D. A. Hejhal, *The Selberg Trace Formula for* PSL(2,$\mathbb{R}$), *vol.2*, Lecture Notes in Mathematics, vol. 1001, Springer-Verlag, Berlin, 1983.
13. D. A. Hejhal, *Eigenvalues of the Laplacian for Hecke triangle groups*, Mem. Amer. Math. Soc. **97** (1992), no. 469, vi+165.
14. D. A. Hejhal, *On eigenfunctions of the Laplacian for Hecke triangle groups*, Emerging applications of number theory (ed. by D. Hejhal and J. Friedman et al), IMA Vol. Math. Appl., vol. 109, Springer, New York, 1999, pp. 291–315.
15. D. A. Hejhal, *On the Calculation of Maass Cusp Forms*, these proceedings.
16. D. A. Hejhal and S. Arno, *On Fourier coefficients of Maass waveforms for* PSL(2,$\mathbb{Z}$), Math. Comp. **61** (1993), no. 203, 245–267, S11–S16.
17. D. A. Hejhal and A. Strömbergsson, *On quantum chaos and Maass waveforms of CM-type*, Foundations of Physics **31** (2001), 519–533.
18. M. N. Huxley, *Scattering matrices for congruence subgroups*, in *Modular forms* (ed. by R. Rankin), Ellis Horwood, Chichester, 1984, pp. 141–156.
19. H. Iwaniec, *Introduction to the Spectral Theory of Automorphic Forms*, Revista Matemática Iberoamericana, 1995.
20. H. Iwaniec, *Topics in Classical Automorphic Forms*, AMS, 1997.
21. S. Katok, *Fuchsian Groups*, The University of Chicago Press, 1992.
22. S.L. Krushkal, B.N. Apanasov, and N.A. Gusevskii, *Kleinian Groups and Uniformization in Examples and Problems*, Translations of Mathematical Monographs, AMS, 1986.
23. H. Maass, *Über eine neue Art von nichtanalytischen automorphen Funktionen und die Bestimmung Dirichletscher Reihen durch Funktionalgleichungen*, Math. Annalen **121** (1949), 141–183.
24. H. Maass, *Lectures on Modular Munctions of One Complex Variable*, second ed., Tata Institute of Fundamental Research Lectures on Mathematics and Physics, vol. 29, Tata Institute of Fundamental Research, Bombay, 1983.
25. B. Maskit, *Kleinian Groups*, Springer-Verlag, 1988.
26. T. Miyake, *On automorphic forms on* GL$_2$ *and Hecke operators*, Ann. of Math. (2) **94** (1971), 174–189.
27. T. Miyake, *Modular Forms*, Springer-Verlag, 1997.
28. A. Orihara, *On the Eisenstein series for the principal congruence subgroups*, Nagoya Math. J. **34** (1969), 129–142.
29. R. S. Phillips and P. Sarnak, *On cusp forms for co-finite subgroups of* PSL(2,$\mathbb{R}$), Invent. Math. **80** (1985), 339–364.
30. H. Rademacher, *Über die Erzeugenden von Kongruenzuntergruppe der Modulgruppe*, Abhandlungen Hamburg **7** (1929), 134–148.
31. R. A. Rankin, *Modular Forms and Functions*, Cambridge University Press, 1976.
32. J. G. Ratcliffe, *Foundations of Hyperbolic Manifolds*, Springer-Verlag, 1994.
33. M. S. Risager, *Asymptotic densities of Maass newforms*, J. Number Theory **109** (2004), no. 1, 96–119.

34. B. Selander, *Arithmetic of three-point covers*, PhD thesis, Uppsala University, 2001, http://urn.kb.se/resolve?urn=urn:nbn:se:uu:diva-7497.

35. G. Shimura, *Introduction to the Arithmetic Theory of Automorphic Forms*, Princeton Univ Press, 1971.

36. A. Strömbergsson, *Studies in the analytical and spectral theory of automorphic forms*, Ph.D. thesis, Uppsala University, Dept. of Math., 2001.

37. F. Strömberg, *Computational aspects of Maass waveforms*, Ph.D. thesis, Uppsala University, Dept. of Math., 2005.

38. A. Terras, *Harmonic Analysis on Symmetric Spaces and Applications*, vol. I, Springer-Verlag, New York, 1985.

39. H. Then, *Maass cusp forms for large eigenvalues*, Math. Comp. **74** (2005), no. 249, 363–381.

VII. Numerical Computation of Maass Waveforms and an Application to Cosmology

Ralf Aurich[1], Frank Steiner[1,2], and Holger Then[3]

[1] Institut für Theoretische Physik, Universität Ulm, Albert-Einstein-Allee 11, 89069 Ulm, Germany
[2] Université Lyon 1, Centre de Recherche Astrophysique de Lyon, CNRS UMR 5574, 9 avenue Charles André, F-69230 Saint-Genis-Laval, France
[3] School of Mathematics, University of Bristol, Bristol BS8 1TW, UK
ralf.aurich@uni-ulm.de, frank.steiner@uni-ulm.de,
holger.then@bristol.ac.uk

Summary. We compute numerically eigenvalues and eigenfunctions of the Laplacian in a three-dimensional hyperbolic space. Applying the results to cosmology, we demonstrate that the methods learned in quantum chaos can be used in other fields of research.

1 Introduction

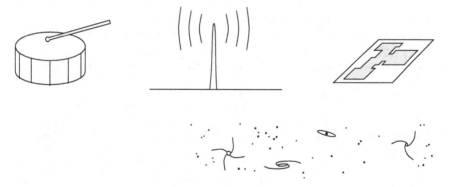

Fig. 1. Applications of the Laplacian. A drum (top left), electromagnetic waves (top middle), semiconductors resp. quantum mechanics (top right), and the universe (bottom).

The Laplacian Δ, a second order differential operator, is of fundamental interest in several fields of mathematics and physics. One of the oldest examples is a drum $\mathcal{D}$, see figure 1 top left. If one hits it, its membrane oscillates and gives a sound. The vibrations of the drum are solutions of the wave equation $(c = 1)$,

$$(\Delta - \frac{\partial^2}{\partial t^2})u(x,t) = 0 \quad \forall x \in \mathcal{D}, \tag{1}$$

where u is the displacement of the membrane. Fixing the membrane to its frame gives rise to Dirichlet boundary conditions,

$$u(x,t) = 0 \quad \forall x \in \partial\mathcal{D}. \tag{2}$$

A Fourier transformation,

$$u(x,t) = \int_{-\infty}^{\infty} v(x,\omega)e^{-i\omega t}\, d\omega, \tag{3}$$

allows us to eliminate the time-dependence. The sound of the drum is determined by the eigenvalues of the time-independent Helmholtz equation,

$$(\Delta + \omega^2)v(x,\omega) = 0 \quad \forall x \in \mathcal{D}, \tag{4}$$

which is nothing else than the eigenvalue equation of the (negative) Laplacian inside the domain $\mathcal{D}$.

Another example is the propagation of electromagnetic waves in a two-dimensional system $\mathcal{D}$, see figure 1 top middle. Each component of the electric and magnetic field is again subject to the wave equation (1). Placing (super)conducting materials around $\mathcal{D}$ yields Dirichlet boundary conditions (2). Electromagnetic waves inside $\mathcal{D}$ can only be transmitted and received, if their frequencies are in the spectrum of the Laplacian.

Of special interest to this proceedings is quantum chaos which yields our next example, see figure 1 top right. A non-relativistic point particle moving freely in a manifold resp. orbifold $\mathcal{D}$ is described by the Schrödinger equation

$$i\hbar\frac{\partial}{\partial t}\Psi(x,t) = -\frac{\hbar^2}{2m}\Delta\Psi(x,t) \quad \forall x \in \mathcal{D}, \tag{5}$$

with appropriate boundary conditions on $\partial\mathcal{D}$. Scaling the units to $\hbar = 2m = 1$, and making the ansatz

$$\Psi(x,t) = \psi(x)e^{-\frac{1}{\hbar}Et}, \tag{6}$$

gives the time-independent Schrödinger equation,

$$(\Delta + E)\psi(x) = 0 \quad \forall x \in \mathcal{D}. \tag{7}$$

The statistical properties of its spectrum and its eigenfunctions are a central subject of study in quantum chaos.

In sections 8 and 9 we present some of the statistical properties of the solutions of the eigenvalue equation (7). But before, in sections 3–5 we introduce the three-dimensional hyperbolic system we are dealing with, and in sections 6 and 7 we develop an efficient algorithm that allows us to compute the solutions numerically. Part of the material was previously presented in [82].

From section 10 on we apply the results of the eigenvalue equation of the Laplacian to the universe, see figure 1 bottom, and compute the temperature fluctuations in the cosmic microwave background (CMB). This final example demonstrates that the methods learned in quantum chaos can be successfully used in other fields of research even though the physical interpretation can differ completely, e.g. metric perturbations and temperature fluctuations in cosmology instead of probability amplitudes in quantum mechanics.

2 General statistical properties in quantum chaos

Concerning the statistical properties of the eigenvalues and the eigenfunctions, we have to emphasise that they depend on the choice of the manifold resp. orbifold $\mathcal{D}$. Depending on whether the corresponding classical system is integrable or not, there are some generally accepted conjectures about the nearest-neighbour spacing distributions of the eigenvalues in the semiclassical limit. The semiclassical limit is the limit of large eigenvalues, $E \to \infty$.

Unless otherwise stated, we use the following assumptions: The quantum mechanical system is desymmetrised with respect to all its unitary symmetries, and whenever we examine the distribution of the eigenvalues we regard them on the scale of the mean level spacings. Moreover, it is generally believed that after desymmetrisation a generic quantum Hamiltonian corresponding to a classically strongly chaotic system possesses no degenerate eigenvalues.

Conjecture 2.1 (Berry, Tabor [15]). If the corresponding classical system is integrable, the eigenvalues behave like independent random variables and the distribution of the nearest-neighbour spacings is in the semiclassical limit close to the Poisson distribution, i.e. there is no level repulsion.

Conjecture 2.2 (Bohigas, Giannoni, Schmit [18, 19]). If the corresponding classical system is chaotic, the eigenvalues are distributed like the eigenvalues of hermitian random matrices [25, 56]. The corresponding ensembles depend only on the symmetries of the system:

- For chaotic systems without time-reversal invariance the distribution of the eigenvalues approaches in the semiclassical limit the distribution of the Gaussian Unitary Ensemble (GUE) which is characterised by a quadratic level repulsion.
- For chaotic systems with time-reversal invariance and integer spin the distribution of the eigenvalues approaches in the semiclassical limit the distribution of the Gaussian Orthogonal Ensemble (GOE) which is characterised by a linear level repulsion.

- For chaotic systems with time-reversal invariance and half-integer spin the distribution of the eigenvalues approaches in the semiclassical limit the distribution of the Gaussian Symplectic Ensemble (GSE) which is characterised by a quartic level repulsion.

These conjectures are very well confirmed by numerical calculations, but several exceptions are known. Here are two examples:

Exception 2.3. The harmonic oscillator is classically integrable, but its spectrum is equidistant.

Exception 2.4. The geodesic motion on surfaces of constant negative curvature provides a prime example for classical chaos. In some cases, however, the nearest-neighbour distribution of the eigenvalues of the Laplacian on these surfaces appears to be Poissonian.

"A strange arithmetical structure of chaos" in the case of surfaces of constant negative curvature that are generated by arithmetic fundamental groups was discovered by Aurich and Steiner [5], see also Aurich, Bogomolny, and Steiner [2]. (For the definition of an arithmetic group we refer the reader to [22]). Deviations from the expected GOE-behaviour in the case of a particular arithmetic surface were numerically observed by Bohigas, Giannoni, and Schmit [19] and by Aurich and Steiner [6]. Computations showed [6, 7], however, that the level statistics on 30 generic (i.e. non-arithmetic) surfaces were in nice agreement with the expected random-matrix theory prediction in accordance with conjecture 2.2. This has led Bogomolny, Georgeot, Giannoni, and Schmit [17], Bolte, Steil, and Steiner [21], and Sarnak [70] to introduce the concept of arithmetic quantum chaos.

Conjecture 2.5 (Arithmetic Quantum Chaos [16, 17, 20, 21, 70]). On surfaces of constant negative curvature that are generated by arithmetic fundamental groups, the distribution of the eigenvalues of the quantum Hamiltonian approaches in the semiclassical limit the Poisson distribution. Due to level clustering small spacings occur comparably often.

In order to carry out some specific numerical computations, we have to specify the manifold resp. orbifold $\mathcal{D}$. We will choose $\mathcal{D}$ to be given by the quotient space $\Gamma \backslash \mathcal{H}$ in the hyperbolic upper half space $\mathcal{H}$ where we choose Γ to be the Picard group, see section 5.

3 The hyperbolic upper half-space

Let

$$\mathcal{H} = \{(x_0, x_1, y) \in \mathbb{R}^3; \quad y > 0\} \tag{8}$$

be the upper half-space equipped with the hyperbolic metric of constant curvature -1

$$ds^2 = \frac{dx_0^2 + dx_1^2 + dy^2}{y^2}. \tag{9}$$

Due to the metric the Laplacian reads

$$\Delta = y^2 \left(\frac{\partial^2}{\partial x_0^2} + \frac{\partial^2}{\partial x_1^2} + \frac{\partial^2}{\partial y^2} \right) - y \frac{\partial}{\partial y}, \tag{10}$$

and the volume element is

$$d\mu = \frac{dx_0 dx_1 dy}{y^3}. \tag{11}$$

The geodesics of a particle moving freely in the upper half-space are straight lines and semicircles perpendicular to the x_0-x_1-plane, respectively, see figure 2.

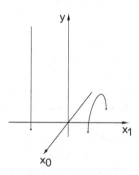

Fig. 2. Geodesics in the upper half-space of constant negative curvature.

Expressing any point $(x_0, x_1, y) \in \mathcal{H}$ as a Hamilton quaternion, $z = x_0 + \mathrm{i}x_1 + \mathrm{j}y$, with the multiplication defined by $\mathrm{i}^2 = -1$, $\mathrm{j}^2 = -1$, $\mathrm{ij} + \mathrm{ji} = 0$, all motions in the upper half-space are given by linear fractional transformations

$$z \mapsto \gamma z = (az + b)(cz + d)^{-1}; \quad a, b, c, d \in \mathbb{C}, \quad ad - bc = 1. \tag{12}$$

The group of these transformations is isomorphic to the group of matrices

$$\gamma = \begin{pmatrix} a & b \\ c & d \end{pmatrix} \in \mathrm{SL}(2, \mathbb{C}) \tag{13}$$

up to a common sign of the matrix entries,

$$\mathrm{SL}(2, \mathbb{C})/\{\pm 1\} = \mathrm{PSL}(2, \mathbb{C}). \tag{14}$$

The motions provided by the elements of $\mathrm{PSL}(2, \mathbb{C})$ exhaust all orientation preserving isometries of the hyperbolic metric on $\mathcal{H}$.

4 Topology

The topology is given by the manifold resp. orbifold. An orbifold is a space that locally looks like a Euclidean space modulo the action of a discrete group, e.g. a rotation group. A manifold is an orbifold that locally resembles Euclidean space.

A hyperbolic three-orbifold can be realised by a quotient

$$\Gamma \backslash \mathcal{H} = \{\Gamma z; \quad z \in \mathcal{H}\}, \tag{15}$$

where $\mathcal{H}$ is the upper-half space and Γ is a discrete subgroup of the isometries on $\mathcal{H}$. The elements of Γ are 2×2-matrices whose determinants equal one. If the trace of an element is real, it is called hyperbolic, parabolic, or elliptic, depending on whether the absolute value of its trace is larger, equal, or smaller than two, respectively. If the trace of an element is not real, the element is called loxodromic.

The action of Γ on $\mathcal{H}$ identifies all Γ-equivalent points with each other. All the Γ-equivalent points of z give an orbit

$$\Gamma z = \{\gamma z; \quad \gamma \in \Gamma\}. \tag{16}$$

The set of all the orbits is the orbifold. If, except of the identity, Γ contains only parabolic and hyperbolic elements, then $\Gamma \backslash \mathcal{H}$ is a manifold.

There exist infinitely many hyperbolic three-orbifolds. But opposed to the two-dimensional case there do not exist any deformations of hyperbolic three-orbifolds, because of the Mostow rigidity theorem [59, 66].

Each hyperbolic three-orbifold has its specific volume which can be finite or infinite. The volumes are always bounded from below by a positive constant.

If the volume of $\Gamma \backslash \mathcal{H}$ is finite and if Γ does not contain any parabolic elements, then the orbifold is compact. If Γ does contain parabolic elements, then the orbifold has cusps and is non-compact.

The hyperbolic three-manifold of smallest volume is unknown. Only a lower limit for the volume is proven to be [67]

$$\mathrm{vol}(\mathcal{M}) > 0.281. \tag{17}$$

The smallest known hyperbolic three-manifold is the Weeks manifold [84] whose volume is

$$\mathrm{vol}(\mathcal{M}) \simeq 0.943. \tag{18}$$

The volume of the Weeks manifold is the smallest among the volumes of arithmetic hyperbolic three-manifolds [24].

In contrast, hyperbolic three-orbifolds are known which have smaller volumes, whereby the one with smallest volume is also unknown. A lower limit for the volume is [57, 58]

$$\mathrm{vol}(\Gamma\backslash\mathcal{H}) > 8.2 \cdot 10^{-4}, \tag{19}$$

and the smallest known hyperbolic orbifold is the twofold extension of the tetrahedral Coxeter group CT(22). With

$$\Gamma = \mathrm{CT}(22)_2^+ = \mathrm{CT}(22)_2 \cap \mathrm{Iso}^+(\mathcal{H}), \tag{20}$$

where $\mathrm{Iso}^+(\mathcal{H})$ are the orientation preserving isometries of $\mathcal{H}$, the volume of the orbifold is

$$\mathrm{vol}(\Gamma\backslash\mathcal{H}) \simeq 0.039. \tag{21}$$

The volume of this orbifold is the smallest among the volumes of arithmetic hyperbolic three-orbifolds [23].

5 The Picard group

In the following we choose the hyperbolic three-orbifold $\Gamma\backslash\mathcal{H}$ of constant negative curvature that is generated by the Picard group [65],

$$\Gamma = \mathrm{PSL}(2, \mathbb{Z}[\mathrm{i}]), \tag{22}$$

where $\mathbb{Z}[\mathrm{i}] = \mathbb{Z} + \mathrm{i}\mathbb{Z}$ are the Gaussian integers.

The Picard group is generated by the cosets of three elements,

$$\begin{pmatrix} 1 & 1 \\ 0 & 1 \end{pmatrix}, \quad \begin{pmatrix} 1 & \mathrm{i} \\ 0 & 1 \end{pmatrix}, \quad \begin{pmatrix} 0 & -1 \\ 1 & 0 \end{pmatrix}, \tag{23}$$

which yield two translations and one inversion,

$$z \mapsto z + 1, \quad z \mapsto z + \mathrm{i}, \quad z \mapsto -z^{-1}. \tag{24}$$

The three motions generating Γ, together with the coset of the element

$$\begin{pmatrix} \mathrm{i} & 0 \\ 0 & -\mathrm{i} \end{pmatrix} \tag{25}$$

that is isomorphic to the symmetry

$$z = x + \mathrm{j}y \mapsto \mathrm{i}z\mathrm{i} = -x + \mathrm{j}y, \tag{26}$$

can be used to construct the fundamental domain.

Definition 5.1. *A fundamental domain of the discrete group Γ is a closed subset $\mathcal{F} \subset \mathcal{H}$ with the following conditions:*
(i) $\mathcal{F}$ meets each orbit Γz at least once,
(ii) if an orbit Γz does not meet the boundary of $\mathcal{F}$ it meets $\mathcal{F}$ at most once,
(iii) the boundary of $\mathcal{F}$ has Lebesgue measure zero.

For the Picard group the fundamental domain of standard shape is

$$\mathcal{F} = \{z = x_0 + ix_1 + jy \in \mathcal{H}; \quad -\frac{1}{2} \le x_0 \le \frac{1}{2}, \quad 0 \le x_1 \le \frac{1}{2}, \quad |z| \ge 1\},$$

$$(27)$$

with the absolute value of z being defined by $|z| = (x_0^2 + x_1^2 + y^2)^{\frac{1}{2}}$, see figure 3. Identifying the faces of the fundamental domain according to the elements

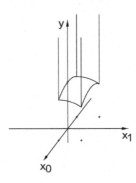

Fig. 3. The fundamental domain of the Picard group.

of the group Γ leads to a realisation of the quotient space $\Gamma \backslash \mathcal{H}$, see figure 4.

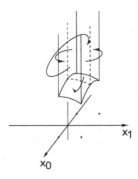

Fig. 4. Identifying the faces of the fundamental domain according to the elements of the Picard group.

With the hyperbolic metric the quotient space $\Gamma \backslash \mathcal{H}$ inherits the structure of an orbifold that has one parabolic and four elliptic fixed-points,

$$z = \mathrm{j}\infty, \quad z = \mathrm{j}, \quad z = \frac{1}{2} + \mathrm{j}\sqrt{\frac{3}{4}}, \quad z = \frac{1}{2} + \mathrm{i}\frac{1}{2} + \mathrm{j}\sqrt{\frac{1}{2}}, \quad z = \mathrm{i}\frac{1}{2} + \mathrm{j}\sqrt{\frac{3}{4}}.$$

(28)

The parabolic fixed-point corresponds to a cusp at $z = \mathrm{j}\infty$ that is invariant under the parabolic elements

$$\begin{pmatrix} 1 & 1 \\ 0 & 1 \end{pmatrix} \quad \text{and} \quad \begin{pmatrix} 1 & \mathrm{i} \\ 0 & 1 \end{pmatrix}.$$

(29)

Because of the hyperbolic metric the volume of the non-compact orbifold $\Gamma\backslash\mathcal{H}$ is finite [43],

$$\mathrm{vol}(\Gamma\backslash\mathcal{H}) = \frac{\zeta_K(2)}{4\pi^2} = 0.30532186472\ldots,$$

(30)

where

$$\zeta_K(s) = \frac{1}{4} \sum_{\nu\in\mathbb{Z}[\mathrm{i}]-\{0\}} (\nu\bar{\nu})^{-s}, \quad \Re s > 1,$$

(31)

is the Dedekind zeta function.

6 Maass waveforms

We are interested in the smooth and square-integrable eigenfunctions of the Laplacian in the orbifold $\Gamma\backslash\mathcal{H}$. A function being defined on the upper half-space that is invariant under all discrete linear fractional transformations,

$$\psi(\gamma z) = \psi(z) \quad \forall \gamma \in \Gamma,$$

(32)

is called an automorphic function. Any automorphic function can be identified with a function living on the quotient space $\Gamma\backslash\mathcal{H}$, and vice versa any function being defined on the quotient space can be identified with an automorphic function living in the upper half-space. The functions we are interested in match the definition of Maass waveforms [53, 54].

Definition 6.1. *Let Γ be a discrete subgroup of the isometries $\Gamma \subset \mathrm{PSL}(2, \mathbb{C})$ containing parabolic elements. A function $\psi(z)$ is called a Maass waveform if it is an automorphic eigenfunction of the Laplacian that is smooth and square-integrable on the fundamental domain,*

$$\psi(z) \in C^\infty(\mathcal{H}),$$

(33)

$$\psi(z) \in L^2(\Gamma\backslash\mathcal{H}),$$

(34)

$$(\Delta + E)\psi(z) = 0 \quad \forall z \in \mathcal{H},$$

(35)

$$\psi(\gamma z) = \psi(z) \quad \forall \gamma \in \Gamma, \; z \in \mathcal{H}.$$

(36)

Since the Maass waveforms are automorphic with respect to a discrete subgroup Γ that contains parabolic elements, they are periodic in the directions perpendicular to the cusp. This allows to expand them into Fourier series.

In the case of the Picard group we have

$$\psi(z) = u(y) + \sum_{\beta \in \mathbb{Z}[i]-\{0\}} a_\beta y K_{ik}(2\pi|\beta|y) e^{2\pi i \Re \beta x}, \tag{37}$$

where $\Re \beta x$ is the real part of the complex scalar product βx, and

$$u(y) = \begin{cases} b_0 y^{1+ik} + b_1 y^{1-ik} & \text{if } k \neq 0, \\ b_2 y + b_3 y \ln y & \text{if } k = 0. \end{cases} \tag{38}$$

$K_{ik}(t)$ is the K-Bessel function [83] whose order is connected with the eigenvalue E by

$$E = k^2 + 1. \tag{39}$$

If a Maass waveform vanishes in the cusp,

$$\lim_{z \to j\infty} \psi(z) = 0, \tag{40}$$

it is called a Maass cusp form.

According to the Roelcke-Selberg spectral resolution of the Laplacian [38, 68, 72], its spectrum contains both a discrete and a continuous part. The discrete part is spanned by the constant eigenfunction ψ_{k_0} and a countable number of Maass cusp forms $\psi_{k_1}, \psi_{k_2}, \psi_{k_3}, \ldots$ which we take to be ordered with increasing eigenvalues, $0 = E_{k_0} < E_{k_1} \leq E_{k_2} \leq E_{k_3} \leq \ldots$. The continuous part of the spectrum $E \geq 1$ is spanned by the Eisenstein series $E(z, 1 + ik)$ which are known analytically [31, 46]. The Fourier coefficients of the function $\Lambda_K(1 + ik)E(z, 1 + ik)$ are given by

$$b_0 = \Lambda_K(1+ik), \quad b_1 = \Lambda_K(1-ik), \quad a_\beta = 2 \sum_{\substack{\lambda,\mu \in \mathbb{Z}[i] \\ \lambda\mu=\beta}} \left|\frac{\lambda}{\mu}\right|^{ik}, \tag{41}$$

where $\beta \in \mathbb{Z}[i] - \{0\}$, and

$$\Lambda_K(s) = 4\pi^{-s}\Gamma(s)\zeta_K(s) \tag{42}$$

has an analytic continuation into the complex plane except for a pole at $s = 1$.

Defining

$$\psi_k^{\text{Eisen}}(z) = \frac{\Lambda_K(1+ik)E(z, 1+ik)}{\sqrt{\pi}|\Lambda_K(1+ik)|}, \tag{43}$$

the Eisenstein series $\psi_k^{\text{Eisen}}(z)$ is real.

Normalising the Maass cusp forms according to

$$\langle \psi_{k_i}, \psi_{k_i} \rangle = 1, \tag{44}$$

we can expand any square integrable function $\phi \in L^2(\Gamma \backslash \mathcal{H})$ in terms of the Maass cusp forms and the Eisenstein series, [32],

$$\phi(z) = \sum_{i \geq 0} \langle \psi_{k_i}, \phi \rangle \psi_{k_i}(z) + \frac{1}{2\pi i} \int_{\Re s = 1} \langle E(\cdot, s), \phi \rangle E(z, s)\, ds, \tag{45}$$

where

$$\langle \psi, \phi \rangle = \int_{\Gamma \backslash \mathcal{H}} \bar{\psi} \phi \, d\mu \tag{46}$$

is the Petersson scalar product.

The discrete eigenvalues and their associated Maass cusp forms are not known analytically. Thus, one has to calculate them numerically. Previous calculations of eigenvalues for the Picard group can be found in [35,44,74,78]. By making use of the Hecke operators [37,41,74,77] and the multiplicative relations among the coefficients, Steil [78] obtained a non-linear system of equations which allowed him to compute 2545 consecutive eigenvalues. Another way is to extend Hejhal's algorithm [39] to three dimensions [82]. Improving the procedure of finding the eigenvalues [81], we computed 13950 consecutive eigenvalues and their corresponding eigenfunctions.

7 Hejhal's algorithm

Hejhal found a linear stable algorithm for computing Maass waveforms together with their eigenvalues which he used for groups acting on the two-dimensional hyperbolic plane [39], see also [10,40,71,80]. We extend this algorithm which is based on the Fourier expansion and the automorphy condition and apply it to the Picard group acting on the three-dimensional hyperbolic space. For the Picard group no small eigenvalues $0 < E = k^2 + 1 < 1$ exist [79]. Therefore, k is real and the term $u(y)$ in the Fourier expansion of Maass cusp forms vanishes. Due to the exponential decay of the K-Bessel function for large arguments,

$$K_{ik}(t) \sim \sqrt{\frac{\pi}{2t}} e^{-t} \quad \text{for } t \to \infty, \tag{47}$$

and the polynomial bound of the coefficients [54],

$$a_\beta = O(|\beta|), \quad |\beta| \to \infty, \tag{48}$$

the absolutely convergent Fourier expansion can be truncated,

$$\psi(z) = \sum_{\substack{\beta \in \mathbb{Z}[\mathrm{i}] - \{0\} \\ |\beta| \leq M}} a_\beta y K_{\mathrm{i}k}(2\pi|\beta|y) e^{2\pi \mathrm{i} \Re \beta x} + [[\varepsilon]], \qquad (49)$$

if we bound y from below. Given $\varepsilon > 0$, k, and y, we determine the smallest $M = M(\varepsilon, k, y)$ such that the inequalities

$$2\pi M y \geq k \quad \text{and} \quad K_{\mathrm{i}k}(2\pi M y) \leq \varepsilon \max_t (K_{\mathrm{i}k}(t)) \qquad (50)$$

hold. Larger y allow smaller M. In all remainder terms,

$$[[\varepsilon]] = \sum_{\substack{\beta \in \mathbb{Z}[\mathrm{i}] - \{0\} \\ |\beta| > M}} a_\beta y K_{\mathrm{i}k}(2\pi|\beta|y) e^{2\pi \mathrm{i} \Re \beta x}, \qquad (51)$$

the K-Bessel function decays exponentially in $|\beta|$, and already the K-Bessel function of the first summand of the remainder terms is smaller than ε times most of the K-Bessel functions in the sum of (49). Thus, the error $[[\varepsilon]]$ does at most marginally exceed ε. The reason why $[[\varepsilon]]$ can exceed ε somewhat is due to the possibility that the summands in (49) cancel each other, or that the coefficients in the remainder terms are larger than in (49). By a finite two-dimensional Fourier transformation the Fourier expansion (49) is solved for its coefficients

$$a_\gamma y K_{\mathrm{i}k}(2\pi|\gamma|y) = \frac{1}{(2Q)^2} \sum_{x \in \mathbb{X}[\mathrm{i}]} \psi(x + \mathrm{j}y) e^{-2\pi \mathrm{i} \Re \gamma x} + [[\varepsilon]], \qquad (52)$$

where $\gamma \in \mathbb{Z}[\mathrm{i}] - \{0\}$, and $\mathbb{X}[\mathrm{i}]$ is a two-dimensional equally distributed set of $(2Q)^2$ numbers,

$$\mathbb{X}[\mathrm{i}] = \{\frac{l_0 + \mathrm{i} l_1}{2Q}; \quad l_j = -Q + \tfrac{1}{2}, -Q + \tfrac{3}{2}, \ldots, Q - \tfrac{3}{2}, Q - \tfrac{1}{2}, \quad j = 0, 1\}, \qquad (53)$$

with $2Q > M + |\gamma|$.

By automorphy we have

$$\psi(z) = \psi(z^*), \qquad (54)$$

where z^* is the Γ-pullback of the point z into the fundamental domain $\mathcal{F}$,

$$z^* = \gamma z, \quad \gamma \in \Gamma, \quad z^* \in \mathcal{F}. \qquad (55)$$

Thus, a Maass cusp form can be approximated by

$$\psi(x + \mathrm{j}y) = \psi(x^* + \mathrm{j}y^*) = \sum_{\substack{\beta \in \mathbb{Z}[\mathrm{i}] - \{0\} \\ |\beta| \leq M_0}} a_\beta y^* K_{\mathrm{i}k}(2\pi|\beta|y^*) e^{2\pi \mathrm{i} \Re \beta x^*} + [[\varepsilon]], \qquad (56)$$

where y^* is always larger or equal than the height y_0 of the lowest points of the fundamental domain $\mathcal{F}$,

$$y_0 = \min_{z \in \mathcal{F}}(y) = \frac{1}{\sqrt{2}}, \qquad (57)$$

allowing us to replace $M(\varepsilon, k, y)$ by $M_0 = M(\varepsilon, k, y_0)$.

Choosing y smaller than y_0 the Γ-pullback $z \mapsto z^*$ of any point into the fundamental domain $\mathcal{F}$ makes at least once use of the inversion $z \mapsto -z^{-1}$, possibly together with the translations $z \mapsto z+1$ and $z \mapsto z+\mathrm{i}$. This is called implicit automorphy, since it guarantees the invariance $\psi(z) = \psi(-z^{-1})$. The conditions $\psi(z) = \psi(z+1)$ and $\psi(z) = \psi(z+\mathrm{i})$ are automatically satisfied because of the Fourier expansion.

Making use of the implicit automorphy by replacing $\psi(x+\mathrm{j}y)$ in (52) by the right-hand side of (56) gives

$$a_\gamma y K_{ik}(2\pi|\gamma|y)$$
$$= \frac{1}{(2Q)^2} \sum_{x \in \mathbf{X}[\mathrm{i}]} \sum_{\substack{\beta \in \mathbf{Z}[\mathrm{i}]-\{0\} \\ |\beta| \leq M_0}} a_\beta y^* K_{ik}(2\pi|\beta|y^*) \mathrm{e}^{2\pi \mathrm{i} \Re \beta x^*} \mathrm{e}^{-2\pi \mathrm{i} \Re \gamma x} + [[2\varepsilon]], \quad (58)$$

which is the central identity in the algorithm.

The symmetry in the Picard group and the symmetries of the fundamental domain imply that the Maass waveforms fall into four symmetry classes [78] named **D**, **G**, **C**, and **H**, satisfying

$$\mathbf{D}: \quad \psi(x+\mathrm{j}y) = \psi(\mathrm{i}x+\mathrm{j}y) = \psi(-\bar{x}+\mathrm{j}y), \qquad (59)$$
$$\mathbf{G}: \quad \psi(x+\mathrm{j}y) = \psi(\mathrm{i}x+\mathrm{j}y) = -\psi(-\bar{x}+\mathrm{j}y), \qquad (60)$$
$$\mathbf{C}: \quad \psi(x+\mathrm{j}y) = -\psi(\mathrm{i}x+\mathrm{j}y) = \psi(-\bar{x}+\mathrm{j}y), \qquad (61)$$
$$\mathbf{H}: \quad \psi(x+\mathrm{j}y) = -\psi(\mathrm{i}x+\mathrm{j}y) = -\psi(-\bar{x}+\mathrm{j}y), \qquad (62)$$

respectively, see figure 5, from which the symmetry relations among the coefficients follow,

$$\mathbf{D}: \quad a_\beta = a_{\mathrm{i}\beta} = a_{\bar{\beta}}, \qquad (63)$$
$$\mathbf{G}: \quad a_\beta = a_{\mathrm{i}\beta} = -a_{\bar{\beta}}, \qquad (64)$$
$$\mathbf{C}: \quad a_\beta = -a_{\mathrm{i}\beta} = a_{\bar{\beta}}, \qquad (65)$$
$$\mathbf{H}: \quad a_\beta = -a_{\mathrm{i}\beta} = -a_{\bar{\beta}}. \qquad (66)$$

Defining

$$\mathrm{cs}(\beta, x) = \sum_{\sigma \in \mathbb{S}_\beta} s_{\sigma\beta} \mathrm{e}^{2\pi \mathrm{i} \Re \sigma x}, \qquad (67)$$

where $s_{\sigma\beta}$ is given by

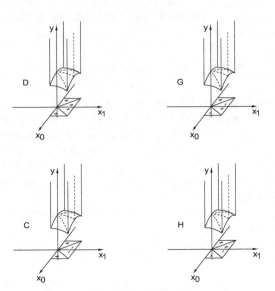

Fig. 5. The symmetries **D**, **G**, **C**, and **H** from top left to bottom right.

$$a_\sigma = s_{\sigma\beta}a_\beta \qquad (68)$$

and

$$\sigma \in \$_\beta = \begin{cases} \{\beta, \mathrm{i}\beta, -\beta, -\mathrm{i}\beta, \bar{\beta}, \mathrm{i}\bar{\beta}, -\bar{\beta}, -\mathrm{i}\bar{\beta}\} & \text{if } \bar{\beta} \notin \{\beta, \mathrm{i}\beta, -\beta, -\mathrm{i}\beta\}, \\ \{\beta, \mathrm{i}\beta, -\beta, -\mathrm{i}\beta\} & \text{else,} \end{cases} \qquad (69)$$

the Fourier expansion (37) of the Maass waveforms can be written

$$\psi(z) = u(y) + \sum_{\beta \in \tilde{\mathbb{Z}}[\mathrm{i}]-\{0\}} a_\beta y K_{\mathrm{i}k}(2\pi|\beta|y)\,\mathrm{cs}(\beta, x), \qquad (70)$$

where the tilde operator on a set of numbers is defined such that

$$\tilde{\mathbb{X}} \subset \mathbb{X}, \quad \bigcup_{x \in \tilde{\mathbb{X}}} \$_x = \mathbb{X}, \quad \text{and} \quad \bigcap_{x \in \tilde{\mathbb{X}}} \$_x = \emptyset \qquad (71)$$

holds.

Forgetting about the error $[[2\varepsilon]]$ the set of equations (58) can be written as

$$\sum_{\substack{\beta \in \tilde{\mathbb{Z}}[\mathrm{i}]-\{0\} \\ |\beta| \le M_0}} V_{\gamma\beta}(k, y)a_\beta = 0, \quad \gamma \in \tilde{\mathbb{Z}}[\mathrm{i}] - \{0\}, \quad |\gamma| \le M_0, \qquad (72)$$

where the matrix $V = (V_{\gamma\beta})$ is given by

$$V_{\gamma\beta}(k,y) = \#\{\sigma \in \$_\gamma\} y K_{ik}(2\pi|\gamma|y)\delta_{\gamma\beta}$$

$$- \frac{1}{(2Q)^2} \sum_{x\in\mathbb{X}[i]} y^* K_{ik}(2\pi|\beta|y^*)\,\mathrm{cs}(\beta,x^*)\,\mathrm{cs}(-\gamma,x). \quad (73)$$

Since $y < y_0$ can always be chosen such that $K_{ik}(2\pi|\gamma|y)$ is not too small, the diagonal terms in the matrix V do not vanish for large $|\gamma|$ and the matrix is well conditioned.

We are now looking for the non-trivial solutions of (72) for $1 \le |\gamma| \le M_0$ that simultaneously give the eigenvalues $E = k^2 + 1$ and the coefficients a_β. Trivial solutions are avoided by setting the first non-vanishing coefficient equal to one, $a_\alpha = 1$, where α is 1, $2+i$, 1, and $1+i$, for the symmetry classes **D**, **G**, **C**, and **H**, respectively.

Since the eigenvalues are unknown, we discretise the k-axis and solve for each k-value on this grid the inhomogeneous system of equations

$$\sum_{\substack{\beta\in\tilde{\mathbb{Z}}[i]-\{0,\alpha\}\\|\beta|\le M_0}} V_{\gamma\beta}(k,y^{\#1})a_\beta = -V_{\gamma\alpha}(k,y^{\#1}), \quad 1 \le |\gamma| \le M_0, \quad (74)$$

where $y^{\#1} < y_0$ is chosen such that $K_{ik}(2\pi|\gamma|y^{\#1})$ is not too small for $1 \le |\gamma| \le M_0$. A good value to try for $y^{\#1}$ is given by $2\pi M_0 y^{\#1} = k$.

It is important to check whether

$$g_\gamma = \sum_{\substack{\beta\in\tilde{\mathbb{Z}}[i]-\{0\}\\|\beta|\le M_0}} V_{\gamma\beta}(k,y^{\#2})a_\beta, \quad 1 \le |\gamma| \le M_0, \quad (75)$$

vanishes where $y^{\#2}$ is another y value independent of $y^{\#1}$. Only if all g_γ vanish simultaneously the solution of (74) is independent of y. In this case $E = k^2+1$ is an eigenvalue and the a_β's are the coefficients of the Fourier expansion of the corresponding Maass cusp form.

The probability to find a k-value such that all g_γ vanish simultaneously is zero, because the discrete eigenvalues are of measure zero in the real numbers. Therefore, we make use of the intermediate value theorem where we look for simultaneous sign changes in g_γ when k is varied. Once we have found them in at least half of the g_γ's, we have found an interval which contains an eigenvalue with high probability. By some bisection and interpolation we can see if this interval really contains an eigenvalue, and by nesting up the interval until its size tends to zero we obtain the eigenvalue.

It is conjectured [17,20,21,70] that the eigenvalues of the Laplacian to cusp forms of each particular symmetry class possess a spacing distribution close to that of a Poisson random process, see conjecture 2.5. One therefore expects that small spacings will occur rather often (due to level clustering). In order not to miss eigenvalues which lie close together, we have to make sure that at least one point of the k-grid lies between any two successive eigenvalues. On the other hand, we do not want to waste CPU time if there are large spacings.

Therefore, we use an adaptive algorithm which tries to predict the next best k-value of the grid. It is based on the observation that the coefficients a_β of two Maass cusp forms of successive eigenvalues must differ. Assume that two eigenvalues lie close together and that the coefficients of the two Maass cusp forms do not differ much. Numerically then both Maass cusp forms would tend to be similar – which contradicts the fact that different Maass cusp forms are orthogonal to each other with respect to the Petersson scalar product

$$\langle \psi_{k_i}, \psi_{k_j} \rangle = 0, \quad \text{if } k_i \neq k_j. \tag{76}$$

Maass cusp forms corresponding to different eigenvalues are orthogonal because the Laplacian is an essentially self-adjoint operator. Thus, if successive eigenvalues lie close together, the coefficients a_β must change fast when varying k. In contrast, if successive eigenvalues are separated by large spacings, numerically it turns out that often the coefficients change only slowly upon varying k. Defining

$$\tilde{a}_\beta = \frac{a_\beta}{\sqrt{\sum_{\substack{\gamma \in \tilde{\mathbb{Z}}[i] - \{0\} \\ |\gamma| \leq M_0}} |a_\gamma|^2}}, \quad 1 \leq |\beta| \leq M_0, \tag{77}$$

our adaptive algorithm predicts the next k-value of the grid such that the change in the coefficients is

$$\sum_{\substack{\beta \in \tilde{\mathbb{Z}}[i] - \{0\} \\ |\beta| \leq M_0}} |\tilde{a}_\beta(k_{\text{old}}) - \tilde{a}_\beta(k_{\text{new}})|^2 \approx 0.04. \tag{78}$$

For this prediction, the last step in the k-grid together with the last change in the coefficients is used to extrapolate linearly the choice for the next k-value of the grid.

However the adaptive algorithm is not a rigorous one. Sometimes the prediction of the next k-value fails so that it is too close or too far away from the previous one. A small number of small steps does not bother us unless the step size tends to zero. But, if the step size is too large, such that the left-hand side of (78) exceeds 0.16, we reduce the step size and try again with a smaller k-value.

Compared to earlier algorithms, our adaptive one tends to miss significantly less eigenvalues per run.

8 Eigenvalues

We have computed 13950 consecutive eigenvalues and their corresponding eigenfunctions of the (negative) Laplacian for the Picard group. The smallest non-trivial eigenvalue is $E = k^2 + 1$ with $k = 6.6221193402528$ which is in agreement with the lower bound $E > \frac{2\pi^2}{3}$ [79]. Table 1 shows the first few

Table 1. The first few eigenvalues of the negative Laplacian for the Picard group. Listed is k, related to the eigenvalues via $E = k^2 + 1$.

D	G	C	H
8.55525104		6.62211934	
11.10856737		10.18079978	
12.86991062		12.11527484	12.11527484
14.07966049		12.87936900	
15.34827764		14.14833073	
15.89184204		14.95244267	14.95244267
17.33640443		16.20759420	
17.45131992	17.45131992	16.99496892	16.99496892
17.77664065		17.86305643	17.86305643
19.06739052		18.24391070	
19.22290266		18.83298996	
19.41119126		19.43054310	19.43054310
20.00754583		20.30030720	20.30030720
20.70798880	20.70798880	20.60686743	
20.81526852		21.37966055	21.37966055
21.42887079		21.44245892	
22.12230276		21.83248972	21.83248972
22.63055256		22.58475297	22.58475297
22.96230105	22.96230105	22.85429195	
23.49617692		23.49768305	23.49768305
23.52784503		23.84275866	
23.88978413	23.88978413	23.89515755	23.89515755
24.34601664		24.42133829	24.42133829
24.57501426		25.03278076	25.03278076
24.70045917		25.42905483	
25.47067539		25.77588591	25.77588591
25.50724616		26.03903968	
25.72392169	25.72392169	26.12361823	26.12361823
25.91864376	25.91864376	26.39170209	
26.42695914		27.07065195	27.07065195
27.03326136		27.16341524	27.16341524

Table 2. Some consecutive large eigenvalues of the negative Laplacian for the Picard group. Listed is k, related to the eigenvalues via $E = k^2 + 1$.

D	G	C	H
139.65419675	139.65419675	139.66399548	139.66399548
139.65434417	139.65434417	139.66785333	139.66785333
139.65783548	139.65783548	139.66922266	139.66922266
139.66104047	139.66104047	139.67870460	139.67870460
139.67694018		139.68234200	139.68234200
139.68162707	139.68162707	139.68424704	139.68424704
139.68657976		139.69369972	139.69369972
139.71803029	139.71803029	139.69413379	139.69413379
139.72166907	139.72166906	139.69657741	139.69657741
139.78322452	139.78322452	139.73723373	139.73723373
139.81928622	139.81928622	139.73828541	139.73828541
139.81985670	139.81985670	139.74467774	139.74467774
139.82826034	139.82826034	139.75178180	139.75178180
139.84250751		139.75260292	139.75260292
139.87781072	139.87781072	139.79620628	139.79620628
139.87805540		139.80138072	139.80138072
139.88211647	139.88211647	139.81243991	139.81243991
139.91782003	139.91782003	139.81312982	139.81312982
139.91893517		139.82871870	139.82871870
139.92397167	139.92397167	139.86401372	139.86401372
139.92721861	139.92721861	139.86461581	139.86461581
139.93117207	139.93117207	139.89407865	139.89407865
139.93149277	139.93149277	139.89914777	139.89914777
139.94067283		139.90090849	139.90090849
139.94396890	139.94396890	139.91635302	139.91635302
139.95074070		139.94071729	139.94071729
139.95124805	139.95124805	139.95080198	139.95080198
139.99098324	139.99098324	139.97043676	139.97043676
140.00011792	140.00011792	140.00409202	140.00409202
140.00109753	140.00109753	140.02733151	140.02733151
140.00626902	140.00626902	140.04198905	140.04198905

eigenvalues of each symmetry class and table 2 shows some larger ones. The eigenvalues listed in table 1 agree with those of Steil [78] up to five decimal places.

One may ask whether we have found all eigenvalues. The answer can be given by comparing our results with Weyl's law. Consider the level counting function (taking all symmetry classes into account),

$$N(k) = \#\{\, i \mid k_i \leq k \},$$
(79)

(where the trivial eigenvalue, $E = 0$, is excluded), and split it into two parts

$$N(k) = \bar{N}(k) + N_{fluc}(k).$$
(80)

Here $\bar{N}$ is a smooth function describing the average increase in the number of levels, and N_{fluc} describes the fluctuations around the mean such that

$$\lim_{K \to \infty} \frac{1}{K} \int_1^K N_{fluc}(k) dk = 0.$$
(81)

The average increase in the number of levels is given by Weyl's law [9, 85] and higher order corrections have been calculated by Matthies [55]. She obtained

$$\bar{N}(k) = \tfrac{\text{vol}(\mathcal{F})}{6\pi^2} k^3 + a_2 k \log k + a_3 k + a_4 + o(1)$$
(82)

with the constants

$$a_2 = -\tfrac{3}{2\pi},$$
(83)

$$a_3 = \tfrac{1}{\pi} [\tfrac{13}{16} \log 2 + \tfrac{7}{4} \log \pi - \log \Gamma(\tfrac{1}{4}) + \tfrac{2}{9} \log(2 + \sqrt{3}) + \tfrac{3}{2}],$$
(84)

$$a_4 = -\tfrac{3}{2}.$$
(85)

We compare our results for $N(k)$ with (82) by defining

$$N_{fluc}(k) = N(k) - \bar{N}(k).$$
(86)

N_{fluc} fluctuates around zero or a negative integer whose absolute value gives the number of missing eigenvalues, see figure 6. Unfortunately, our algorithm does not find all eigenvalues in one single run. In the first run it finds about 97% of the eigenvalues. Apart from very few exceptions the remaining eigenvalues are found in the third run. To be more specific, we plotted N_{fluc} decreased by $\tfrac{1}{2}$, because $N(k) - \bar{N}(k)$ is approximately $\tfrac{1}{2}$ whenever $E = k^2 + 1$ is an eigenvalue. A plot indicating that N_{fluc} fluctuates around zero is shown in figure 7 where we plotted the integral

$$I(K) = \frac{1}{K} \int_1^K N_{fluc}(k) dk.$$
(87)

Desymmetrising the spectrum yields Weyl's law to be

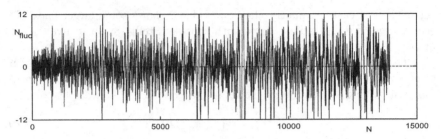

Fig. 6. $N_{fluc}(k_i)$ as a function of $N(k_i) \equiv i$ fluctuating around zero.

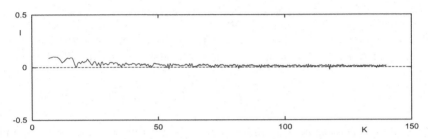

Fig. 7. I as a function of K showing that $I \xrightarrow{K \to \infty} 0$.

$$\bar{N}(k) = \frac{\mathrm{vol}(\mathcal{F})}{24\pi^2}k^3 + O(k^2) \tag{88}$$

for each symmetry class. Looking at table 1 it seems somehow surprising that there are not equally many eigenvalues listed for each symmetry class. Especially in the symmetry classes **G** and **H** there seem to be much less eigenvalues than in the symmetry classes **D** and **C**. Indeed, as was shown by Steil [78], there occur systematic degenerated eigenvalues between different symmetry classes.

Theorem 8.1 (Steil [78]). *If* $E = k^2 + 1$ *is an eigenvalue corresponding to an eigenfunction of the symmetry class* **G** *resp.* **H***, then there exists an eigenfunction of the symmetry class* **D** *resp.* **C** *corresponding to the same eigenvalue.*

Based on our numerical results we conjecture [82]:

Conjecture 8.2. Taking all four symmetry classes together, there are no degenerate eigenvalues other than those explained by Steil's theorem. Furthermore, the degenerate eigenvalues which are explained by Steil's theorem occur only in pairs of two degenerate eigenvalues. They never occur in sets of three or more degenerate eigenvalues.

Looking at the semiclassical limit, $E \to \infty$, we finally find that almost all eigenvalues are two-fold degenerated, see e.g. table 2, which is an immediate consequence of Weyl's law, Steil's theorem, and conjecture 8.2. This means that as $E \to \infty$

$$\frac{\#\{\text{non-degenerate eigenvalues} \leq E\}}{\#\{\text{all eigenvalues} \leq E\}} \to 0. \tag{89}$$

Finally, we remark that the distribution of the eigenvalues of each individual symmetry class agrees numerically with conjecture 2.5, see [78, 82].

9 Eigenfunctions

Concerning the eigenfunctions of the Laplacian, it is believed that they behave like random waves. The conjecture of Berry [14] predicts for each eigenfunction in the semiclassical limit, $E \to \infty$, a Gaussian value distribution,

$$d\rho(u) = \frac{1}{\sqrt{2\pi}\sigma} e^{-\frac{u^2}{2\sigma^2}} \, du, \tag{90}$$

inside any compact regular subregion F of $\mathcal{F}$. This means that

$$\lim_{E \to \infty} \frac{\frac{1}{\text{vol}(F)} \int_F \chi_{[a,b]}(\psi(z)) \, d\mu}{\int_a^b d\rho(u)} = 1 \tag{91}$$

holds with variance

$$\sigma^2 = \frac{1}{\text{vol}(F)} \int_F |\psi(z)|^2 \, d\mu \tag{92}$$

for any $-\infty < a < b < \infty$, where $\chi_{[a,b]}$ is the indicator function of the interval $[a, b]$. Figure 8 shows the value distribution of the 80148th[4] eigenfunction corresponding to the eigenvalue $E = k^2 + 1$ with $k = 250.0018575195$ inside a small subregion

$$F = \{z = x + iy; \quad -0.43750 < x_0 < -0.31435,$$
$$0.06250 < x_1 < 0.18565, \tag{93}$$
$$1.10000 < y < 1.34881\}.$$

Our numerical data agree quite well with Berry's conjecture, providing numerical evidence that the conjecture holds. A plot of the eigenfunction inside the region F is given in figure 9.

10 An application to cosmology

In the remaining sections we apply the eigenvalues and eigenfunctions of the Laplacian to a perturbed Friedmann-Lemaître universe and compute the temperature fluctuations in the cosmic microwave background (CMB).

250   Ralf Aurich, Frank Steiner, and Holger Then

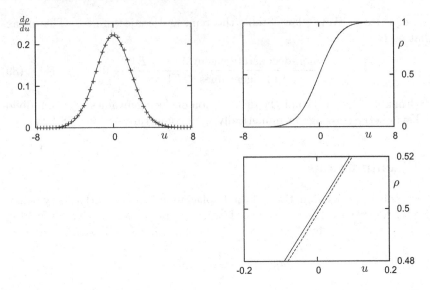

Fig. 8. In the top left figure the value distribution of the eigenfunction corresponding to the eigenvalue $E = k^2 + 1$ with $k = 250.0018575195$ inside the region F is shown as ($^+$). The solid line is the conjectured Gaussian. In the figure on the top right the dashed line is the integrated value distribution of the eigenfunction which is nearly indistinguishable from the integrated Gaussian (solid line). The figure on the bottom right displays a detailed magnification of the integrated value distribution showing that it lies slightly below the integrated Gaussian.

The CMB is a relic from the primeval fireball of the early universe. It is the light that comes from the time when the universe was 379 000 years old [12]. It was predicted by Gamow in 1948 and explained in detail by Peebles [63]. In 1978, Penzias and Wilson [64] won the Nobel Prize for Physics for first measuring the CMB at a wavelength of 7.35 cm. Within the resolution of their experiment they found the CMB to be completely isotropic over the whole sky. Later with the much better resolution of the NASA satellite mission Cosmic Background Explorer (COBE), Smoot et al. [73] found fluctuations in the CMB which are of amplitude 10^{-5} relative to the mean background temperature of $T_0 = 2.725$ K, except for the large dipole moment, see figure 10. These small fluctuations serve as a fingerprint of the early universe, since the temperature fluctuations are related to the density fluctuations at the time of last scattering. They show how isotropic the universe was at early times. In the inflationary scenario the fluctuations originate from quantum fluctuations which are inflated to macroscopic scales. Due to gravitational instabilities the fluctuations grow steadily and give rise to the formation of stars and galaxies.

The theoretical framework in which the CMB and its fluctuations are explained is Einstein's general theory of relativity [26–30]. Thereby a ho-

---

[4] The number 80148 was determined approximately using Weyl's law (82).

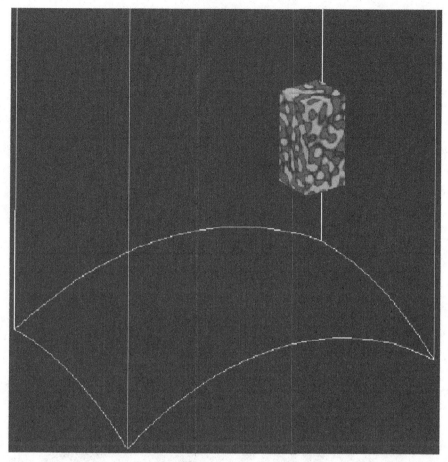

Fig. 9. A plot of the eigenfunction corresponding to the eigenvalue $E = k^2 + 1$ with $k = 250.0018575195$ inside the region F.

mogeneous and isotropic background given by a Friedmann-Lemaître universe [33, 34, 49] is perturbed. The time-evolution of the perturbations can be computed in the framework of linear perturbation theory [11, 52, 69].

An explanation for the presence of the CMB is the following, see also figure 11: We live in an expanding universe [42]. At early enough times the universe was so hot and dense that it was filled with a hot plasma consisting of ionised atoms, unbounded electrons, and photons. Due to Thomson scattering of photons with electrons, the hot plasma was in thermal equilibrium and the mean free path of the photons was small, hence the universe was opaque. Due to its expansion the universe cooled down and became less dense. When the universe was around 379 000 years old, its temperature T has dropped down to approximately 3000 K. At this time, called the time of last scattering, the electrons got bound to the nuclei forming a gas of neutral atoms, mainly hy-

Fig. 10. Sky maps of the temperature fluctuations in the CMB as observed by the NASA satellite mission COBE. The sky map on the top shows the dipole anisotropy after the mean background temperature of $T_0 = 2.725$ K has been subtracted. The amplitude of the dipole anisotropy is about 3 mK. Also subtracting the dipole yields the sky map in the middle. One sees the small temperature fluctuations whose amplitude is roughly $30\,\mu$K. But one also sees a lot of foreground contamination along the equator that comes from nearby stars in our galaxy. After removing the foregrounds one finally gets the sky map on the bottom showing the temperature fluctuations in the CMB. Downloaded from LAMBDA [48].

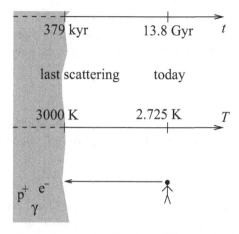

Fig. 11. The expanding universe. At the time of last scattering occured a phase transition from an opaque to a transparent universe.

drogen and helium, and the universe became transparent. Since this time the photons travel freely on their geodesics through the universe. At the time of last scattering the photons had an energy distribution according to a Planck spectrum with temperature of nearly 3000 K. The further expansion of the universe redshifted the photons such that they nowadays have an energy distribution according to a Planck spectrum with temperature of $T_0 = 2.725$ K. This is what we observe as the CMB.

Due to the thermal equilibrium before the time of last scattering the CMB is nearly perfectly isotropic, but small density fluctuations lead to small temperature fluctuations. The reason for the small temperature fluctuations comes from a variety of effects. Most dominant on large scales are the Sachs-Wolfe and the integrated Sachs-Wolfe effect. The former results from density fluctuations at the time of last scattering, the latter from time varying gravitational potentials along the line of sight. In these lecture notes we restrict our attention to just these two effects. The inclusion of further physical effects is done in [3, 4].

11 Friedmann-Lemaître universes

Assume a universe whose spatial part is locally homogeneous and isotropic. Its metric is given by the Robertson-Walker metric,

$$ds^2 = dt^2 - \tilde{a}^2(t)\gamma_{ij}dx^i dx^j, \tag{94}$$

where we use the Einstein summation convention. Notice that we have changed the notation slightly. Instead of the quaternion z for the spatial variables, we now write x. γ_{ij} is the metric of a homogeneous and isotropic three-dimensional space, and the units are rescaled such that the speed of light is $c = 1$. Introducing the conformal time $d\eta = \frac{dt}{\tilde{a}(t)}$ we have

$$ds^2 = a^2(\eta)\left[d\eta^2 - \gamma_{ij}dx^i dx^j\right], \tag{95}$$

where $a(\eta) = \tilde{a}(t(\eta))$ is the cosmic scale factor.

With the Robertson-Walker metric the Einstein equations simplify to the Friedmann equations [33, 34, 49]. One of the two Friedmann equations reads

$$a'^2 + \kappa a^2 = \frac{8\pi G}{3}T_0^0 a^4 + \frac{1}{3}\Lambda a^4, \tag{96}$$

and the other Friedmann equation is equivalent to local energy conservation. a' is the derivative of the cosmic scale factor with respect to the conformal time η. κ is the curvature parameter which we choose to be negative, $\kappa = -1$. G is Newton's gravitational constant, T_ν^μ is the energy-momentum tensor, and Λ is the cosmological constant.

Assuming the energy and matter in the universe to be a perfect fluid consisting of radiation, non-relativistic matter, and a cosmological constant, the time-time component of the energy-momentum tensor reads

$$T_0^0 = \varepsilon_r(\eta) + \varepsilon_m(\eta), \tag{97}$$

where the energy densities of radiation and matter scale like

$$\varepsilon_r(\eta) = \varepsilon_r(\eta_0)\left(\frac{a(\eta_0)}{a(\eta)}\right)^4 \quad \text{and} \quad \varepsilon_m(\eta) = \varepsilon_m(\eta_0)\left(\frac{a(\eta_0)}{a(\eta)}\right)^3. \tag{98}$$

Here η_0 denotes the conformal time at the present epoch.

Specifying the initial conditions (Big Bang!) $a(0) = 0$, $a'(0) > 0$, the Friedmann equation (96) can be solved analytically [8],

$$a(\eta) = \frac{-\left(\frac{\Omega_r}{\Omega_c}\right)^{\frac{1}{2}}\mathcal{P}'(\eta) + \frac{1}{2}\left(\frac{\Omega_m}{\Omega_c}\right)\left(\mathcal{P}(\eta) - \frac{1}{12}\right)}{2\left(\mathcal{P}(\eta) - \frac{1}{12}\right)^2 - \frac{1}{2}\frac{\Omega_\Lambda\Omega_r}{\Omega_c^2}}a(\eta_0), \tag{99}$$

where $\mathcal{P}(\eta)$ denotes the Weierstrass $\mathcal{P}$-function which can numerically be evaluated very efficiently, see [1], by

$$\mathcal{P}(\eta) = \mathcal{P}(\eta; g_2, g_3) = \frac{1}{\eta^2} + \sum_{n=2}^{\infty} c_n \eta^{2n-2} \tag{100}$$

with

$$c_2 = \frac{g_2}{20}, \quad c_3 = \frac{g_3}{28}, \quad \text{and} \quad c_n = \frac{3}{(2n+1)(n-3)}\sum_{m=2}^{n-2} c_m c_{n-m} \quad \text{for } n \geq 4. \tag{101}$$

The so-called invariants g_2 and g_3 are determined by the cosmological parameters,

$$g_2 = \frac{\Omega_\Lambda \Omega_r}{\Omega_c^2} + \frac{1}{12}, \quad g_3 = \frac{1}{6}\frac{\Omega_\Lambda \Omega_r}{\Omega_c^2} - \frac{1}{16}\frac{\Omega_\Lambda \Omega_m^2}{\Omega_c^3} - \frac{1}{216}, \qquad (102)$$

with

$$\Omega_r = \frac{8\pi G \varepsilon_r(\eta_0)}{3H^2(\eta_0)}, \qquad (103)$$

$$\Omega_m = \frac{8\pi G \varepsilon_m(\eta_0)}{3H^2(\eta_0)}, \qquad (104)$$

$$\Omega_c = \frac{1}{H^2(\eta_0)a^2(\eta_0)}, \qquad (105)$$

$$\Omega_\Lambda = \frac{\Lambda}{3H^2(\eta_0)}, \qquad (106)$$

where

$$H(\eta) = \frac{a'(\eta)}{a^2(\eta)} \qquad (107)$$

is the Hubble parameter.

Because of the homogeneity and isotropy, nowhere in the equations of a Friedmann-Lemaître universe appears the Laplacian.

12 Perturbed Friedmann-Lemaître universes

The idealisation to an exact homogeneous and isotropic universe was essential to derive the spacetime of the Robertson-Walker metric. But obviously, we do not live in a universe which is perfectly homogeneous and isotropic. We see individual stars, galaxies, and in between large empty space. Knowing the spacetime of the Robertson-Walker metric, we can study small perturbations around the homogeneous and isotropic background. Since the amplitude of the large scale fluctuations in the universe is of relative size 10^{-5} [73], we can use linear perturbation theory. In longitudinal gauge the most general scalar perturbation of the Robertson-Walker metric reads

$$ds^2 = a^2(\eta)\big[(1 + 2\Phi)d\eta^2 - (1 - 2\Psi)\gamma_{ij}dx^i dx^j\big], \qquad (108)$$

where $\Phi = \Phi(\eta, x)$ and $\Psi = \Psi(\eta, x)$ are functions of spacetime.

Assuming that the energy and matter density in the universe can be described by a perfect fluid, consisting of radiation, non-relativistic matter, and a cosmological constant, and neglecting possible entropy perturbations, the Einstein equations reduce in first order perturbation theory [60] to

$$\Phi = \Psi, \tag{109}$$

$$\Phi'' + 3\hat{H}(1 + c_s^2)\Phi' - c_s^2 \Delta\Phi + \left(2\hat{H}' + (1 + 3c_s^2)(\hat{H}^2 + 1)\right)\Phi = 0, \tag{110}$$

where $\hat{H} = \frac{a'}{a}$ and $c_s^2 = (3 + \frac{9}{4}\frac{\varepsilon_m}{\varepsilon_r})^{-1}$ are given by the solution of the non-perturbed Robertson-Walker universe. In the partial differential equation (110) the Laplacian occurs. If the initial and the boundary conditions of Φ are specified, the time-evolution of the metric perturbations can be computed.

With the separation ansatz

$$\Phi(\eta, x) = \sum_k f_k(\eta)\psi_k(x) + \int dk\, f_k(\eta)\psi_k(x), \tag{111}$$

where the ψ_k are the eigenfunctions of the negative Laplacian, and the E_k are the corresponding eigenvalues,

$$-\Delta\psi_k(x) = E_k\psi_k(x), \tag{112}$$

(110) simplifies to

$$f_k''(\eta) + 3\hat{H}(1 + c_s^2)f_k'(\eta) + \left(c_s^2 E_k + 2\hat{H}' + (1 + 3c_s^2)(\hat{H}^2 + 1)\right)f_k(\eta) = 0. \tag{113}$$

The ODE (113) can be computed numerically in a straightforward way, and we finally obtain the metric of the whole universe. This gives the input to the Sachs-Wolfe formula [69] which connects the metric perturbations with the temperature fluctuations,

$$\frac{\delta T}{T_0}(\hat{n}) = 2\Phi(\eta_{\mathrm{SLS}}, x(\eta_{\mathrm{SLS}})) - \frac{3}{2}\Phi(0, x(0)) + 2\int_{\eta_{\mathrm{SLS}}}^{\eta_0} d\eta\, \frac{\partial}{\partial\eta}\Phi(\eta, x(\eta)), \tag{114}$$

where $\hat{n}$ is a unit vector in the direction of the observed photons. $x(\eta)$ is the geodesic along which the light travels from the surface of last scattering (SLS) towards us, and η_{SLS} is the time of last scattering.

If we choose the topology of the universe to be the orbifold of the Picard group, we can use in (111) the Maass cusp forms and the Eisenstein series computed in sections 6–9. Let us further choose the initial conditions to be

$$f_k(0) = \frac{\sigma_k \alpha}{\sqrt{E_k}\sqrt{E_k - 1}} \quad \text{and} \quad f_k'(0) = \frac{-\Omega_m f_k(0)}{16(\Omega_c \Omega_r)^{\frac{1}{2}}}, \tag{115}$$

[8, 45, 50, 51], which carry over to a Harrison-Zel'dovich spectrum having a spectral index $n = 1$ and selecting only the non-decaying modes. α is a constant independent of k which is fitted to the amplitude of the observed temperature fluctuations. The quantities σ_k are random signs, $\sigma_k \in \{-1, 1\}$.

The following cosmological parameters are used $\Omega_m = 0.3, \Omega_\Lambda = 0.6, \Omega_c = 1 - \Omega_{\mathrm{tot}} = 1 - \Omega_r - \Omega_m - \Omega_\Lambda, H(\eta_0) = 100\, h_0\, \mathrm{km\, s^{-1}\, Mpc^{-1}}$ with $h_0 = 0.65$. The density $\Omega_r \approx 10^{-4}$ is determined by the current temperature $T_0 = 2.725\,\mathrm{K}$. The point of the observer is chosen to be at $x_{\mathrm{obs}} = 0.2 + 0.1\mathrm{i} + 1.6\mathrm{j}$. The sky map of figure 12 is computed with the expansion (111) using cusp forms and Eisenstein series. For numerical reasons the infinite spectrum is cut such that only the eigenvalues with $E = k^2 + 1 \leq 19601$ and their corresponding eigenfunctions are taken into account. In addition the integration of the continuous spectrum is approximated numerically using a Gauss quadrature with 16 integration points per unit interval. The resulting map in figure 12 is not in agreement with the cosmological observations [12] which are shown in figure 14. Especially in the direction of the cusp, the temperature fluctuations of our calculated model show a strong peak. Clearly, such a pronounced hot spot is unphysical, since it is not observed. This peak comes from the contribution of the Eisenstein series. The sky map in figure 13, which results from taking only the cusp forms into account, yields a much better agreement with the cosmological observations.

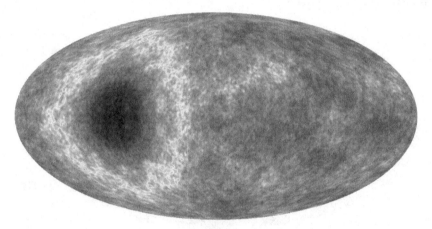

Fig. 12. The sky map of the calculated temperature fluctuations of the CMB for $\Omega_{\mathrm{tot}} = 0.9, \Omega_{\mathrm{m}} = 0.3, \Omega_\Lambda = 0.6, h_0 = 0.65$, and $x_{\mathrm{obs}} = 0.2 + 0.1\mathrm{i} + 1.6\mathrm{j}$, where the Eisenstein series and the cusp forms are taken into account. The coordinate system is oriented such that the cusp is located in the left part of the figure where the hot spot can be seen.

The hot spot in the direction of the cusp, which results from the contribution of the Eisenstein series, is sometimes used, see e.g. [75], as an argument that the Eisenstein series should not be taken into account. Some papers even ignore the existence of the Eisenstein series completely, e.g. [51].

Another possibility is to take the Eisenstein series into account, but to choose for them initial conditions which differ from (115) such that the intensity towards the cusp is not increasing. To see whether this is possible, let us consider the completeness relation of Maass waveforms,

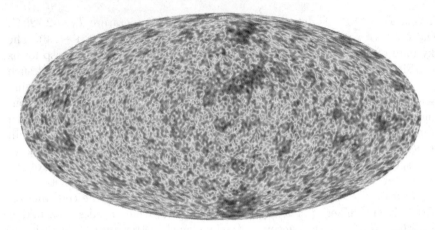

Fig. 13. The sky map of the calculated temperature fluctuations of the CMB for $\Omega_{\text{tot}} = 0.9, \Omega_{\text{m}} = 0.3, \Omega_{\Lambda} = 0.6, h_0 = 0.65$, and $x_{\text{obs}} = 0.2 + 0.1\text{i} + 1.6\text{j}$, if only the cusp forms are taken into account.

Fig. 14. The temperature fluctuations of the CMB observed by WMAP. Downloaded from LAMBDA [48].

$$\Phi(\eta, x) = \sum_{n \geq 0} \langle \psi_{k_n}, \Phi(\eta, \cdot) \rangle \psi_{k_n}(x) + \int_0^\infty \langle \psi_k^{\text{Eisen}}, \Phi(\eta, \cdot) \rangle \psi_k^{\text{Eisen}}(x) \, dk, \quad (116)$$

cf. (43) and (45), saying that within the fundamental cell any desired metric perturbation $\Phi(\eta_{\text{SLS}}, x)$ at the time of last scattering can be expressed via the conditions

$$f_{k_n}(\eta_{\text{SLS}}) = \langle \psi_{k_n}, \Phi(\eta_{\text{SLS}}, \cdot) \rangle \qquad \text{for the discrete spectrum,}$$
$$f_k^{\text{Eisen}}(\eta_{\text{SLS}}) = \langle \psi_k^{\text{Eisen}}, \Phi(\eta_{\text{SLS}}, \cdot) \rangle \qquad \text{for the continuous spectrum.}$$

E.g., choosing the metric perturbation to be

$$\Phi(\eta_{\mathrm{SLS}}, x) = \cos(k' \ln x_2) \tag{117}$$

would neither result in a hot spot in the direction of the cusp nor would $f_k^{\mathrm{Eisen}}(\eta)$ vanish identically. A more general ansatz would be

$$\Phi(\eta_{\mathrm{SLS}}, x) = \int_0^\infty dk' \left(A_{k'}(\eta_{\mathrm{SLS}}, x) \cos(k' \ln x_2) + B_{k'}(\eta_{\mathrm{SLS}}, x) \sin(k' \ln x_2) \right), \tag{118}$$

where $A_{k'}(\eta_{\mathrm{SLS}}, x)$ and $B_{k'}(\eta_{\mathrm{SLS}}, x)$ are some amplitudes.

Concerning the initial conditions we found that (115) has to be modified for the continuous part of the spectrum. Hence, the important question occurs: How do the correct initial conditions look like?

Not knowing the correct initial conditions, we use (115) for the discrete part and neglect the continuous part of the spectrum. We proceed with the discussion of the properties that come from the discrete part of the spectrum, only.

13 Comparison with the cosmological observations

The key experiments that observed the fluctuations of the CMB are: COBE, Smoot et al. [73]; Boomerang, de Bernardis et al. [13]; MAXIMA-1, Hanany et al. [36]; ACBAR, Kuo et al. [47]; CBI, Pearson et al. [62]; WMAP, Bennett et al. [12].

Concerning the topology of the universe which manifests itself in the low multipole moments, there exist only the cosmological observations from COBE and WMAP. These two experiments have scanned a full sky map of the temperature fluctuations. The NASA satellite mission of the Cosmic Background Explorer (COBE) was the first experiment that detected the temperature fluctuations of the CMB in 1992. In addition, it found that the quadrupole moment of the temperature fluctuations is suppressed. In 2003 the NASA satellite mission of the Wilkinson Microwave Anisotropy Probe (WMAP) scanned a full sky map of the CMB with a much higher resolution confirming and improving the results of COBE. All the other experiments were either ground based or balloon flights that could only scan parts of the sky and thus did not measure the first multipole moments. Their strength was to measure the finer-scale anisotropy manifested in the higher multipole moments.

In order to quantitatively compare our results with the observations, we introduce the angular power spectrum and the correlation function. Expanding the temperature fluctuations in the CMB into spherical harmonics,

$$\delta T\big(\hat{n}(\theta, \phi)\big) = \sum_{l=2}^\infty \sum_{m=-l}^{l} a_{lm} Y_l^m(\theta, \phi), \tag{119}$$

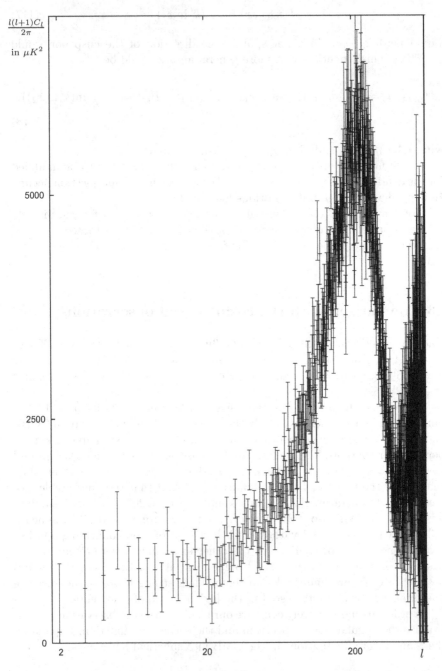

Fig. 15. The angular power spectrum of the temperature fluctuations observed by WMAP. Downloaded from LAMBDA [48].

yields the expansion coefficients a_{lm}. We remark that in (119) the sum over l starts at $l = 2$, i.e. the monopole and dipole have been subtracted, because the monopole term does not give rise to an anisotropy and the dipole term is unobservable due to the peculiar velocity of our solar system relative to the background. From the expansion coefficients we obtain the multipole moments of the CMB anisotropies,

$$C_l = \frac{1}{2l+1} \sum_{m=-l}^{l} |a_{lm}|^2, \tag{120}$$

the angular power spectrum,

$$\frac{l(l+1)}{2\pi} C_l, \tag{121}$$

and the two-point correlation function,

$$C(\vartheta) = \langle \delta T(\hat{n})\delta T(\hat{n}') \rangle_{\cos\vartheta = \hat{n}\cdot\hat{n}'} = \frac{1}{8\pi^2 \sin\vartheta} \int\int_{\substack{S^2 \times S^2 \\ \hat{n}\cdot\hat{n}' = \cos\vartheta}} d\hat{n}\, d\hat{n}'\, \delta T(\hat{n})\delta T(\hat{n}'). \tag{122}$$

Assuming a Gaussian value distribution for the expansion coefficients a_{lm}, the two-point correlation function is related to the multipole moments via

$$C(\vartheta) = \frac{1}{4\pi} \sum_{l=2}^{\infty} (2l+1)C_l P_l(\cos\vartheta). \tag{123}$$

When determining the correlation function numerically, we consider (123) as its definition irrespectively whether the expansion coefficients a_{lm} are Gaussian distributed or not.

Figures 15–17 show the angular power spectrum and the correlation function, respectively, measured by WMAP [76]. In figure 15 we see that the first multipole moments resulting from the cosmological observations are suppressed, especially the quadrupole, C_2. The next few multipole moments up to $l \approx 25$ give rise to a plateau in the angular power spectrum, before the larger multipole moments with $l > 25$ start to increase until they reach the first acoustic peak. Page et al. [61] find that the first acoustic peak is at $l = 220.1 \pm 0.8$. The trough following this peak is at $l = 411.7 \pm 3.5$, and the second peak is at $l = 546 \pm 10$.

Figures 16 and 17 show the angular power spectrum and the correlation function, respectively, corresponding to the calculated sky map for the Picard model of figure 13 in comparison with the results of the cosmological observations and with the concordance model [76]. Regarding our results of the calculated temperature fluctuations, see figure 16, we see that the first multipole moments are suppressed in accordance with the cosmological observations. Especially the quadrupole term, C_2, computed in our model, is suppressed.

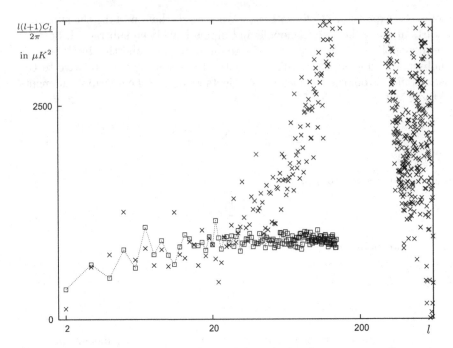

Fig. 16. The angular power spectrum of the calculated temperature fluctuations for $\Omega_{\text{tot}} = 0.9, \Omega_{\text{m}} = 0.3, \Omega_\Lambda = 0.6, h_0 = 0.65$, and $x_{\text{obs}} = 0.2 + 0.1\text{i} + 1.6\text{j}$ ($\Box$) in comparison with the angular power spectrum of the WMAP observations ($\times$).

However the suppression of the observed one is still stronger. Note, that in the case of the concordance model, the first multipoles are increased in contrast to our results. This points to a non-trivial topology of our universe.

The next multipole moments give rise to a plateau in the angular power spectrum. We use this plateau to fit the constant α of the initial conditions (115) such that the sum of the first 19 computed multipole moments matches with the cosmological observations,

$$\sum_{l=2}^{20} C_{l,\text{computed}} \overset{!}{=} \sum_{l=2}^{20} C_{l,\text{observed}}. \tag{124}$$

But concerning the fluctuations on finer scales (smaller angular resolution), i.e. higher multipoles, we do not obtain the acoustic peaks in our numerical calculations. This is due to the fact that we neglect several physical effects which perturb the temperature on finer scales. We ignore all the physical effects that dominate on scales smaller than those corresponding to $l \gg 25$, since our main interest is in the influence of the non-trivial topology of the universe due to the Picard group. See [3] for the inclusion of further physical effects.

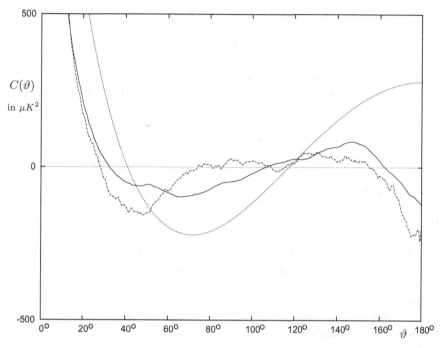

Fig. 17. The solid line is the correlation function of the calculated temperature fluctuations for $\Omega_{\text{tot}} = 0.9, \Omega_{\text{m}} = 0.3, \Omega_\Lambda = 0.6, h_0 = 0.65$, and $x_{\text{obs}} = 0.2 + 0.1\text{i} + 1.6\text{j}$. The dashed line represents the correlation function of the WMAP observations, and the dotted line shows the concordance model [76].

Another comparison of our calculated temperature fluctuations with the cosmological observations can be done via the correlation function (123) which emphasises large scales, i.e. the behaviour of low multipoles. Concerning the correlation function $C(\vartheta)$, we find quite good agreement of our calculated temperature fluctuations with the cosmological observations, see figure 17. The reason for this good agreement of our model with the observations is due to the suppressed quadrupole moment. We also show in figure 17 (dotted line) the concordance model [76] which is not in good agreement with the data for $\vartheta \gtrsim 7°$ in contrast to our model (full curve). Especially for large angular separations, $\vartheta \gtrsim 160°$, the concordance model is not able to describe the observed anticorrelation in $C(\vartheta)$. This anticorrelation constitutes a fingerprint in the CMB that favours a non-trivial topology for the universe.

14 Conclusion

We pointed out the importance of the eigenvalue equation of the Laplacian and gave some physical applications. In the framework of quantum chaos, we focused on the eigenvalue equation of the Picard orbifold and computed the solutions numerically. Our main goal was to determine the statistical properties of the solutions, but also to apply the solutions to a completely different problem, namely the temperature fluctuations in the CMB. Here, we demonstrate that a model of the universe with the topology of the Picard group matches the large-scale anisotropy in the CMB better than the current concordance model. Thereby, we show that the methods developed in quantum chaos can be successfully applied to other fields of research although the physical interpretation can differ completely.

15 Acknowledgments

H. T. gratefully acknowledges the encouraging advice of Prof. Dennis A. Hejhal. This work has been supported by the European Commission under the Research Training Network (Mathematical Aspects of Quantum Chaos) no HPRN-CT-2000-00103 of the IHP Programme and by the Deutsche Forschungsgemeinschaft under the contract no. DFG Ste 241/16-1. The computations were run on the computers of the Universitäts-Rechenzentrum Ulm. We acknowledge the use of the Legacy Archive for Microwave Background Data Analysis (LAMBDA). Support for LAMBDA is provided by the NASA Office of Space Science.

References

1. M. Abramowitz and I. A. Stegun. *Handbook of mathematical functions with formulas, graphs, and mathematical tables*. National Bureau of Standards Applied Mathematics Series, 55. U.S. Government Printing Office, Washington, D.C., 1964.
2. R. Aurich, E. B. Bogomolny, and F. Steiner. Periodic orbits on the regular hyperbolic octagon. *Physica D*, 48:91–101, 1991.
3. R. Aurich, S. Lustig, F. Steiner, and H. Then. Hyperbolic universes with a horned topology and the CMB anisotropy. *Class. Quant. Grav.* 21:4901–4925, 2004.
4. R. Aurich, S. Lustig, F. Steiner, and H. Then. Indications about the shape of the universe from the WMAP data. *Phys. Rev. Lett.* 94:021301, 2005.
5. R. Aurich and F. Steiner. On the periodic orbits of a strongly chaotic system. *Physica D*, 32:451–460, 1988.
6. R. Aurich and F. Steiner. Periodic-orbit sum rules for the Hadamard-Gutzwiller model. *Physica D*, 39:169–193, 1989.
7. R. Aurich and F. Steiner. Energy-level statistics of the Hadamard-Gutzwiller ensemble. *Physica D*, 43:155–180, 1990.
8. R. Aurich and F. Steiner. The cosmic microwave background for a nearly flat compact hyperbolic universe. *Mon. Not. Roy. Astron. Soc.*, 323:1016–1024, 2001.

9. V. G. Avakumović. Über die Eigenfunktionen auf geschlossenen Riemannschen Mannigfaltigkeiten. (German). *Math. Z.*, 65:327–344, 1956.

10. H. Avelin. On the deformation of cusp forms (Licentiate Thesis). *UUDM report 2003:8*, Uppsala 2003.

11. J. Bardeen. Gauge-invariant cosmological perturbations. *Phys. Rev. D*, 22:1882–1905, 1980.

12. C. L. Bennett, M. Halpern, G. Hinshaw, N. Jarosik, A. Kogut, M. Limon, S. S. Meyer, L. Page, D. N. Spergel, G. S. Tucker, E. Wollack, E. L. Wright, C. Barnes, M. R. Greason, R. S. Hill, E. Komatsu, M. R. Nolta, N. Odegard, H. V. Peirs, L. Verde, and J. L. Weiland. First year Wilkinson microwave anisotropy probe (WMAP) observations: Preliminary maps and basic results. *Astroph. J. Suppl.*, 148:1–27, 2003.

13. P. de Bernardis, P. A. R. Ade, J. J. Bock, J. R. Bond, J. Borrill, A. Boscaleri, K. Coble, B. P. Crill, G. De Gasperis, P. C. Farese, P. G. Ferreira, K. Ganga, M. Giacometti, E. Hivon, V. V. Hristov, A. Iacoangeli, A. H. Jaffe, A. E. Lange, L. Martinis, S. Masi, P. Mason, P. D. Mauskopf, A. Melchiorri, L. Miglio, T. Montroy, C. B. Netterfield, E. Pascale, F. Piacentini, D. Pogosyan, S. Prunet, S. Rao, G. Romeo, J. E. Ruhl, F. Scaramuzzi, D. Sforna, and N. Vittorio. A flat universe from high-resolution maps of the cosmic microwave background radiation. *Nature*, 404:955–959, 2000.

14. M. V. Berry. Regular and irregular semiclassical wavefunctions. *J. Phys. A*, 10:2083–2091, 1977.

15. M. V. Berry and M. Tabor. Closed orbits and the regular bound spectrum. *Proc. Roy. Soc. London Ser. A*, 349:101–123, 1976.

16. E. Bogomolny, Quantum and arithmetical chaos. In P. Cartier, B. Julia, P. Moussa, and P. Vanhove, editors, *Frontiers in Number Theory, Physics, and Geometry I*. Springer, 2006.

17. E. B. Bogomolny, B. Georgeot, M.-J. Giannoni, and C. Schmit. Chaotic billiards generated by arithmetic groups. *Phys. Rev. Lett.*, 69:1477–1480, 1992.

18. O. Bohigas, M.-J. Giannoni, and C. Schmit. Characterization of chaotic quantum spectra and universality of level fluctuation laws. *Phys. Rev. Lett.*, 52:1–4, 1984.

19. O. Bohigas, M.-J. Giannoni, and C. Schmit. Spectral fluctuations, random matrix theories and chaotic motion. Stochastic processes in classical and quantum systems. *Lecture Notes in Phys.*, 262:118–138, 1986.

20. J. Bolte. Some studies on arithmetical chaos in classical and quantum mechanics. *Int. J. Mod. Phys. B*, 7:4451–4553, 1993.

21. J. Bolte, G. Steil, and F. Steiner. Arithmetical chaos and violation of universality in energy level statistics. *Phys. Rev. Lett.*, 69:2188–2191, 1992.

22. A. Borel. *Introduction aux groupes arithmétiques*. (French). Hermann, 1969.

23. T. Chinburg and E. Friedman. The smallest arithmetic hyperbolic three-orbifold. *Invent. Math.*, 86:507–527, 1986.

24. T. Chinburg, E. Friedman, K. N. Jones, and A. W. Reid. The arithmetic hyperbolic 3-manifold of smallest volume. *Ann. Scuola Norm. Sup. Pisa Cl. Sci.*, 30:1–40, 2001.

25. F. J. Dyson. Correlations between the eigenvalues of a random matrix. *Commun. Math. Phys.*, 19:235–250, 1970.

26. A. Einstein. Zur allgemeinen Relativitätstheorie. (German). *Preuß. Akad. Wiss., Sitzungsber.*, pp. 778–786, 1915.

27. A. Einstein. Zur allgemeinen Relativitätstheorie (Nachtrag). (German). *Preuß. Akad. Wiss., Sitzungsber.*, pp. 799–801, 1915.

28. A. Einstein. Die Feldgleichungen der Gravitation. (German). *Preuß. Akad. Wiss., Sitzungsber.*, pp. 844–847, 1915.

29. A. Einstein. Die Grundlage der allgemeinen Relativitätstheorie. (German). *Ann. Phys.*, 49:769–822, 1916.

30. A. Einstein. Kosmologische Betrachtungen zur Allgemeinen Relativitätstheorie. (German). *Preuß. Akad. Wiss., Sitzungsber.*, pp. 142–152, 1917.

31. J. Elstrodt, F. Grunewald, and J. Mennicke. Eisenstein series on three-dimensional hyperbolic space and imaginary quadratic number fields. *J. Reine Angew. Math.*, 360:160–213, 1985.

32. J. Elstrodt, F. Grunewald, and J. Mennicke. *Groups Acting on Hyperbolic Space.* Springer, 1998.

33. A. Friedmann. Über die Krümmung des Raumes. (German). *Z. Phys.*, 10:377–386, 1922.

34. A. Friedmann. Über die Möglichkeit einer Welt mit konstanter negativer Krümmung des Raumes. (German). *Z. Phys.*, 21:326–332, 1924.

35. F. Grunewald and W. Huntebrinker. A numerical study of eigenvalues of the hyperbolic Laplacian for polyhedra with one cusp. *Experiment. Math.*, 5:57–80, 1996.

36. S. Hanany, P. Ade, A. Balbi, J. Bock, J. Borrill, A. Boscaleri, P. de Bernardis, P. G. Ferreira, V. V. Hristov, A. H. Jaffe, A. E. Lange, A. T. Lee, P. D. Mauskopf, C. B. Netterfield, S. Oh, E. Pascale, B. Rabii, P. L. Richards, G. F. Smoot, R. Stompor, C. D. Winant, and J. H. P. Wu. MAXIMA-1: A measurement of the cosmic microwave background anisotropy on angular scales of 10 arcminutes to 5 degrees. *Astroph. J.*, 545:L5, 2000.

37. D. Heitkamp. Hecke-Theorie zur SL(2; o). (German). *Schriftenreihe des Mathematischen Instituts der Universität Münster, 3. Serie*, 5, 1992.

38. D. A. Hejhal. *The Selberg trace formula for* PSL(2, ℝ). Lecture Notes in Math. 1001. Springer, 1983.

39. D. A. Hejhal. On eigenfunctions of the Laplacian for Hecke triangle groups. In D. A. Hejhal, J. Friedman, M. C. Gutzwiller, and A. M. Odlyzko, editors, *Emerging applications of number theory*, IMA Series No. 109, pp. 291–315. Springer, 1999.

40. D. A. Hejhal. On the calculation of Maass cusp forms, *these proceedings.*

41. D. A. Hejhal and S. Arno. On Fourier coefficients of Maass waveforms for PSL(2, ℤ). *Math. Comp.*, 61:245–267, 1993.

42. E. P. Hubble. A relation between distance and radial velocity among extragalactic nebulae. *Proc. Nat. Acad. Sci. (USA)*, 15:168–173, 1929.

43. G. Humbert. Sur la mesure des classes d'Hermite de discriminant donné dans un corps quadratique imaginaire, et sur certaines volumes non euclidiens. (French). *C. R. Acad. Sci. Paris*, 169:448–454, 1919.

44. W. Huntebrinker. Numerical computation of eigenvalues of the Laplace-Beltrami operator on three-dimensional hyperbolic spaces by finite-element methods. *Diss. Summ. Math.*, 1:29–36, 1996.

45. K. T. Inoue, K. Tomita, and N. Sugiyama. Temperature correlations in a compact hyperbolic universe. *Mon. Not. Roy. Astron. Soc.*, 314:L21, 2000.

46. T. Kubota. *Elementary Theory of Eisenstein Series.* Kodansha, Tokyo and Halsted Press, 1973.

47. C. L. Kuo, P. A. R. Ade, J. J. Bock, C. Cantalupo, M. D. Daub, J. Goldstein, W. L. Holzapfel, A. E. Lange, M. Lueker, M. Newcomb, J. B. Peterson, J. Ruhl, M. C. Runyan, and E. Torbet. High resolution observations of the CMB power spectrum with ACBAR. *Astroph. J.*, 600:32–51, 2004.

48. The Legacy Archive for Microwave Background Data Analysis (LAMBDA), http://lambda.gsfc.nasa.gov/

49. G. Lemaître. Un univers homogène de masse constante et de rayon croissant, rendant compte de la vitesse radiale de nébuleuses extragalactiques. (French). *Ann. Soc. Sci. Bruxelles*, 47A:47–59, 1927.

50. J. Levin. Topology and the cosmic microwave background. *Phys. Rep.*, 365:251–333, 2002.

51. J. J. Levin, J. D. Barrow, E. F. Bunn, and J. Silk. Flat spots: Topological signatures of an open universe in cosmic background explorer sky maps. *Phys. Rev. Lett.*, 79:974–977, 1997.

52. E. Lifshitz. On the gravitational stability of the expanding universe. *Zh. Eksp. Teor. Fiz.*, 16:587–602, 1946.

53. H. Maaß. Über eine neue Art von nichtanalytischen automorphen Funktionen und die Bestimmung Dirichletscher Reihen durch Funktionalgleichungen. (German). *Math. Ann.*, 121:141–183, 1949.

54. H. Maaß. Automorphe Funktionen von mehreren Veränderlichen und Dirichletsche Reihen. (German). *Abh. Math. Semin. Univ. Hamb.*, 16:72–100, 1949.

55. C. Matthies. *Picards Billard. Ein Modell für Arithmetisches Quantenchaos in drei Dimensionen.* (German). PhD thesis, Universität Hamburg, 1995.

56. M. L. Mehta. *Random matrices.* Academic Press, second edition, 1991.

57. R. Meyerhoff. A lower bound for the covolume of hyperbolic 3-orbifolds. *Duke Math. J.*, 57:185–203, 1988.

58. R. Meyerhoff. Sphere packing and volume in hyperbolic 3-space. *Comment. Math. Helv.*, 61:271–278, 1988.

59. G. D. Mostow. *Strong rigidity of locally symmetric spaces.* Annals of Mathematics Studies, Princeton University Press, 1973.

60. V. F. Mukhanov, H. A. Feldman, and R. H. Brandenberger. Theory of cosmological perturbations. *Phys. Reports*, 215:203–333, 1992.

61. L. Page, M. R. Nolta, C. Barnes, C. L. Bennett, M. Halpern, G. Hinshaw, N. Jarosik, A. Kogut, M. Limon, S. S. Meyer, H. V. Peiris, D. N. Spergel, G. S. Tucker, E. Wollack, and E. L. Wright. First year Wilkinson Microwave Anisotropy Probe (WMAP) observations: Interpretation of the TT and TE angular power spectrum peaks. *Astroph. J. Suppl.*, 148:233–241, 2003.

62. T. J. Pearson, B. S. Mason, A. C. S. Readhead, M. C. Shepherd, J. L. Sievers, P. S. Udomprasert, J. K. Cartwright, A. J. Farmer, S. Padin, S. T. Myers, J. R. Bond, C. R. Contaldi, U.-L. Pen, S. Prunet, D. Pogosyan, J. E. Carlstrom, J. Kovac, E. M. Leitch, C. Pryke, N. W. Halverson, W. L. Holzapfel, P. Altamirano, L. Bronfman, S. Casassus, J. May, and M. Joy. The anisotropy of the microwave background to l = 3500: Mosaic observations with the Cosmic Background Imager (CBI). *Astroph. J.*, 591:556–574, 2003.

63. P. J. E. Peebles. The black-body radiation content of the universe and the formation of galaxies. *Astroph. J.*, 142:1317–1326, 1965.

64. A. A. Penzias and R. W. Wilson. A measurement of excess antenna temperature at 4080 Mc/s. *Astroph. J.*, 142:419–421, 1965.

65. E. Picard. Sur un groupe de transformations des points de l'espace situé du même coté d'un plan. *Bull. Soc. Math. France* 12:43–47, 1884.

66. G. Prasad. Strong rigidity of Q-rank 1 lattices. *Invent. Math.*, 21:255–286, 1973.

67. A. Przeworski. Cones embedded in hyperbolic manifolds. *J. Differential Geom.*, 58:219–232, 2001.

68. W. Roelcke. Das Eigenwertproblem der automorphen Formen in der hyperbolischen Ebene, I and II. (German). *Math. Ann.*, 167:292–337, 1966 and 168:261–324, 1967.

69. R. K. Sachs and A. M. Wolfe. Perturbations of a cosmological model and angular variations of the microwave background. *Astroph. J.*, 147:73–90, 1967.

70. P. Sarnak. Arithmetic quantum chaos. *Israel Math. Conf. Proc.*, 8:183–236, 1995.

71. B. Selander and A. Strömbergsson. Sextic coverings of genus two which are branched at three points. *UUDM report 2002:16*, Uppsala 2002.

72. A. Selberg. Harmonic analysis and discontinuous groups in weakly symmetric Riemannian spaces with applications to Dirichlet series. *J. Indian Math. Soc.*, 20:47–87, 1956.

73. G. F. Smoot, C. L. Bennett, A. Kogut, E. L. Wright, J. Aymon, N. W. Boggess, E. S. Cheng, G. De Amici, S. Gulkis, M. G. Hauser, G. Hinshaw, P. D. Jackson, M. Janssen, E. Kaita, T. Kelsall, P. Keegstra, C. Lineweaver, K. Loewenstein, P. Lubin, J. Mather, S. S. Meyer, S. H. Moseley, T. Murdock, L. Rokke, R. F. Silverberg, L. Tenorio, R. Weiss, and D. T. Wilkinson. Structure in the COBE Differential Microwave Radiometer first-year maps. *Astroph. J.*, 396:L1–L5, 1992.

74. M. N. Smotrov and V. V. Golovčanskiĭ. Small eigenvalues of the Laplacian on $\Gamma \backslash H_3$ for $\Gamma = PSL_2(\mathbb{Z}[i])$. *Preprint*, 91-040, Bielefeld 1991.

75. D. D. Sokolov and A. A. Starobinskii. Globally inhomogeneous "spliced" universes. *Sov. Astron.*, 19:629–633, 1976.

76. D. N. Spergel, L. Verde, H. V. Peiris, E. Komatsu, M. R. Nolta, C. L. Bennett, M. Halpern, G. Hinshaw, N. Jarosik, A. Kogut, M. Limon, S. S. Meyer, L. Page, G. S. Tucker, J. L. Weiland, E. Wollack, and E. L. Wright. First year Wilkinson microwave anisotropy probe (WMAP) observations: Determination of cosmological parameters. *Astroph. J. Suppl.*, 148:175–194, 2003.

77. H. M. Stark. Fourier coefficients of Maass waveforms. In R. A. Rankin, editor, *Modular Forms*, pp. 263–269. Ellis Horwood, 1984.

78. G. Steil. Eigenvalues of the Laplacian for Bianchi groups. In D. A. Hejhal, J. Friedman, M. C. Gutzwiller, and A. M. Odlyzko, editors, *Emerging applications of number theory*, IMA Series No. 109, pp. 617–641. Springer, 1999.

79. K. Stramm. Kleine Eigenwerte des Laplace-Operators zu Kongruenzgruppen. (German). *Schriftenreihe des Mathematischen Instituts der Universität Münster, 3. Serie*, 11, 1994.

80. F. Strömberg. Maass waveforms on $(\Gamma_0(N), \chi)$, *this volume*.

81. H. Then. Maass cusp forms for large eigenvalues. *Math. Comp.*, 74:363–381, 2005.

82. H. Then. Arithmetic quantum chaos of Maass waveforms. In P. Cartier, B. Julia, P. Moussa, and P. Vanhove, editors, *Frontiers in Number Theory, Physics, and Geometry I*, pp 183–212. Springer, 2006.

83. G. N. Watson. *A treatise on the theory of Bessel functions*. Cambridge University Press, 1944.

84. J. Weeks. *Hyperbolic structures on 3-manifolds*. PhD thesis, Princeton University, 1985.

85. H. Weyl. Das asymptotische Verteilungsgesetz der Eigenwerte linearer partieller Differentialgleichungen. (German). *Math. Ann.*, 71:441–479, 1912.

Index

Printed in the United States
by Baker & Taylor Publisher Services